Dimensions Math®
Teacher's Guide KB

Authors and Reviewers

Cassandra Turner

Elizabeth Curran

Allison Coates

Tricia Salerno

Pearly Yuen

Jenny Kempe

Singapore Math Inc.

Published by Singapore Math Inc.

19535 SW 129th Avenue
Tualatin, OR 97062
www.singaporemath.com

Dimensions Math® Teacher's Guide Kindergarten B
ISBN 978-1-947226-31-9

First published 2018
Reprinted 2019, 2020 (twice), 2021 (twice)

Printed in China

Acknowledgments

Editing by the Singapore Math Inc. team.
Design and illustration by Cameron Wray with Carli Bartlett.

Contents

Dimensions Math® Curriculum

The **Dimensions Math®** series is a Pre-Kindergarten to Grade 5 series based on the pedagogy and methodology of math education in Singapore. The main goal of the **Dimensions Math®** series is to help students develop competence and confidence in mathematics.

The series follows the principles outlined in the Singapore Mathematics Framework below.

Pedagogical Approach and Methodology

- Through Concrete-Pictorial-Abstract development, students view the same concepts over time with increasing levels of abstraction.
- Thoughtful sequencing creates a sense of continuity. The content of each grade level builds on that of preceding grade levels. Similarly, lessons build on previous lessons within each grade.
- Group discussion of solution methods encourages expansive thinking.
- Interesting problems and activities provide varied opportunities to explore and apply skills.
- Hands-on tasks and sharing establish a culture of collaboration.
- Extra practice and extension activities encourage students to persevere through challenging problems.
- Variation in pictorial representation (number bonds, bar models, etc.) and concrete representation (straws, linking cubes, base ten blocks, discs, etc.) broaden student understanding.

Each topic is introduced, then thoughtfully developed through the use of a variety of learning experiences, problem solving, student discourse, and opportunities for mastery of skills. This combination of hands-on practice, in-depth exploration of topics, and mathematical variability in teaching methodology allows students to truly master mathematical concepts.

Singapore Mathematics Framework

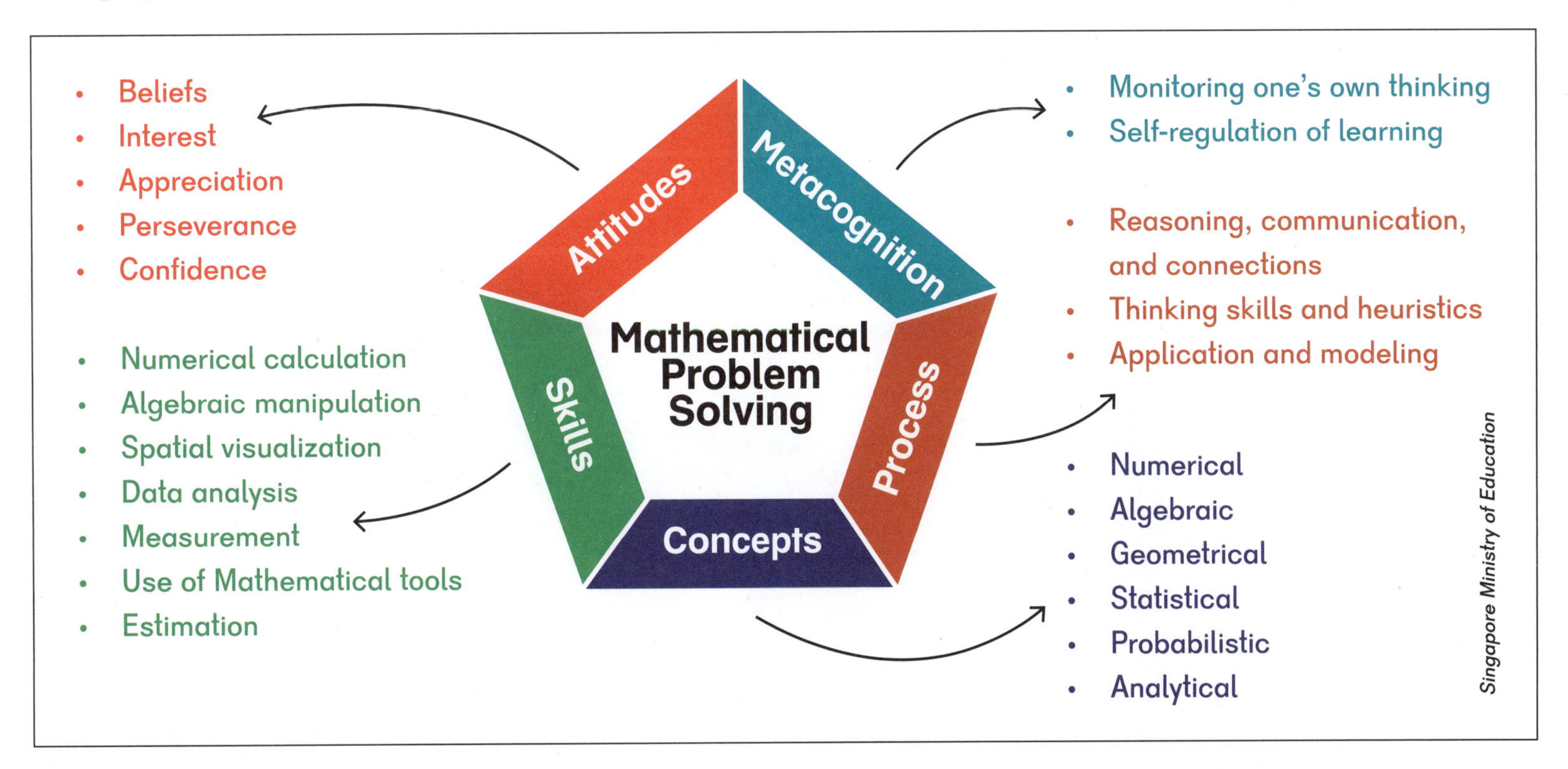

Dimensions Math® Program Materials

Textbooks

Textbooks are designed to help students build a solid foundation in mathematical thinking and efficient problem solving. Careful sequencing of topics, well-chosen problems, and simple graphics foster deep conceptual understanding and confidence. Mental math, problem solving, and correct computation are given balanced attention in all grades. As skills are mastered, students move to increasingly sophisticated concepts within and across grade levels.

Students work through the textbook lessons with the help of five friends: Emma, Alex, Sofia, Dion, and Mei. The characters appear throughout the series and help students develop metacognitive reasoning through questions, hints, and ideas.

A pencil icon at the end of the textbook lessons links to exercises in the workbooks.

Workbooks

Workbooks provide additional problems that range from basic to challenging. These allow students to independently review and practice the skills they have learned.

Teacher's Guides

Teacher's Guides include lesson plans, mathematical background, games, helpful suggestions, and comprehensive resources for daily lessons.

Tests

Tests contained differentiated assessments to systematically evaluate student progress.

Online Resources

The following can be downloaded from dimensionsmath.com.

- **Blackline Masters** used for various hands-on tasks.
- **Letters Home** to be emailed or sent home with students for continued exploration. These outline what the student is learning in math class and offer suggestions for related activities at home. Reinforcement at home supports deep understanding of mathematical concepts.
- **Videos** of popular children songs used for singing activities.
- **Material Lists** for each chapter and lesson, so teachers and classroom helpers can prepare ahead of time.
- **Activities** that can done with students who need more practice or a greater challenge, organized by concept, chapter, and lesson.
- **Standards Alignments** for various states.

Using the Teacher's Guide

This guide is designed to assist in planning daily lessons, and should be considered a helping hand between the curriculum and the classroom. It provides introductory notes on mathematical content, key points, and suggestions for activities. It also includes ideas for differentiation within each lesson, and answers and solutions to textbook and workbook problems.

Each chapter of the guide begins with the following.

- **Overview**

  Includes objectives and suggested number of class periods for each chapter.

- **Notes**

  Highlights key learning points, provides background on math concepts, explains the purpose of certain activities, and helps teachers understand the flow of topics throughout the year.

- **Materials**

  Lists materials, manipulatives, and Blackline Masters used in the Explore and Learn sections of the guide. It also includes suggested storybooks. Blackline Masters can be found at dimensionsmath.com.

The guide goes through the Chapter Openers, Daily Lessons, and Practices of each chapter in the following general format.

- **Explore**

  Introduces students to math concepts through hands-on activities that can be done in small or whole group formats.

- **Learn**

  Summarizes the main concepts of the lesson, including exercises from Look and Talk pages. Small and whole group activities are often utilized in this section.

Whole Group & Small Group Activities

Allows students to practice concepts through hands-on tasks and games, including suggestions for outdoor play (most of which can be modified for a gymnasium or classroom).

Level of difficulty in the games and activities are denoted by the following symbols.

● Foundational activities
▲ On-level activities
★ Challenge or extension activities

Extend

This expands on **Explore**, **Learn**, and **Activities** and provides opportunities for students to deepen their understanding and build confidence.

Discussion is a critical component of each lesson. Have students share their ideas with a partner, small group, or the class as often as possible. As each classroom is different, this guide does not anticipate all situations. Teachers are encouraged to elicit higher level thinking and discussion through questions like these:

- Why? How do you know?
- Can you explain that?
- Can you draw a picture of that?
- Does your answer make sense? How do you know?
- How is this task like the one we did before? How is it different?
- What did you learn before that can help you to solve this problem?
- What is alike and what is different about this?
- Can you solve that a different way?
- How do you know it's true?
- Can you restate or say in your own words what your classmate shared?

Lesson structures and activities do not have to conform exactly to what is shown in the guide. Teachers are encouraged to exercise their discretion in using this material in a way that best suits their classes.

Dimensions Math® Scope & Sequence

PKA

Chapter 1
Match, Sort, and Classify

Red and Blue
Yellow and Green
Color Review
Soft and Hard
Rough, Bumpy, and Smooth
Sticky and Grainy
Size — Part 1
Size — Part 2
Sort Into Two Groups
Practice

Chapter 2
Compare Objects

Big and Small
Long and Short
Tall and Short
Heavy and Light
Practice

Chapter 3
Patterns

Movement Patterns
Sound Patterns
Create Patterns
Practice

Chapter 4
Numbers to 5 — Part 1

Count 1 to 5 — Part 1
Count 1 to 5 — Part 2
Count Back
Count On and Back
Count 1 Object
Count 2 Objects
Count Up to 3 Objects
Count Up to 4 Objects
Count Up to 5 Objects
How Many? — Part 1
How Many? — Part 2
How Many Now? — Part 1
How Many Now? — Part 2
Practice

Chapter 5
Numbers to 5 — Part 2

1, 2, 3
1, 2, 3, 4, 5 — Part 1
1, 2, 3, 4, 5 — Part 2
How Many? — Part 1
How Many? — Part 2
How Many Do You See?
How Many Do You See Now?
Practice

Chapter 6
Numbers to 10 — Part 1

0
Count to 10 — Part 1
Count to 10 — Part 2
Count Back
Order Numbers
Count Up to 6 Objects
Count Up to 7 Objects
Count Up to 8 Objects
Count Up to 9 Objects
Count Up to 10 Objects — Part 1
Count Up to 10 Objects — Part 2
How Many?
Practice

Chapter 7
Numbers to 10 — Part 2

6
7
8
9
10
0 to 10
Count and Match — Part 1
Count and Match — Part 2
Practice

PKB

Chapter 8
Ordinal Numbers

First
Second and Third
Fourth and Fifth
Practice

Chapter 9
Shapes and Solids

Cubes, Cylinders, and Spheres
Cubes
Positions
Build with Solids
Rectangles and Circles
Squares
Triangles

Dimensions Math® Scope & Sequence

Count Up to 10 Things — Part 2
Recognize the Numbers 6 to 10
Write the Numbers 6 and 7
Write the Numbers 8, 9, and 10
Write the Numbers 6 to 10
Count and Write the Numbers 1 to 10
Ordinal Positions
One More Than
Practice

Chapter 4
Shapes and Solids

Curved or Flat
Solid Shapes
Closed Shapes
Rectangles
Squares
Circles and Triangles
Where is It?
Hexagons
Sizes and Shapes
Combine Shapes
Graphs
Practice

Chapter 5
Compare Height, Length, Weight, and Capacity

Comparing Height
Comparing Length
Height and Length — Part 1
Height and Length — Part 2
Weight — Part 1
Weight — Part 2
Weight — Part 3
Capacity — Part 1
Capacity — Part 2
Practice

Chapter 6
Comparing Numbers Within 10

Same and More
More and Fewer
More and Less
Practice — Part 1
Practice — Part 2

KB

Chapter 7
Numbers to 20

Ten and Some More
Count Ten and Some More
Two Ways to Count
Numbers 16 to 20
Number Words 0 to 10
Number Words 11 to 15
Number Words 16 to 20
Number Order
1 More Than or Less Than
Practice — Part 1
Practice — Part 2

Chapter 8
Number Bonds

Putting Numbers Together — Part 1
Putting Numbers Together — Part 2
Parts Making a Whole
Look for a Part
Number Bonds for 2, 3, and 4
Number Bonds for 5
Number Bonds for 6
Number Bonds for 7
Number Bonds for 8
Number Bonds for 9
Number Bonds for 10
Practice — Part 1
Practice — Part 2
Practice — Part 3

Chapter 9
Addition

Introduction to Addition — Part 1
Introduction to Addition — Part 2
Introduction to Addition — Part 3
Addition
Count On — Part 1
Count On — Part 2
Add Up to 3 and 4
Add Up to 5 and 6
Add Up to 7 and 8
Add Up to 9 and 10
Addition Practice
Practice

Chapter 10
Subtraction

Take Away to Subtract — Part 1

Dimensions Math® Scope & Sequence

Compare Numbers to 20
Addition
Subtraction
Practice

Chapter 6
Addition to 20

Add by Making 10 — Part 1
Add by Making 10 — Part 2
Add by Making 10 — Part 3
Addition Facts to 20
Practice

Chapter 7
Subtraction Within 20

Subtract from 10 — Part 1
Subtract from 10 — Part 2
Subtract the Ones First
Word Problems
Subtraction Facts Within 20
Practice

Chapter 8
Shapes

Solid and Flat Shapes
Grouping Shapes
Making Shapes
Practice

Chapter 9
Ordinal Numbers

Naming Positions
Word Problems
Practice
Review 2

1B

Chapter 10
Length

Comparing Lengths Directly
Comparing Lengths Indirectly
Comparing Lengths with Units
Practice

Chapter 11
Comparing

Subtraction as Comparison
Making Comparison
Subtraction Stories
Picture Graphs
Practice

Chapter 12
Numbers to 40

Numbers to 40
Tens and Ones
Counting by Tens and Ones
Comparing
Practice

Chapter 13
Addition and Subtraction Within 40

Add Ones
Subtract Ones
Make the Next Ten
Use Addition Facts
Subtract from Tens
Use Subtraction Facts
Add Three Numbers
Practice

Chapter 14
Grouping and Sharing

Adding Equal Groups
Sharing
Grouping
Practice

Chapter 15
Fractions

Halves
Fourths
Practice
Review 3

Chapter 16
Numbers to 100

Numbers to 100
Tens and Ones
Count by Ones or Tens
Compare Numbers to 100
Practice

Chapter 17
Addition and Subtraction Within 100

Add Ones — Part 1
Add Tens
Add Ones — Part 2
Add Tens and Ones — Part 1
Add Tens and Ones — Part 2
Subtract Ones — Part 1
Subtract from Tens
Subtract Ones — Part 2
Subtract Tens

Dimensions Math® Scope & Sequence

Dividing by 5 and 10
Practice C
Word Problems
Review 2

2B

Chapter 8
Mental Calculation

Adding Ones Mentally
Adding Tens Mentally
Making 100
Adding 97, 98, or 99
Practice A
Subtracting Ones Mentally
Subtracting Tens Mentally
Subtracting 97, 98, or 99
Practice B
Practice C

Chapter 9
Multiplication and Division of 3 and 4

The Multiplication Table of 3
Multiplication Facts of 3
Dividing by 3
Practice A
The Multiplication Table of 4
Multiplication Facts of 4
Dividing by 4
Practice B
Practice C

Chapter 10
Money

Making $1
Dollars and Cents
Making Change
Comparing Money
Practice A
Adding Money
Subtracting Money
Practice B

Chapter 11
Fractions

Halves and Fourths
Writing Unit Fractions
Writing Fractions
Fractions that Make 1 Whole
Comparing and Ordering Fractions
Practice
Review 3

Chapter 12
Time

Telling Time
Time Intervals
A.M. and P.M.
Practice

Chapter 13
Capacity

Comparing Capacity
Units of Capacity
Practice

Chapter 14
Graphs

Picture Graphs
Bar Graphs
Practice

Chapter 15
Shapes

Straight and Curved Sides
Polygons
Semicircles and Quarter-circles
Patterns
Solid Shapes
Practice
Review 4
Review 5

3A

Chapter 1
Numbers to 10,000

Numbers to 10,000
Place Value — Part 1
Place Value — Part 2
Comparing Numbers
The Number Line
Practice A
Number Patterns
Rounding to the Nearest Thousand
Rounding to the Nearest Hundred
Rounding to the Nearest Ten
Practice B

Dimensions Math® Scope & Sequence

The Multiplication Table of 9
Multiplying by 8 and 9
Dividing by 8 and 9
Practice B

Chapter 9
Fractions — Part 1

Fractions of a Whole
Fractions on a Number Line
Comparing Fractions with Like Denominators
Comparing Fractions with Like Numerators
Practice

Chapter 10
Fractions — Part 2

Equivalent Fractions
Finding Equivalent Fractions
Simplifying Fractions
Comparing Fractions — Part 1
Comparing Fractions — Part 2
Practice A
Adding and Subtracting Fractions — Part 1
Adding and Subtracting Fractions — Part 2
Practice B

Chapter 11
Measurement

Meters and Centimeters
Subtracting from Meters
Kilometers
Subtracting from Kilometers
Liters and Milliliters
Kilograms and Grams
Word Problems
Practice
Review 3

Chapter 12
Geometry

Circles
Angles
Right Angles
Triangles
Properties of Triangles
Properties of Quadrilaterals
Using a Compass
Practice

Chapter 13
Area and Perimeter

Area
Units of Area
Area of Rectangles
Area of Composite Figures
Practice A
Perimeter
Perimeter of Rectangles
Area and Perimeter
Practice B

Chapter 14
Time

Units of Time
Calculating Time — Part 1
Practice A
Calculating Time — Part 2
Calculating Time — Part 3
Calculating Time — Part 4
Practice B

Chapter 15
Money

Dollars and Cents
Making $10
Adding Money
Subtracting Money
Word Problems
Practice
Review 4
Review 5

4A

Chapter 1
Numbers to One Million

Numbers to 100,000
Numbers to 1,000,000
Number Patterns
Comparing and Ordering Numbers
Rounding 5-Digit Numbers
Rounding 6-Digit Numbers
Calculations and Place Value
Practice

Chapter 2
Addition and Subtraction

Addition
Subtraction
Other Ways to Add and Subtract — Part 1
Other Ways to Add and Subtract — Part 2
Word Problems

Chapter 3
Multiples and Factors

Chapter 4
Multiplication

Chapter 5
Division

Chapter 6
Fractions

Chapter 7
Adding and Subtracting Fractions

Chapter 8
Multiplying a Fraction and a Whole Number

Chapter 9
Line Graphs and Line Plots

4B

Chapter 10
Measurement

Dimensions Math® Scope & Sequence

Chapter 11
Area and Perimeter

Area of Rectangles — Part 1
Area of Rectangles — Part 2
Area of Composite Figures
Perimeter — Part 1
Perimeter — Part 2
Practice

Chapter 12
Decimals

Tenths — Part 1
Tenths — Part 2
Hundredths — Part 1
Hundredths — Part 2
Expressing Decimals as Fractions in Simplest Form
Expressing Fractions as Decimals
Practice A
Comparing and Ordering Decimals
Rounding Decimals
Practice B

Chapter 13
Addition and Subtraction of Decimals

Adding and Subtracting Tenths
Adding Tenths with Regrouping
Subtracting Tenths with Regrouping
Practice A
Adding Hundredths
Subtracting from 1 and 0.1
Subtracting Hundredths
Money, Decimals, and Fractions
Practice B
Review 3

Chapter 14
Multiplication and Division of Decimals

Multiplying Tenths and Hundredths
Multiplying Decimals by a Whole Number — Part 1
Multiplying Decimals by a Whole Number — Part 2
Practice A
Dividing Tenths and Hundredths
Dividing Decimals by a Whole Number — Part 1
Dividing Decimals by a Whole Number — Part 2
Dividing Decimals by a Whole Number — Part 3
Practice B

Chapter 15
Angles

The Size of Angles
Measuring Angles
Drawing Angles
Adding and Subtracting Angles
Reflex Angles
Practice

Chapter 16
Lines and Shapes

Perpendicular Lines
Parallel Lines
Drawing Perpendicular and Parallel Lines
Quadrilaterals
Lines of Symmetry
Symmetrical Figures and Patterns
Practice

Chapter 17
Properties of Cuboids

Cuboids
Nets of Cuboids
Faces and Edges of Cuboids
Practice
Review 4
Review 5

5A

Chapter 1
Whole Numbers

Numbers to One Billion
Multiplying by 10, 100, and 1,000
Dividing by 10, 100, and 1,000
Multiplying by Tens, Hundreds, and Thousands
Dividing by Tens, Hundreds, and Thousands
Practice

Chapter 2
Writing and Evaluating Expressions

Expressions with Parentheses
Order of Operations — Part 1
Order of Operations — Part 2

Dimensions Math® Scope & Sequence

Conversion of Measures
Mental Calculation
Practice B

Chapter 10
The Four Operations of Decimals

Adding Decimals to Thousandths
Subtracting Decimals
Multiplying by 0.1 or 0.01
Multiplying by a Decimal
Practice A
Dividing by a Whole Number — Part 1
Dividing by a Whole Number — Part 2
Dividing a Whole Number by 0.1 and 0.01
Dividing a Whole Number by a Decimal
Practice B

Chapter 11
Geometry

Measuring Angles
Angles and Lines
Classifying Triangles
The Sum of the Angles in a Triangle
The Exterior Angle of a Triangle
Classifying Quadrilaterals
Angles of Quadrilaterals — Part 1
Angles of Quadrilaterals — Part 1
Drawing Triangles and Quadrilaterals
Practice

Chapter 12
Data Analysis and Graphs

Average — Part 1
Average — Part 2
Line Plots
Coordinate Graphs
Line Graphs
Practice

Chapter 13
Ratio

Finding Ratios
Equivalent Ratios
Finding a Quantity
Comparing Three Quantities
Word Problems
Practice

Chapter 14
Rate

Unit Rate
Finding the Total Amount Given the Rate
Finding the Number of Units Given the Rate
Word Problems
Practice

Chapter 15
Percentage

Meaning of Percentage
Writing Percentages as Fractions in Simplest Form
Writing Decimals as Percentages
Writing Fractions as Percentages
Practice A
Percentage of a Quantity
Word Problems
Practice B
Review 3

Suggested number of class periods: 11–12

	Lesson	Page	Resources	Objectives
	Chapter Opener	p. 5	TB: p. 1	
1	Ten and Some More	p. 6	TB: p. 2 WB: p. 1	Identify groups containing 10 objects. Count 10 objects within groups containing more than 10 objects.
2	Count Ten and Some More	p. 8	TB: p. 4 WB: p. 3	Count the number of objects in sets containing between 10 and 20 objects.
3	Two Ways to Count	p. 10	TB: p. 6 WB: p. 5	Identify and write the number of tens and ones in a set of 10 to 15 objects.
4	Numbers 16 to 20	p. 12	TB: p. 8 WB: p. 7	Identify and write the number of tens and ones in a set of 16 to 20 objects.
5	Number Words 0 to 10	p. 14	TB: p. 10 WB: p. 9	Read and write number words 0 to 10.
6	Number Words 11 to 15	p. 16	TB: p. 13 WB: p. 11	Read number words 11 to 15.
7	Number Words 16 to 20	p. 18	TB: p. 15 WB: p. 13	Read number words 16 to 20.
8	Number Order	p. 20	TB: p. 17 WB: p. 15	Order numbers from 11 to 20.
9	1 More Than or Less Than	p. 22	TB: p. 19 WB: p. 17	Identify numbers 1 more than and 1 less than a given number within 20.
10	Practice — Part 1	p. 24	TB: p. 21 WB: p. 19	Practice concepts from the chapter.
11	Practice — Part 2	p. 26	TB: p. 25 WB: p. 21	Practice concepts from the chapter.
	Workbook Solutions	p. 28		

In **Dimensions Math® Kindergarten A**, students focused on the numbers 0 to 10. This chapter extends their understanding to numbers between 10 and 20 with special emphasis on the teen numbers. While students may be able to count by rote to 20 and beyond, this chapter emphasizes number sense and understanding of 10 and some more.

Our system for numeration is a base-10, or decimal system. In a base-10 system, there are 10 digits, 0 to 9, and the position of a number is based on powers of ten (ones, tens, hundreds, thousands, etc.). To write the number that is one more than 9, we write a 1 in the tens place and a 0 in the ones place, indicating there is one group of ten and no ones. The next number, 11, means 1 ten and 1 one. A digit in the tens place indicates a multiple of ten. For example, 35 means the quantity counted is 3 groups of ten, and 5 ones.

This system allows for easy computation of large numbers by applying the same strategies students will learn for numbers to 20 to digits of greater place values. 8 + 9 can be computed the same way whether it is 8 ones + 9 ones or 8 thousands + 9 thousands.

In this chapter, students will learn to count and write numerals from 10 to 20 by focusing on the idea of tens and ones to relate the numerical representation of the number to the quantity. The English number words for teen numbers, e.g., thirteen, can cause some students to write 31 for "thirteen" even into second grade. For that reason, students will begin counting past 10 using words that emphasize place value. The quantity 1 more than 10 will be counted as "ten and one" before using the standard word "eleven."

Standard number language will be introduced in **Lesson 3: Two Ways to Count**.

For students to be successful in counting beyond ten, they need to have moved beyond the strategy of counting all of the items. Counting ten and some more is a strategy that helps students who are not yet proficient at counting on.

Although it is not included in this chapter, students may like to expand their Book of Numbers created in **Dimensions Math® Kindergarten A Chapter 2: Numbers to 5** and **Chapter 3: Numbers to 10**.

It is assumed that all students will have access to recording tools. When a lesson refers to a whiteboard, any writing materials can be used.

Materials

- Chalk
- Counters
- Crayons, markers, paint daubers, or stickers
- Dice
- Dominoes
- Finger paint
- Gelatin or pudding
- Linking cubes
- Painter's tape
- Paper plates
- Sand, rice, or salt
- Sandpaper
- Scrabble letters
- Shallow pan
- Shaving cream
- Small objects such as beads, crayons, pasta, shells, or small erasers
- Whiteboards
- Wikki Stix
- Wooden blocks

Tactile methods of writing numerals

- Sandpaper: Cut numerals out of sandpaper with a die-cut machine or by hand and glue them to cardstock. Students can either cover the numerals with paper and make numeral rubbings or just trace the numerals with their fingers.
- Whiteboards: Have students write the numerals, then use their index fingers to erase.
- Writing in a shallow pan:
  - Filled with sand, rice, or salt
  - Shaving cream writing
  - Gelatin or pudding
- Finger paint writing
- Wikki Stix

Blackline Masters

- 10 and More Number Word Cards
- Blank Double Ten-frames
- Blank Ten-frame
- Double Ten-frame Cards
- Grid
- I Have, Who Has? Cards
- Letter Cards
- Number Cards — Large
- Number Cards
- Number Grid Puzzle
- Number Word Cards
- Ten-frame Cards
- Tens and Ones Worksheet

Storybooks

- *More, Fewer, Less* by Tana Hoban
- *Piglets Playing: Counting from 11 to 20* by Megan Atwood
- *Twenty Big Trucks in the Middle of the Street* by Mark Lee
- *Math for All Seasons* by Greg Tang
- *1 to 20, Animals Aplenty* by Katie Viggers
- *Ten Apples Up on Top!* by Dr. Suess

Letters Home

- Chapter 7 Letter

Notes

This page can be used to assess students' prior knowledge of numbers to 20 by rote counting.

Have students look at page 1.

Ask students what they notice about the caterpillar. They might see that:

- The body has a pattern: green, lighter green, green, lighter green.
- Some circles have 1 number and some have 2 numbers.
- The numbers go in order from 1 to 20, just like we count.

Ask a student to choose a circle on the caterpillar. If she chooses 14, lead her to count as she "hops" her fingers to 14.

Continue asking students to choose circles. Students can clap and count, march and count, or touch toes while they rote count.

If time allows, expand on the **Chapter Opener** by going directly to Lesson 1, which starts on the following page.

Extend

★ 20 Wins!

Players take turns counting from 1. On each turn, a player counts on from the previous player's last number. Players can say 1, 2, or 3 more numbers on their turn. The player that says 20 is the winner.

Example play:

Player 1 counts: 1, 2, 3
Player 2 counts on: 4
Player 1 continues: 5, 6
Player 2 says: 7, 8, 9

The game continues until a player says 20.

Students could use markers and a number chart to 20 as an alternative. Repeat the game multiple times until players can come up with a clear strategy.

Lesson 1 Ten and Some More

Objectives

- Identify groups containing 10 objects.
- Count 10 objects within groups containing more than 10 objects.

Lesson Materials

- Counters, 8 to 15 per student
- Blank Ten-frame (BLM), 1 per student

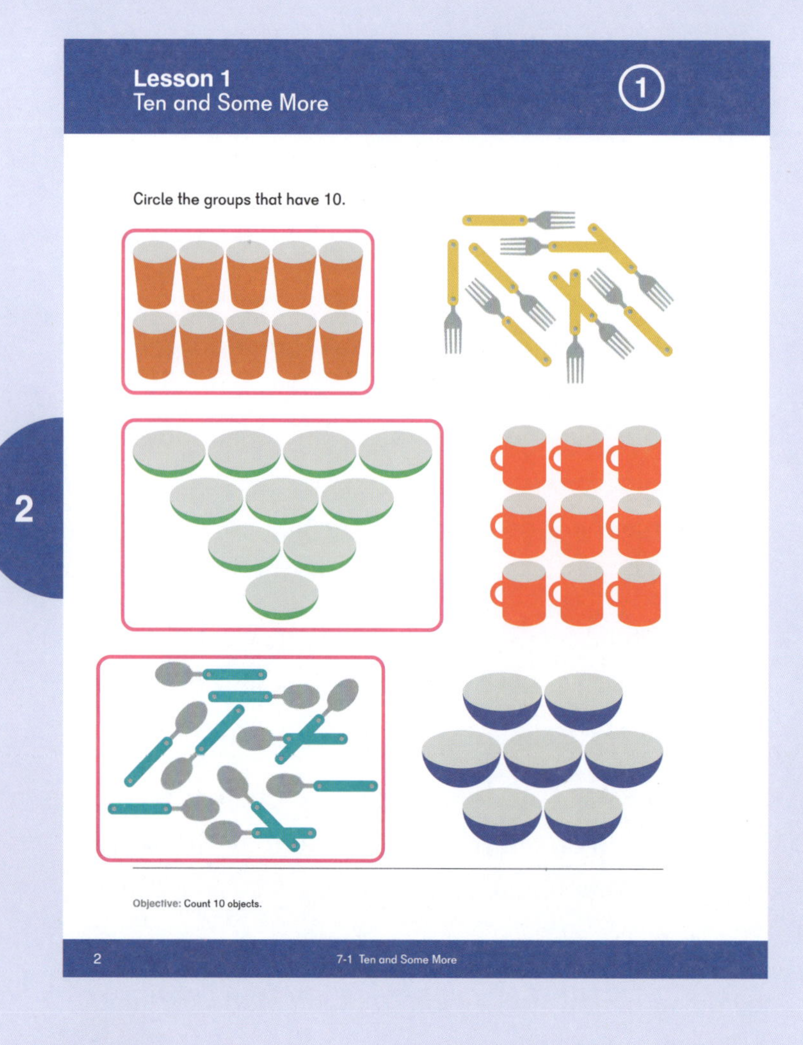

Explore

Provide students with between 8 and 15 counters and ask them if they have more or fewer than 10 counters.

Observe how students are counting.

- **Counting by touching each counter with a finger:** Students touching each counter should be encouraged to move counters to one side as they count.
- **Counting by moving counters to the side:** Students moving counters in this manner should be encouraged to put counters into small groups.
- **Counting by making groups:** Students should be encouraged to recognize when they have 10 counters in a group.
- **Counting by arranging the counters in ten-frame format**

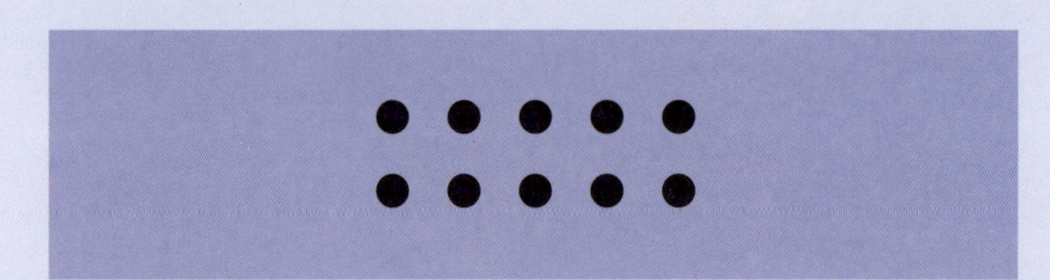

Provide students with a Blank Ten-frame (BLM) and have students use it to organize their counters. They should be able to easily see that when the ten-frame is full, they have 10 counters. Ask students if they have fewer than 10, more than 10, or exactly 10 counters.

Learn

Pages 2–3 can be used to lead students in a "Look and Talk" exercise. Have students represent the quantity of each object on the page with counters on their ten-frames to determine which groups have 10.

Small Group Activities

Textbook Pages 2–3. On page 2, discuss the sets of objects. Students should circle the groups that have 10:

- Glasses
- Green bowls
- Spoons

On page 3, students could cross off 10 and notice there are 10 and some more of each item.

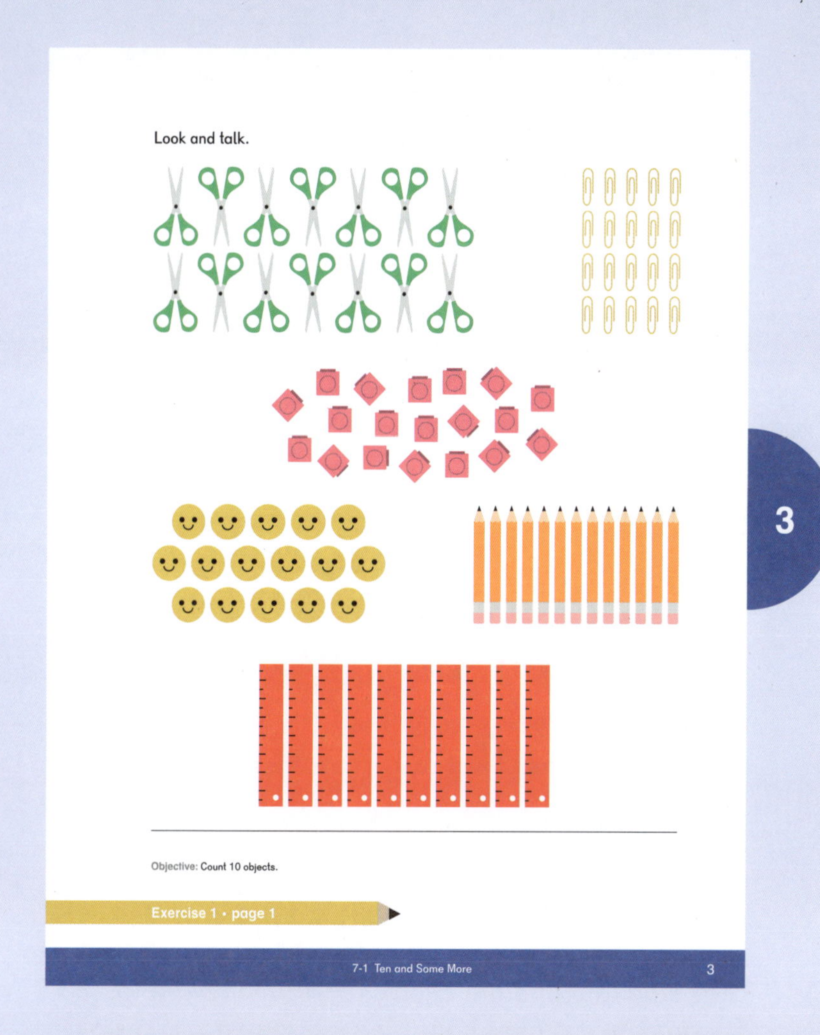

▲ How Tall is the Tower?

Materials: 3–5 wooden blocks and 20 linking cubes per pair of students (a rod of 10 in one color and 10 additional loose cubes)

Students work with partners and take turns building a tower using 3 to 5 wooden building blocks. (The wooden block tower should not exceed the height of 20 linking cubes.) Player 1 builds and Player 2 estimates the height in linking cubes. Player 1 then measures using a rod of 10 linking cubes of one color and some more of another color. If the estimate is correct, the player scores a point.

▲ Make Tens

Materials: Linking cubes

Organize linking cubes for future lessons in this book by having students put ten cubes of the same color into a ten-rod.

Exercise 1 • page 1

Extend

★ Ten-Frame Nim

Materials: Blank Ten-Frame (BLM), 10 counters per pair of students

The classic 2-player game of Nim modified for younger learners can be played throughout the year.

On each turn, players place 1, 2, or 3 counters on the open squares of a Blank Ten-frame (BLM). The first player who places the last counter to fill up the ten-frame wins. Encourage students to play the game multiple times to look for a pattern or strategy for winning.

Objective

- Count the number of objects in sets containing between 10 and 20 objects.

Lesson Materials

- Per student or pair of students:
  - 10 to 20 Counters
  - Blank Ten-frame (BLM)

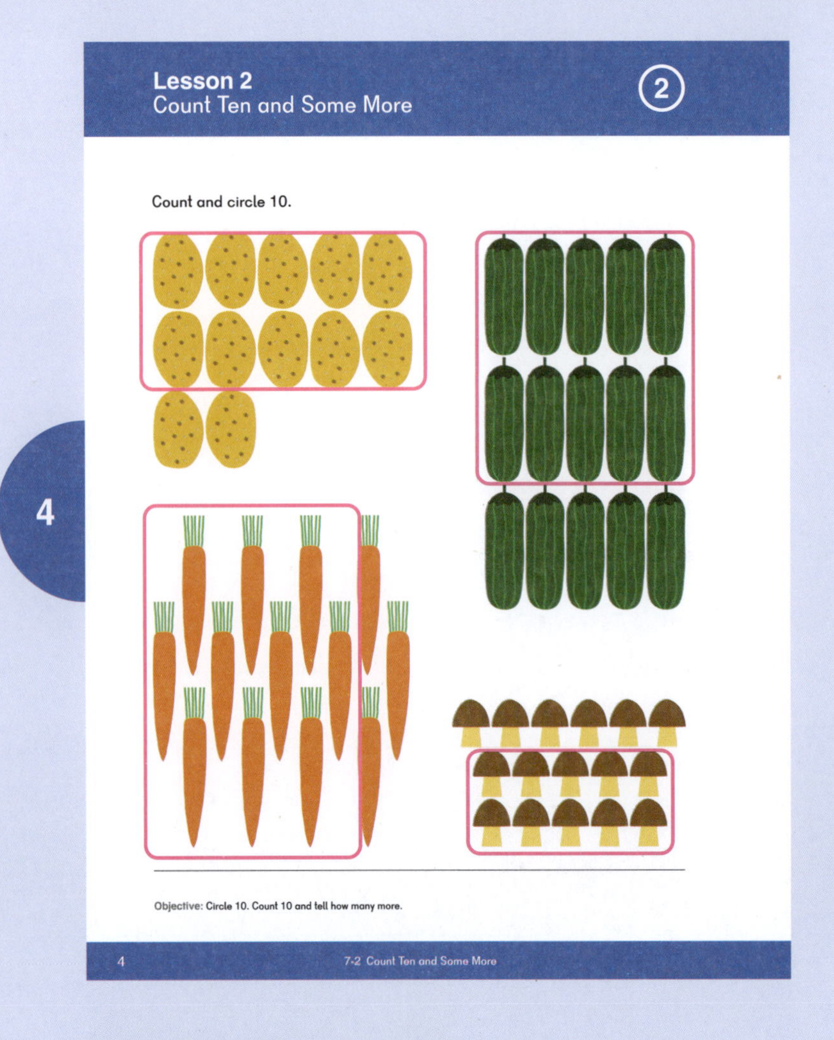

Explore

Provide students or pairs of students with between 10 and 20 counters and ask them to count them. Give each pair of students one Blank Ten-frame (BLM), and have each pair begin by filling one. They should notice that there are more than 10 counters.

Ask students what might be helpful to organize the counters that did not fit. They may suggest using a second Blank Ten-frame (BLM) to organize.

Give students another Blank Ten-frame (BLM) to organize the counters that did not fit in the first ten-frame. Discuss the "ten and some more" structure.

Learn

Look and talk about the illustrations on page 4. Ask how we can find the number of potatoes.

Have students put counters on each of the potatoes and then transfer the counters to the Blank Ten-frame (BLM) to count how many potatoes there are in all (10 and 2).

Repeat with the cucumbers. Ask students if they can determine the numbers of carrots and mushrooms without transferring counters from the page.

Ask if some groups of vegetables are easier to count than others, and have students explain their thinking.

Whole Group Activity

▲ Teen Frame Flash

Materials: Double Ten-frame Cards (BLM)

Hold up a Double Ten-frame Card (BLM) showing a number from the range specified in the lesson. Students will call out the number shown, "10 and 1," or, "11", etc.

Small Group Activities

Textbook Pages 4–5

▲ Double Ten-frame Cards

Materials: 10 Blank Double Ten-frames (BLM) per student, and crayons, markers, paint daubers, or stickers

Have students create a set of cards with numbers from 11 to 20. Students can color in the correct number on each card, or they could use stickers or a paint dauber to create a set.

▲ Match

Materials: 2 sets of Ten-frame Cards (BLM) 11 to 20

Students arrange the cards faceup in a grid. Students take turns finding two cards that go together.

★ Memory

Materials: 2 sets of Ten-frame Cards (BLM) 11 to 20

Students arrange the cards facedown in a grid. Students take turns finding two cards that go together.

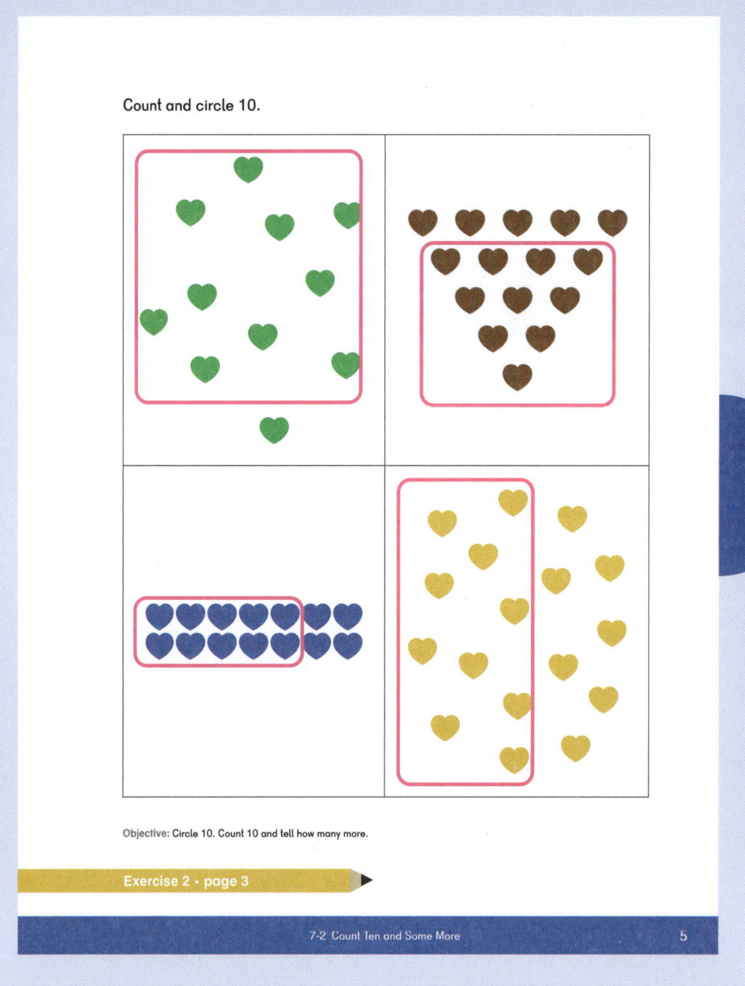

5

Exercise 2 • page 3

Extend

★ High Ten!

Students work with partners. Each player holds out 5 fingers on one hand and 0 to 5 fingers on the other hand.

Players clap the two hands showing 5 (as though giving each other a high five) and say "10," then count the remaining fingers. The first player to say the correct total number of fingers on both hands wins the round.

Lesson 3 Two Ways to Count

Objective

- Identify and write the number of tens and ones in a set of 10 to 15 objects.

Lesson Materials

- Wooden blocks, up to 15 per pair of students
- Counters, 15 per student
- Blank Double Ten-frames (BLM)

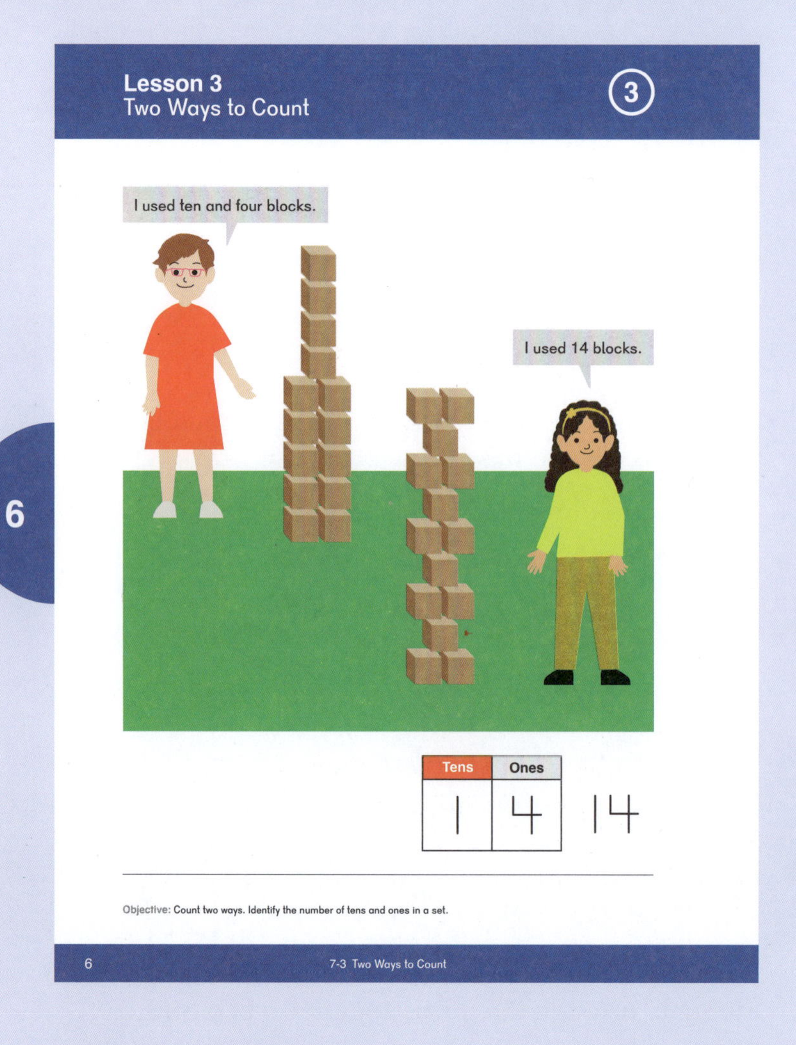

Explore

Have pairs of students make towers of up to 15 blocks. After their structures are complete, have students use counters to show the amount of blocks used on a Blank Double Ten-frame (BLM). Ask students to tell the number of blocks used. Some may say, "Ten and some more" and some may use the number word.

Learn

Have students look at page 6. Ask students to use counters on a Blank Double Ten-frame (BLM) to show the number of blocks each friend has. Ask students to notice how Sofia and Emma have organized their blocks. Emma has organized her blocks into a group of 10 and 4 more.

Have students see that a complete ten-frame doesn't need to be counted, but can be called a ten. The leftover blocks are called ones. We can write that number as 1 ten and 4 ones. 1 ten and 4 ones is the same as 14. Look at and explain how the place value box on page 6 can help us write numbers higher than 9.

Whole Group Activities

▲ Teen Frame Flash

Materials: Double Ten-frame Cards (BLM) 10 to 15

Hold up a Double Ten-frame Card (BLM) showing a number from 10 to 15. Students will call out the number shown, "10 and 1," or, "11", etc.

▲ Magic Thumb

Pointing your thumb up or down, have students chorally count up and down within 15 by ones. For example, start out by saying, "Let's count by ones starting at 10. First number?" The class responds, "11." Then, point your thumb up, and the class responds, "12." Then point your thumb down, and the class responds, "11." Point down again, and the class responds, "10," and so on.

▲ Say and Write

Students can practice saying the number words and writing the numerals from 11 to 15 on whiteboards.

- Write a number on the board and have students say the number.
- Say a number and have students write the number on a whiteboard. For example, say, "Write the number that is 10 and 2." Students write, "12."

Small Group Activities

Textbook Page 7. Have students look at the first problem and say, "10 and 1 more makes 11" as they write the numbers in the place value chart. Repeat with the other problems on the page.

▲ Tactile Writing

Materials: Select from the list below:

- Sandpaper: Cut numerals out of sandpaper with a die-cut machine or by hand and glue them to cardstock. Students can either cover the numerals with paper and make numeral rubbings or just trace the numerals with their fingers.
- Whiteboards: Have students write the numerals, then use their index fingers to erase.
- Writing in a shallow pan:
    - Filled with sand, rice, or salt
    - Shaving cream writing
    - Gelatin or pudding
- Finger paint writing
- Wikki Stix

Have students write with their fingers or create the numbers 11 to 20 out of various materials.

▲ Group 10 and Some More

Materials: Tens and Ones Worksheet (BLM), and groups of 10–15 small objects such as beads, crayons, pasta, shells, or small erasers

Provide students with objects to count. Have them group objects in tens and some more, and record the number on the Tens and Ones Worksheet (BLM).

Exercise 3 • page 5

Extend

★ Ten and More Face-Off

Materials: Deck of Ten-frame Cards (BLM) 0 to 10 and Ten-frame Card (BLM) showing 10 for each player

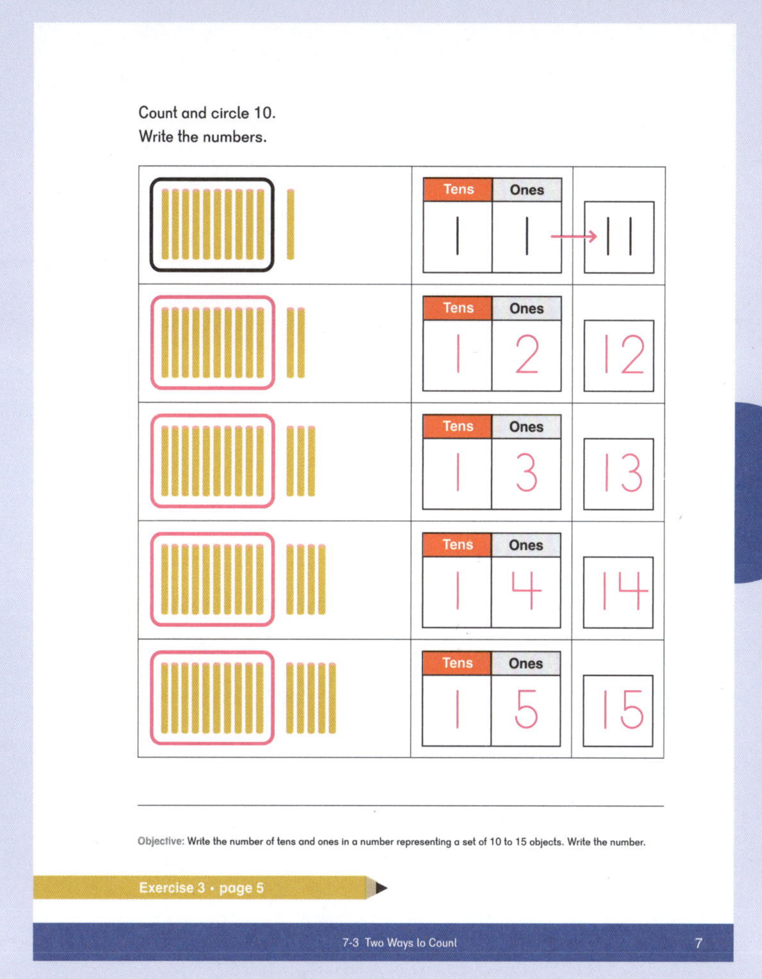

Play in pairs or groups of up to 4 players. Each player receives a full Ten-frame Card (BLM) and places it faceup in front of herself. This 10 card becomes one of the addends in each face off.

Players mix their remaining cards and then turn over the top card of the deck. They say ten and their new numbers. For example, for these cards:

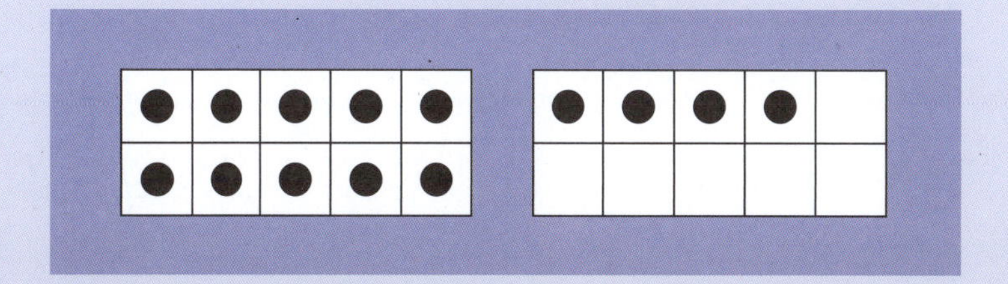

Students say, "Ten and four," or, "Fourteen."

The player with the greatest number wins all of the non-ten cards. All players retain their original 10 cards for their next face off. The player with the most cards at the end wins.

Objective

- Identify and write the number of tens and ones in a set of 16 to 20 objects.

Lesson Materials

- Double Ten-frame Cards (BLM) 11 to 20, 1 per pair of students

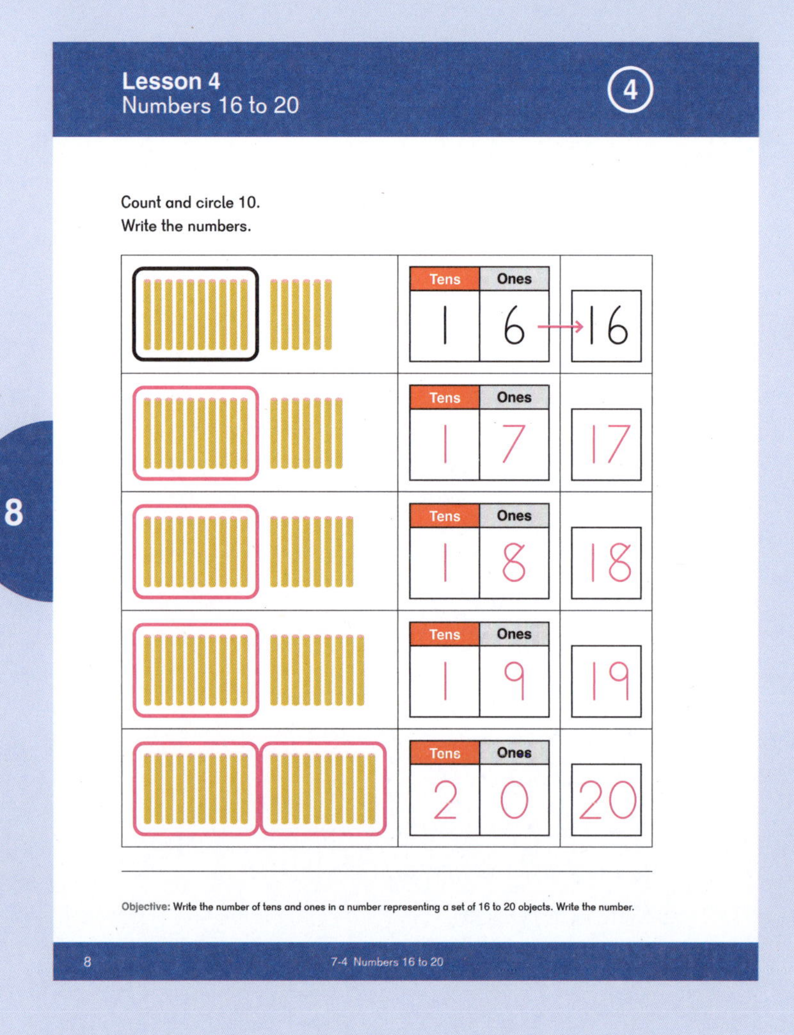
Lesson 4
Numbers 16 to 20

4

Count and circle 10.
Write the numbers.

	Tens	Ones	
	1	6	16
	1	7	17
	1	8	18
	1	9	19
	2	0	20

Objective: Write the number of tens and ones in a number representing a set of 16 to 20 objects. Write the number.

8 7-4 Numbers 16 to 20

8

Explore

Provide pairs of students with Double Ten-frame Cards (BLM) 11 to 20 and have them put them in order. Discuss the two terms used in the previous lesson and how students could name these ten-frame cards similarly. For 16, we could say, "Sixteen," or, "Ten and six."

Learn

Students can practice saying and writing the numbers from 16 to 20 on whiteboards.

- Write a number on the board and have students say the number both ways: "10 and 7" and "seventeen."
- Say a number and have students write the number on a whiteboard. For example, say, "Write the number that is 10 and 9." Students write, "19."

Whole Group Activities

▲ Put Yourself on the Line

Materials: Number Cards — Large (BLM) 0 to 20

Shuffle cards and pass them out so that each student has one card. Invite students to stand up and put themselves in order in front of the class without talking. Repeat as needed, so all students get a chance to participate.

▲ Put Yourself on the Line: Team

Materials: Number Cards — Large (BLM) 0 to 20

Create teams of up to 5 students. Give each student a random card between 11 to 20, so that each team member has a different number. The members in each team put themselves in order from least to greatest, or vice versa. The team that gets itself in the correct order first wins.

Small Group Activities

Textbook Pages 8–9

▲ Tactile Writing

Materials: Select from the list below:

- Sandpaper: Cut numerals out of sandpaper with a die-cut machine or by hand and glue them to cardstock. Students can either cover the numerals with paper and make numeral rubbings or just trace the numerals with their fingers.
- Whiteboards: Have students write the numerals, then use their index fingers to erase.
- Writing in a shallow pan:
  - Filled with sand, rice, or salt
  - Shaving cream writing
  - Gelatin or pudding
- Finger paint writing
- Wikki Stix

Have students write with their fingers or create the numbers 11 to 20 out of various materials.

▲ Group 10

Materials: Blank Double Ten-frame (BLM), Number Cards (BLM) 11 to 20, small objects (up to 20 of each)

Provide each student with Number Cards (BLM) 11 to 20 and a Blank Double Ten-frame (BLM). Students draw a Number Card (BLM) and count out the corresponding number of objects (beads, crayons, pasta, shells, small erasers), organizing them on the Blank Double Ten-frame (BLM).

Exercise 4 • page 7

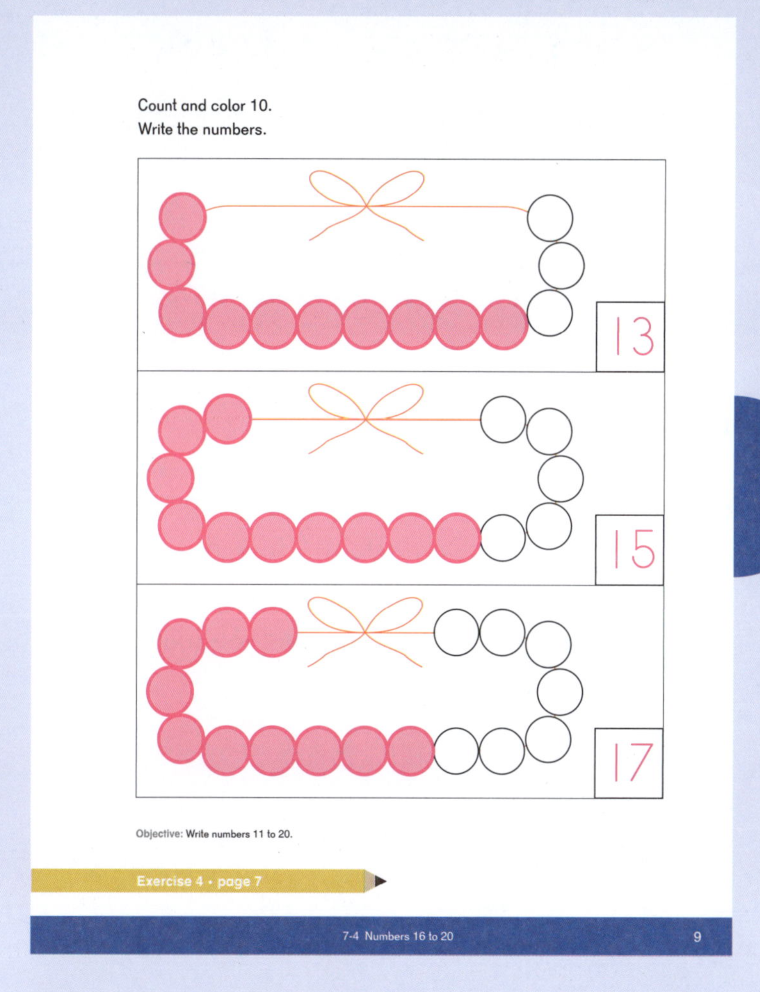

9

Extend

★ Claps and Taps

Materials: Number Cards (BLM) 11 to 19

Students play in small groups. One player draws a Number Card (BLM), reads it, then claps the number of tens and taps the number of ones on the card.

For example, if the number 15 is chosen, the player claps 1 time and taps 5 times. The other players listen and guess the number.

Objective

- Read and write number words 0 to 10.

Lesson Materials

- Ten-frame Cards (BLM) 0 to 10
- Number Cards (BLM) 0 to 10
- Number Word Cards (BLM) 0 to 10

10

Lesson 5
Numbers Words 0 to 10

5

Objective: Match numerals 0 to 10 to number words.

10 7-5 Number Words 0 to 10

Explore

Provide each student with either a Ten-frame Card (BLM) or a Number Card (BLM). Have them find another student holding a card with a matching quantity. Students can line up with their match in order from greatest to least.

Learn

Introduce the Number Word Cards (BLM), reading each aloud.

Mix in the Number Word Cards (BLM) with the Ten-frame Cards (BLM) and Number Cards (BLM). Pass out cards to students. Have students mingle and find other cards that represent matching quantities. Remove some of the Ten-frame Cards (BLM) if there are too many cards.

Whole Group Activity

▲ Group Up

Materials: Number Word Cards (BLM) 1 to 10

Hold up a Number Word Card (BLM) and have students arrange themselves in groups of that number.

Based on the number word shown, not all students will always be in a group, so repeat the activity several times to include all students.

Small Group Activities

Textbook Pages 11–12

▲ Domino Match

Materials: Dominoes with a total of 10 pips, Number Word Cards (BLM) 0 to 10

Have students lay out the Number Word Cards (BLM), then sort and place dominoes in groups by the number word.

▲ Match

Materials: Use two or all three types of the following cards: Ten-frame Cards (BLM) 0 to 10, Number Cards (BLM) 0 to 10, Number Word Cards (BLM) 0 to 10

Students arrange the cards faceup in a grid. Students take turns finding two cards that go together.

★ Memory

Materials: Use two or all three types of the following cards: Ten-frame Cards (BLM) 0 to 10, Number Cards (BLM) 0 to 10, Number Word Cards (BLM) 0 to 10

Students arrange the cards facedown in a grid. Students take turns finding two cards that go together.

Exercise 5 • page 9

Extend

★ Find the Letters

Materials: Number Cards (BLM) 0 to 10, Number Word Cards (BLM) 0 to 10, Letter Cards (BLM) or Scrabble letters

Students find a matching Number Word Card (BLM) and Number Card (BLM). They then find the Letter Cards (BLM) needed and put them in order to spell the number word.

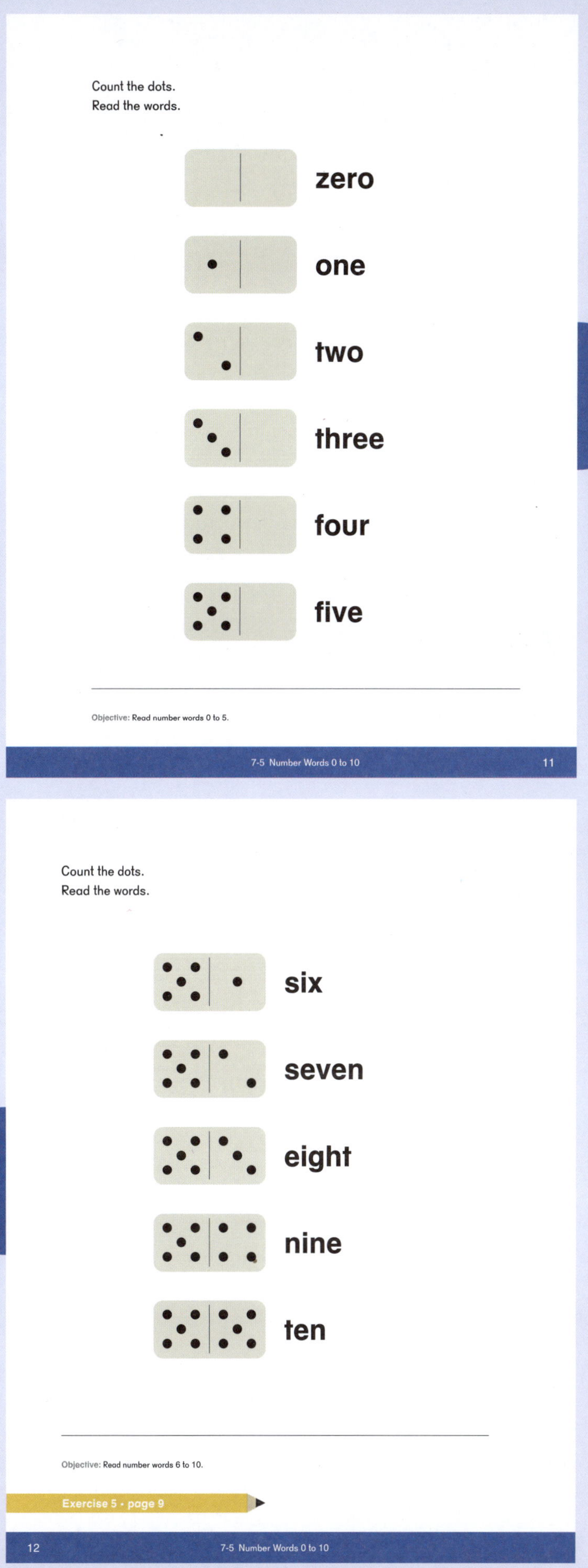

Lesson 6 Number Words 11 to 15

Objective

- Read number words 11 to 15.

Lesson Materials

- Double Ten-frame Cards (BLM) 11 to 15
- Number Cards (BLM) 11 to 15
- Number Word Cards (BLM) 11 to 15
- 10 and More Number Word Cards (BLM) 11 to 15
- Ten-frame Card (BLM) 10, 1 per student

Explore

Provide students with either a Double Ten-frame Card (BLM) or a Number Card (BLM) 11 to 15. Similar to the previous lesson, have students mingle and find another student with a card with a matching quantity. Students can line up with their match in order from greatest to least.

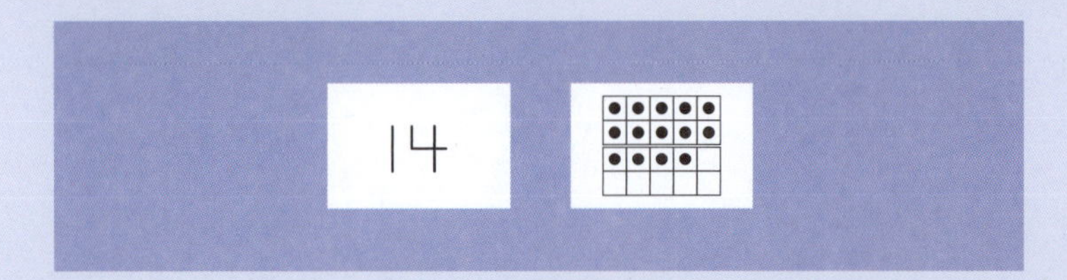

Learn

Look and talk about page 13. Introduce the 10 and More Number Word Cards (BLM) and Number Word Cards (BLM) for the numbers 11 to 15, reading each aloud.

Add the Number Word Cards (BLM) to the Double Ten-frame Cards (BLM) and number cards, and repeat the activity in **Explore** with a mix of all three types of cards. Remove some of the Double Ten-frame Cards (BLM) if there are too many cards.

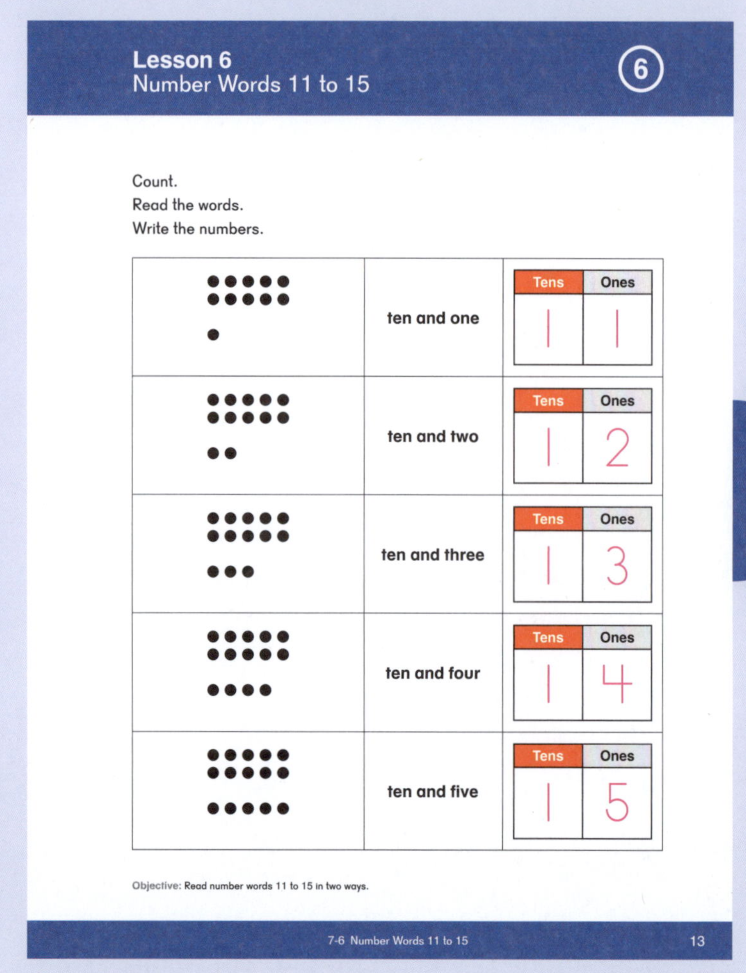

Lesson 6
Number Words 11 to 15 (6)

Count.
Read the words.
Write the numbers.

Count	Words	Tens	Ones
●●●●● ●●●●● ●	ten and one	1	1
●●●●● ●●●●● ●●	ten and two	1	2
●●●●● ●●●●● ●●●	ten and three	1	3
●●●●● ●●●●● ●●●●	ten and four	1	4
●●●●● ●●●●● ●●●●●	ten and five	1	5

Objective: Read number words 11 to 15 in two ways.

7-6 Number Words 11 to 15 13

Whole Group Activity

▲ Flash

Materials: 10 and More Number Word Cards (BLM) 11 to 15, Number Word Cards (BLM) 11 to 15

Show students a 10 and More Number Word Card (BLM) or Number Word Card (BLM), and have them read or say the number word. Alternatively, students can write the number on personal whiteboards.

Small Group Activities

Textbook Pages 13–14

▲ Domino Match

Materials: Dominoes with a total of 15 pips, Number Word Cards (BLM) 0 to 15

Have students lay out the Number Word Cards (BLM), then sort and place dominoes in groups by the number word.

▲ Roll and Race

Materials: Grid (BLM) labeled 10 to 15, Ten-frame Card (BLM) 10, 50 counters per student, die with modified sides: 0, 1, 2, 3, 4, 5

Give each student a complete set of materials as listed above. Have students roll the die and add that number to the 10 on the Ten-frame Card (BLM). Students determine the sum of their die roll and Ten-frame Card (BLM), and place a counter in the lowest possible box on the Grid (BLM) in the column labeled with that number. Play continues until one bar reaches the top of the graph.

▲ Match

Materials: Use two or all three types of the following cards: Ten-frame Cards (BLM) 0 to 15, Number Cards (BLM) 0 to 15, Number Word Cards (BLM) 0 to 15

Students arrange the cards faceup in a grid. Students take turns finding two cards that go together.

★ Memory

Materials: Use two or all three types of the following cards: Ten-frame Cards (BLM) 0 to 15, Number Cards (BLM) 0 to 15, Number Word Cards (BLM) 0 to 15

Students arrange the cards facedown in a grid. Students take turns finding two cards that go together.

Exercise 6 • page 11

14

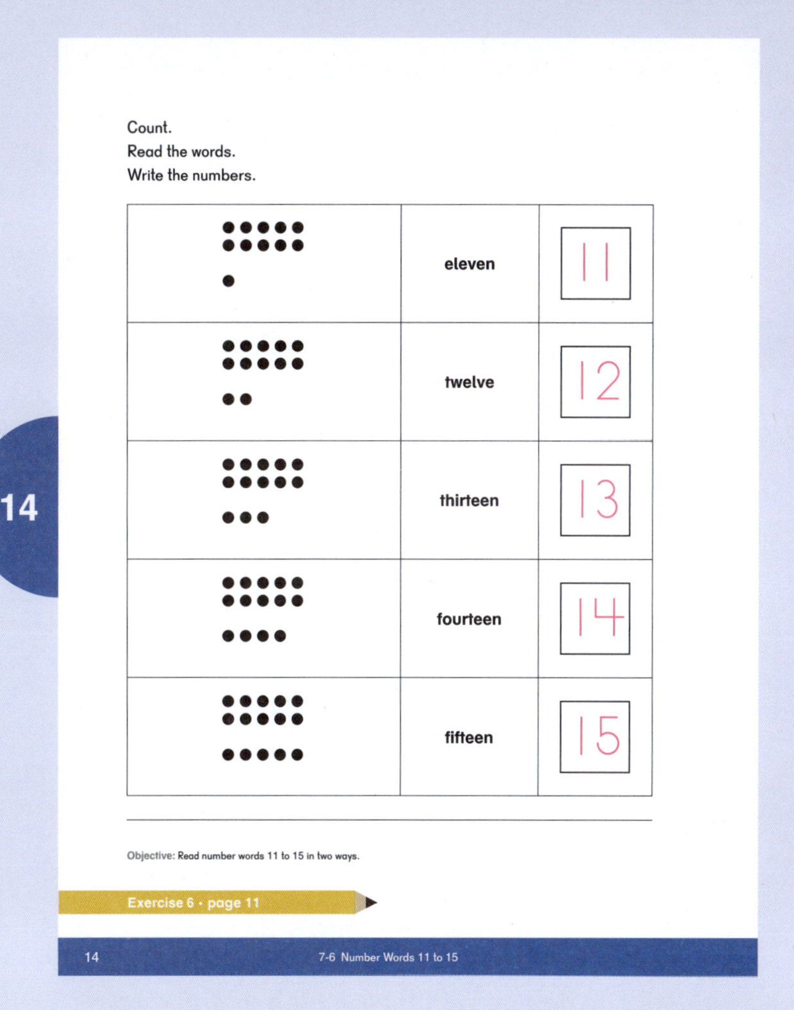
Count.
Read the words.
Write the numbers.

●●●●● ●●●●● ●	eleven	11
●●●●● ●●●●● ●●	twelve	12
●●●●● ●●●●● ●●●	thirteen	13
●●●●● ●●●●● ●●●●	fourteen	14
●●●●● ●●●●● ●●●●●	fifteen	15

Objective: Read number words 11 to 15 in two ways.

Exercise 6 • page 11

14 7-6 Number Words 11 to 15

Extend

★ How Long or Tall?

Materials: Linking cubes (rods of 10 and singles)

Ask students to measure objects around the room using linking cubes and record the length or height in terms of ten and some more. Students may discover that 2 tens is 20 and 3 tens is 30 and extend their understanding of a ten and some more to write numbers greater than 20. Use objects shorter than 20 linking cubes long for students not counting beyond 20.

Objective

- Read number words 16 to 20.

Lesson Materials

- 10 and More Number Word Cards
- Double Ten-frame Cards (BLM) 16 to 20
- Number Cards (BLM) 16 to 20
- Number Word Cards (BLM) 16 to 20

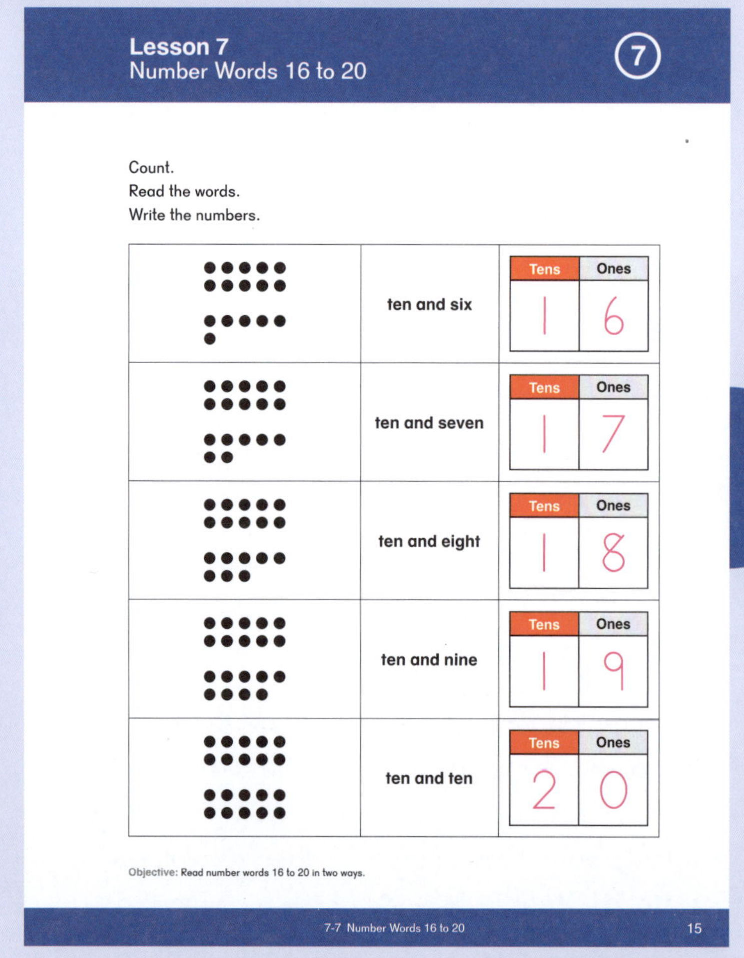

Lesson 7
Number Words 16 to 20 ⑦

Count.
Read the words.
Write the numbers.

		Tens	Ones
	ten and six	1	6
	ten and seven	1	7
	ten and eight	1	8
	ten and nine	1	9
	ten and ten	2	0

Objective: Read number words 16 to 20 in two ways.

7-7 Number Words 16 to 20 15

Explore

Similar to previous lessons, provide students with either a Double Ten-frame Card (BLM) 16 to 20 or a Number Card (BLM) 16 to 20. Have them find another student holding a card with a matching quantity.

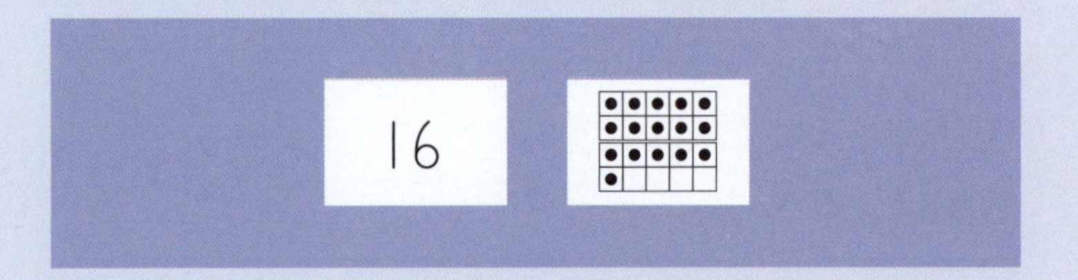

Students can line up with their match in order from greatest to least.

Learn

Introduce the 10 and More Number Word Cards (BLM) and Number Word Cards (BLM) for the numbers 16 to 20, reading each aloud.

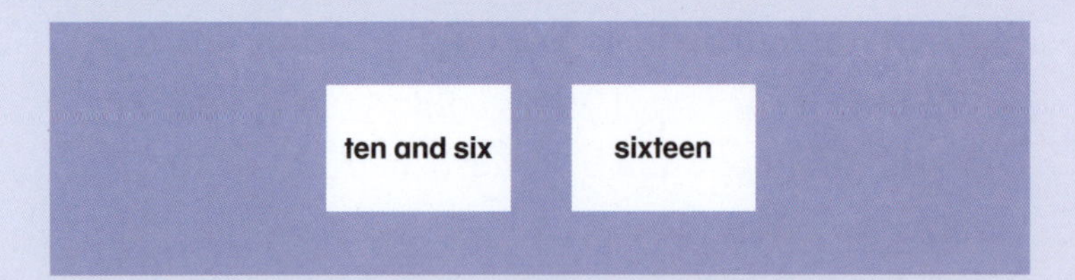

Add the Number Word Cards (BLM) to the sets of Double Ten-frame Cards (BLM) and Number Cards (BLM), and repeat the activity from **Explore** with a mix of all three types of cards. Remove some of the Double Ten-frame Cards (BLM) if there are too many cards for the class to use.

Whole Group Activity

▲ Flash

Materials: 10 and More Number Word Cards (BLM) 16 to 20, Number Word Cards (BLM) 16 to 20

Show students a 10 and More Number Word Card (BLM) or Number Word Card (BLM), and have them read or say the number word. Alternatively, students can write the number on personal whiteboards.

Small Group Activities

Textbook Pages 15–16

▲ Domino Sort

Materials: Set of double 9 dominoes, Number Word Cards (BLM) 0 to 20, Double Ten-frame Cards (BLM), Number Cards (BLM)

Provide students with dominoes and Number Word Cards (BLM). Students work in small groups to sort the dominoes by number word. Extend by asking students to sort Double Ten-frame Cards (BLM) and Number Cards (BLM), as well.

▲ Match

Materials: Use two or all three types of the following cards: Ten-frame Cards (BLM) 0 to 20, Number Cards (BLM) 0 to 20, Number Word Cards (BLM) 0 to 20

Students arrange the cards faceup in a grid. Students take turns finding two cards that go together.

★ Memory

Materials: Use two or all three types of the following cards: Ten-frame Cards (BLM) 0 to 20, Number Cards (BLM) 0 to 20, Number Word Cards (BLM) 0 to 20

Students arrange the cards facedown in a grid. Students take turns finding two cards that go together.

Exercise 7 • page 13

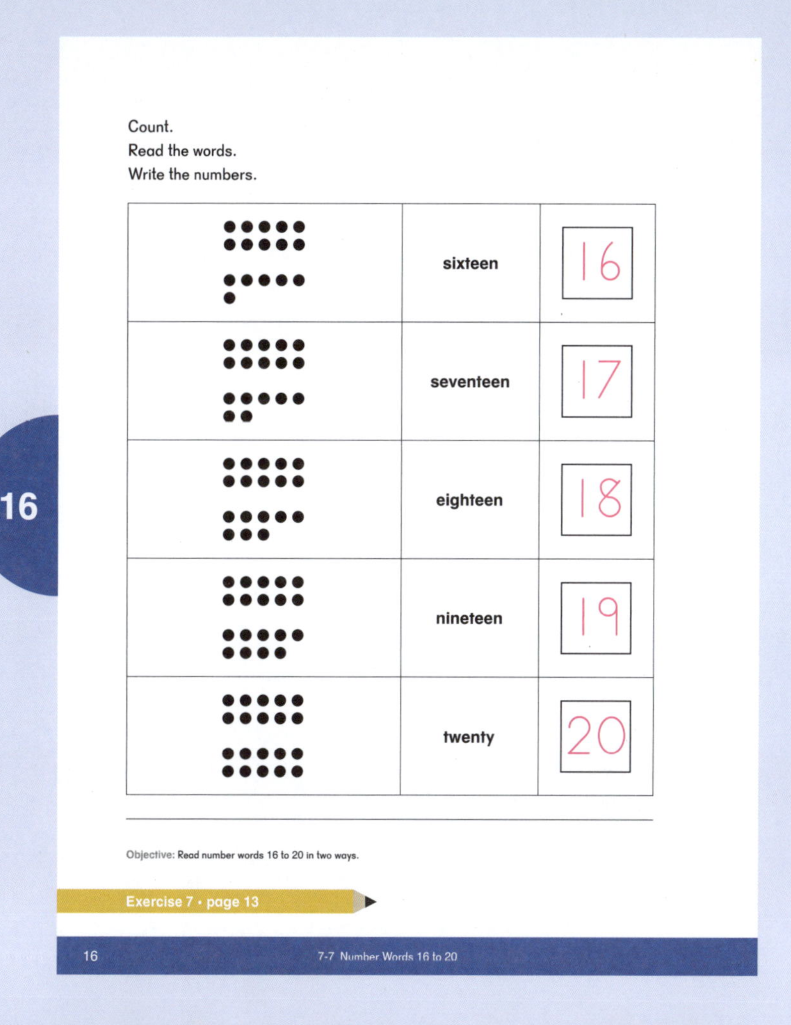
16

Count.
Read the words.
Write the numbers.

●●●●● ●●●●● ●●●●● ●	sixteen	16
●●●●● ●●●●● ●●●●● ●●	seventeen	17
●●●●● ●●●●● ●●●●● ●●●	eighteen	18
●●●●● ●●●●● ●●●●● ●●●●	nineteen	19
●●●●● ●●●●● ●●●●● ●●●●●	twenty	20

Objective: Read number words 16 to 20 in two ways.

Exercise 7 • page 13

16 7-7 Number Words 16 to 20

Extend

★ I Have, Who Has?

Materials: I Have, Who Has? Cards (BLM)

Up to 20 students can play this game. If there are fewer than 20, some students can start with more than one card.

One student starts by reading, "I have 16. Who has 5?" The student with 5 says, "I have 5. Who has 10 and 1?" Play continues until all students have read a card.

Objective

- Order numbers from 11 to 20.

Lesson Materials

- Linking cubes
- Number Cards (BLM) 11 to 20

17

Explore

Provide pairs of students with some linking cubes and a Number Cards (BLM).

Have them make a tower with linking cubes to match the number on their Number Cards (BLM).

Students can be called in order, beginning with 11, to put their cubes in order and line the towers up.

Learn

Have students look and talk about page 17. Ask students what ten-frames or numbers are showing and what is missing.

Whole Group Activity

▲ Magic Thumb

Pointing your thumb up or down, have students chorally count up and down within 15 by ones. For example, start out by saying, "Let's count by ones starting at 10. First number?" The class responds, "11." Then, point your thumb up, and the class responds, "12." Then point your thumb down, and the class responds, "11." Point down again, and the class responds, "10," and so on.

Small Group Activities

Textbook Page 18

▲ What is Missing?

Materials: Ten-frame Cards (BLM) 11 to 20 or Number Cards (BLM) 11 to 20

Shuffle cards. Player 1 pulls one number from the deck, then gives the remaining deck of cards to Player 2. Player 2 puts the cards in order and finds the missing number. Extend the activity by having students remove 2 or 3 cards.

▲ Number Grid Puzzles

Materials: Number Grid Puzzle (BLM)

Cut out the blank puzzle shapes from the second page of the Number Grid Puzzle (BLM).

Provide students with a Number Grid Puzzle (BLM) and some blank puzzle shapes. Player 1 covers some numbers with a puzzle piece without letting Player 2 see the board.

Player 2 then looks at the board and says which numbers he thinks are covered and Player 1 checks if he is right. Players switch roles and play continues.

Number Grid Puzzle

1	2	3	4	5
6	7	8	9	10
11	12	13	14	15
16	17	18	19	20

Exercise 8 • page 15

18

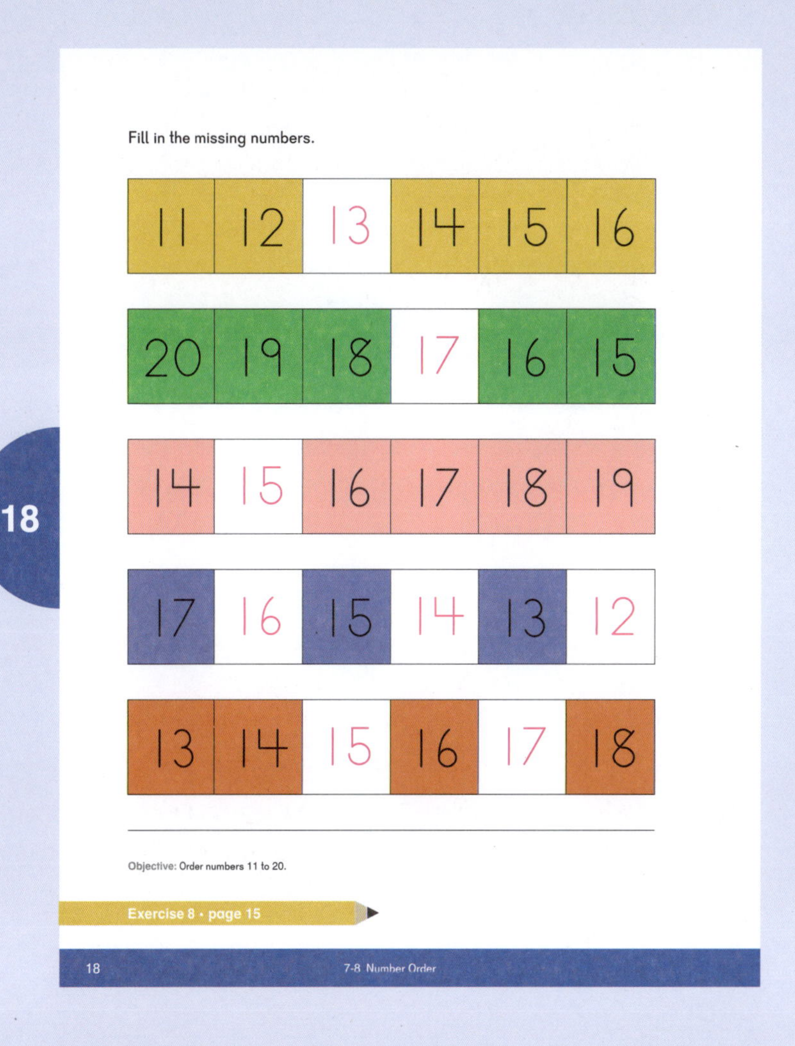
Fill in the missing numbers.

11	12	13	14	15	16
20	19	18	17	16	15
14	15	16	17	18	19
17	16	15	14	13	12
13	14	15	16	17	18

Objective: Order numbers 11 to 20.

Exercise 8 • page 15

18 7-8 Number Order

Extend

★ What's My Number?

Have students choose a number and give clues to a partner or the whole class.

Examples:

- I am more than 10. I have a 3 in the ones place. What number am I?
- I am the first number you say after 17 if you count back from 20. What number am I?
- I am less than 20 and more than 8. My ones place is a 6. What number am I?
- I am less than 20 and more than 8. If you add my ones number and my tens number, they make 2. What number am I?

Objective

- Identify numbers 1 more than and 1 less than a given number within 20.

Lesson Materials

- Linking cubes, 20 per student

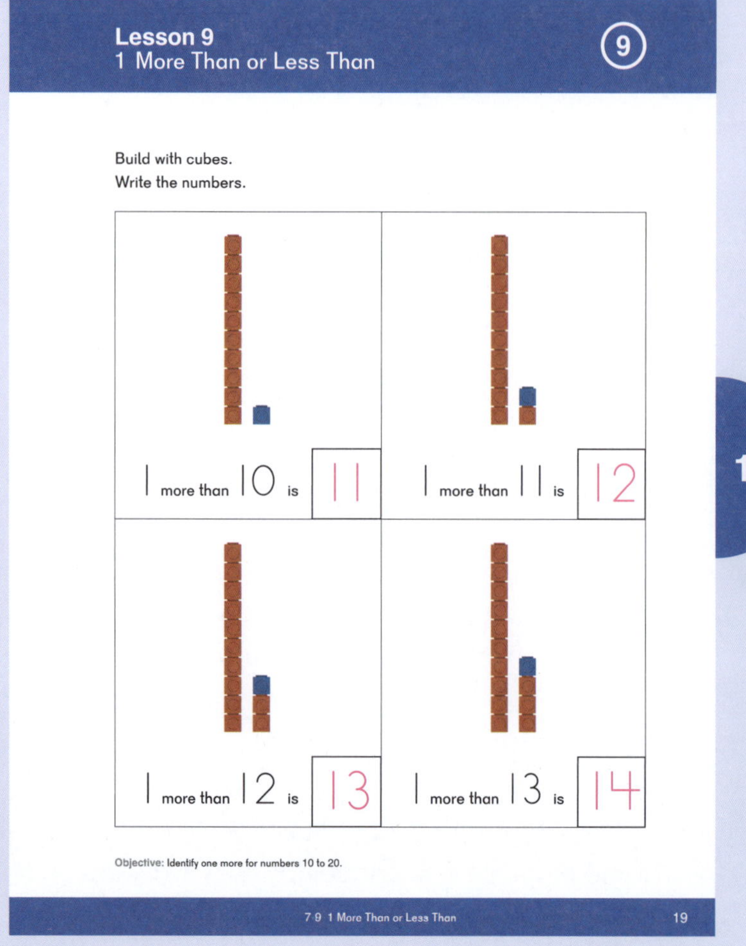

Explore

Give each student linking cubes and have them make one rod of 10 cubes.

Have them add a single cube next to the ten rod and count: "11. 1 more than 10 is 11."

Have students add cubes and continue counting and saying, "1 more than (the previous number) is ____." When they have two 10-rods, or 20 cubes, work back to 10 with, "1 less than." For example, "1 less than 20 is 19."

Learn

Discuss the pattern students are creating. Ask, "How can I find the number that is one more than 14? 1 less than 17?"

Small Group Activities

Textbook Pages 19–20

▲ One More/One Less

Materials: Double Ten-frame Cards (BLM) 11 to 20 or Number Cards (BLM) 11 to 20

Player 1 shows a number card. Player 2 has to say the number that is one more than and one less than Player 1's card. Extend by finding the cards that are two more than or two less than Player 1's card.

▲ More, Less, or Same

Materials: Deck of 4 each of Number Cards (BLM) 10 to 20

Play in groups of up to 4. Each player is dealt 5 Number Cards (BLM). One card is placed faceup in the center. Players take turns either placing a card from their hands that is one more, one less, or the same as the card in the middle. If a player doesn't have a card to play, she must draw a card. The first player to play all of her cards wins.

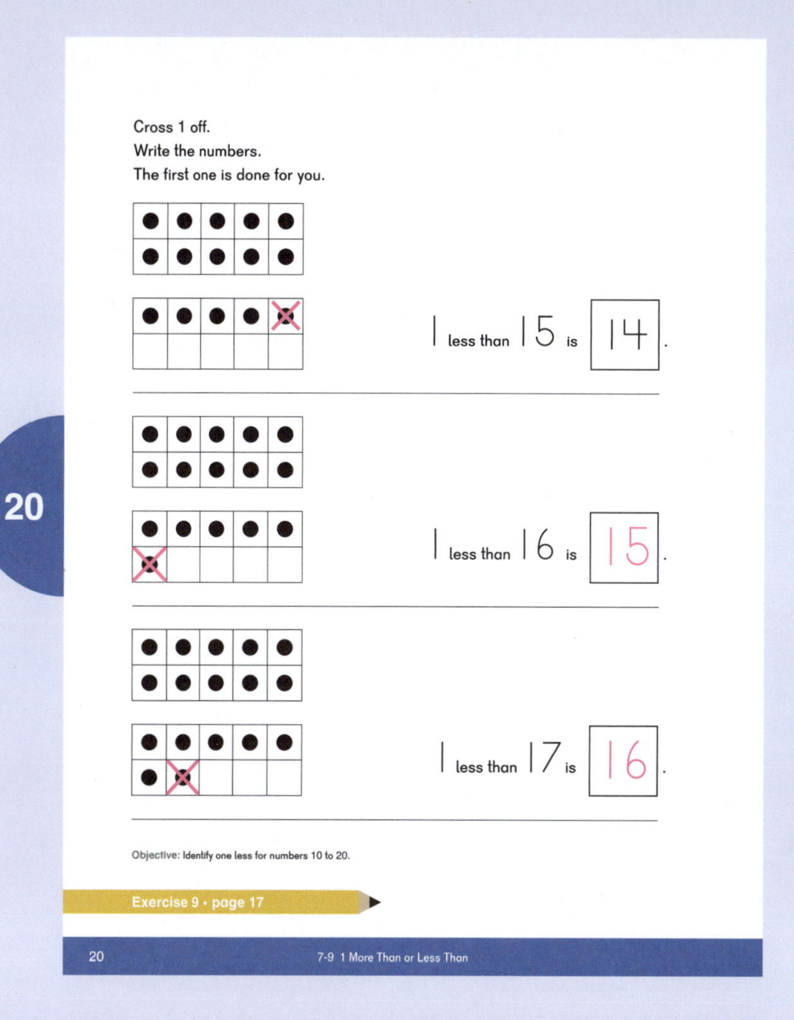

Exercise 9 • page 17

Extend

★ Next Number Snap

Materials: Ten-Frame Cards (BLM) 10 to 20

Deal out all cards to players facedown. Use 4 sets of Ten-Frame Cards (BLM) 10 to 20 for 3 players, and 5 sets of Ten-Frame Cards (BLM) 10 to 20 for 4 players.

Players take turns turning over their top cards and saying the number aloud. They put those cards into a discard pile.

If the new card is one more than the top card on the discard pile, players say, "Snap." The first person to do so collects the discard pile.

The game ends when a player is out of cards. The player with the most cards wins.

Lesson 10 Practice — Part 1

Objective

- Practice concepts from the chapter.

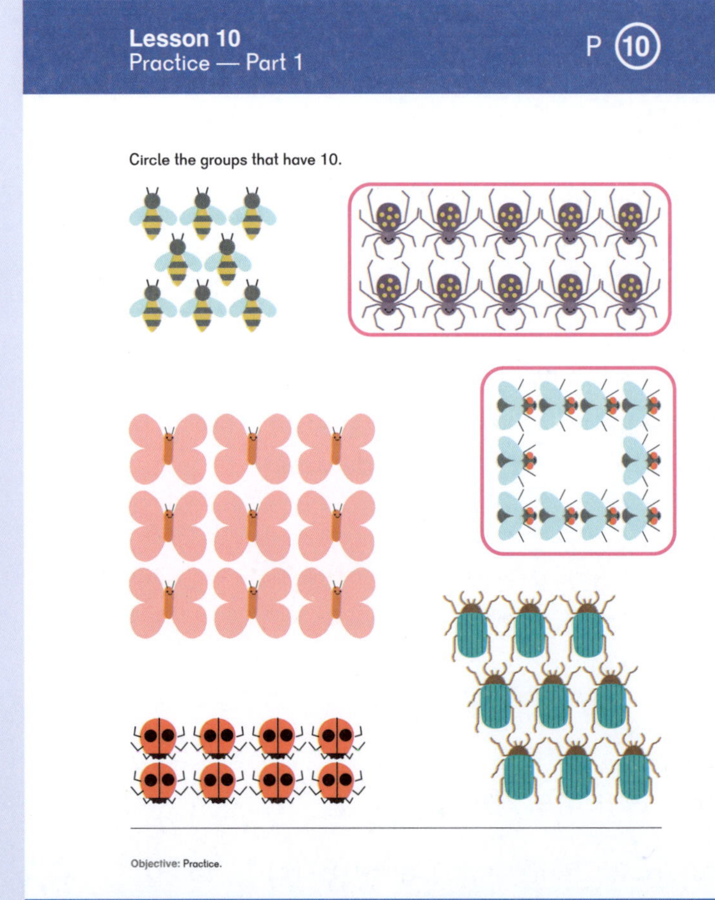

Practice lessons are designed for further practice and assessment as needed.

Students can complete the textbook pages and workbook pages as practice and/or as assessment.

Use activities and extensions from the chapter for additional review and practice.

Exercise 10 • page 19

Extend

★ Number Path Hop

Materials: Chalk, painter's tape, or paper plates

Create a large number path showing numbers 10 to 20. Draw one outside with chalk, or make one inside with masking tape. Paper plates with numbers on them could also be used.

Have students hop through the number path, saying each number as they hop on it.

Create individual number paths. Call a number and have students "hop" (using their fingers) to the number that is 1 more than or 1 less than the number called.

Examples:

- Hop to the number that is 1 less than 12.
- Hop to the number that is 1 more than 10 and 4.

Extend further:

- Hop to a number that is 2 or 3 more or less than the number called.
- Put the numbers out of order so that students must search for the number that is 1 less or 1 more.

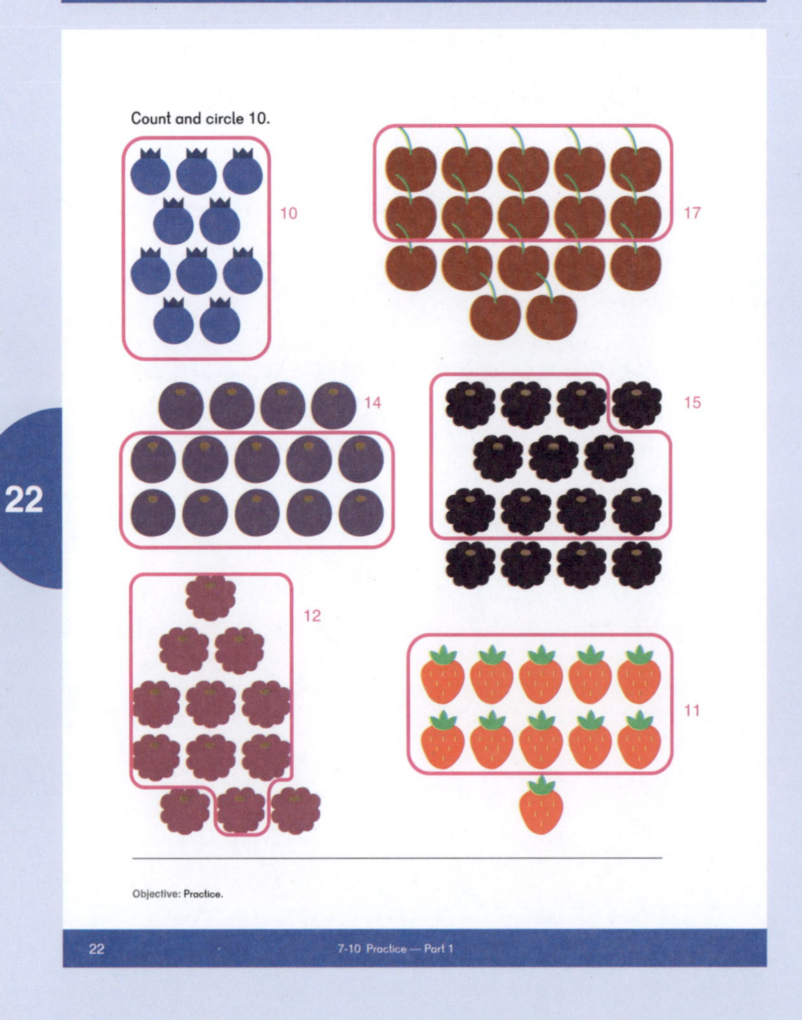

▲ Face-Off

Materials: Number Cards (BLM) 0 to 20, counters or linking cubes

Deal out all Number Cards (BLM) to players equally. Players each flip over a card at the same time. The student with the greatest number shown (or least) wins.

Students can use linking cubes or counters to see whose number is greater.

★ How Many More? Face-Off

Materials: Blank Double Ten-frames (BLM), Number Cards (BLM) 11 to 20 or Number Word Cards (BLM) 11 to 20, counters

Players each draw a Number Card (BLM) or Number Word Card (BLM). Players use counters to build their numbers on a Blank Double Ten-frame (BLM). The player with more keeps the difference as her score for the round. Players may need to line up their counters to determine how many more. The player with the most counters at the end of the game wins.

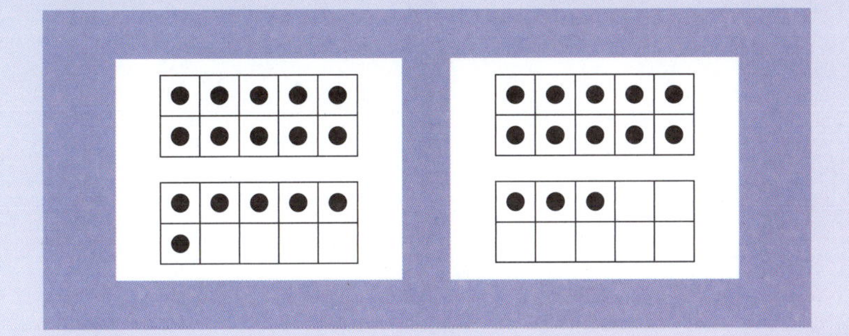

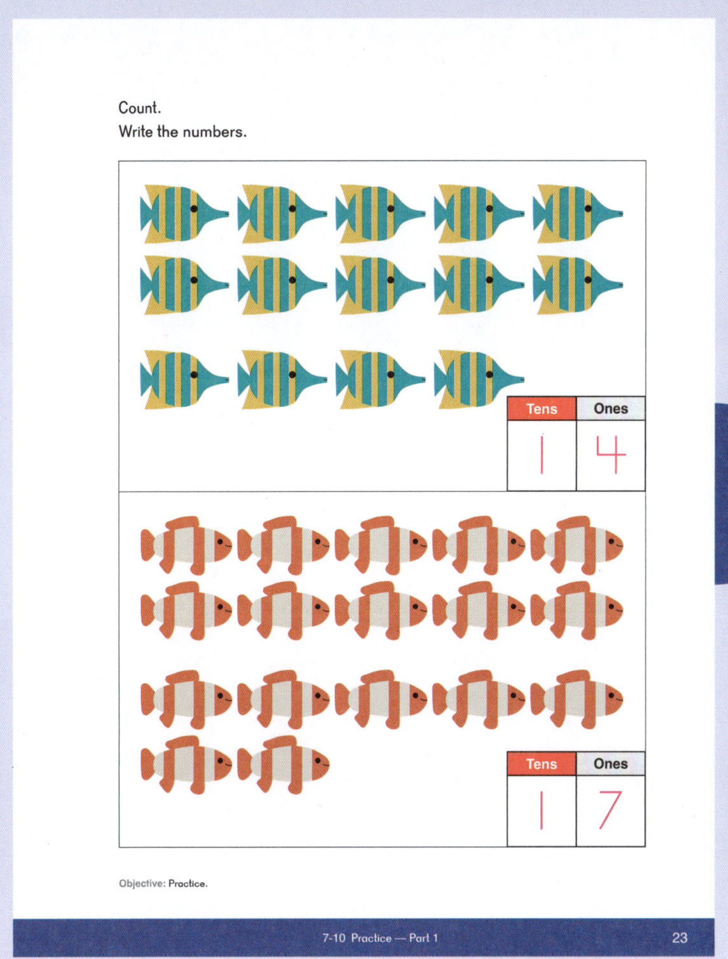

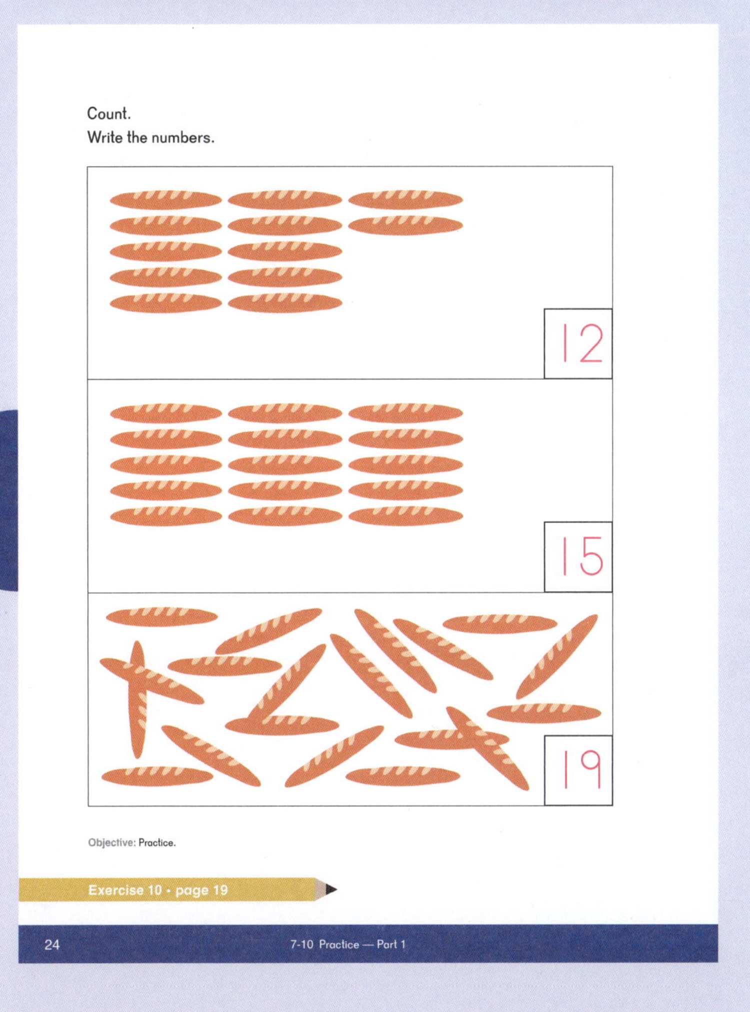

Objective

- Practice concepts from the chapter.

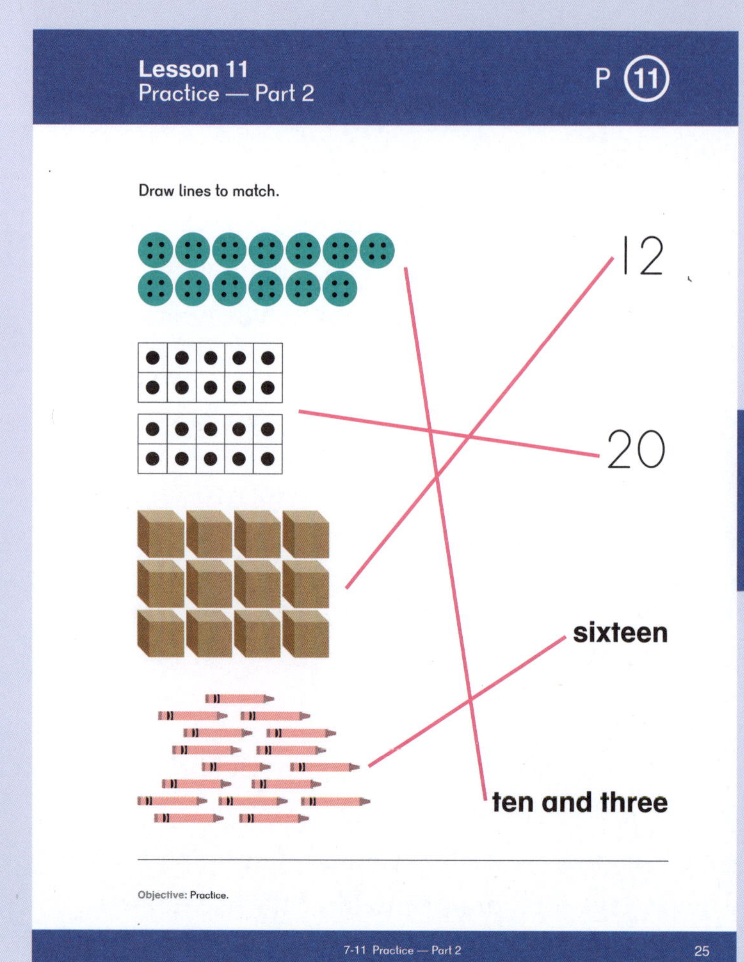

25

26

Exercise 11 • page 21

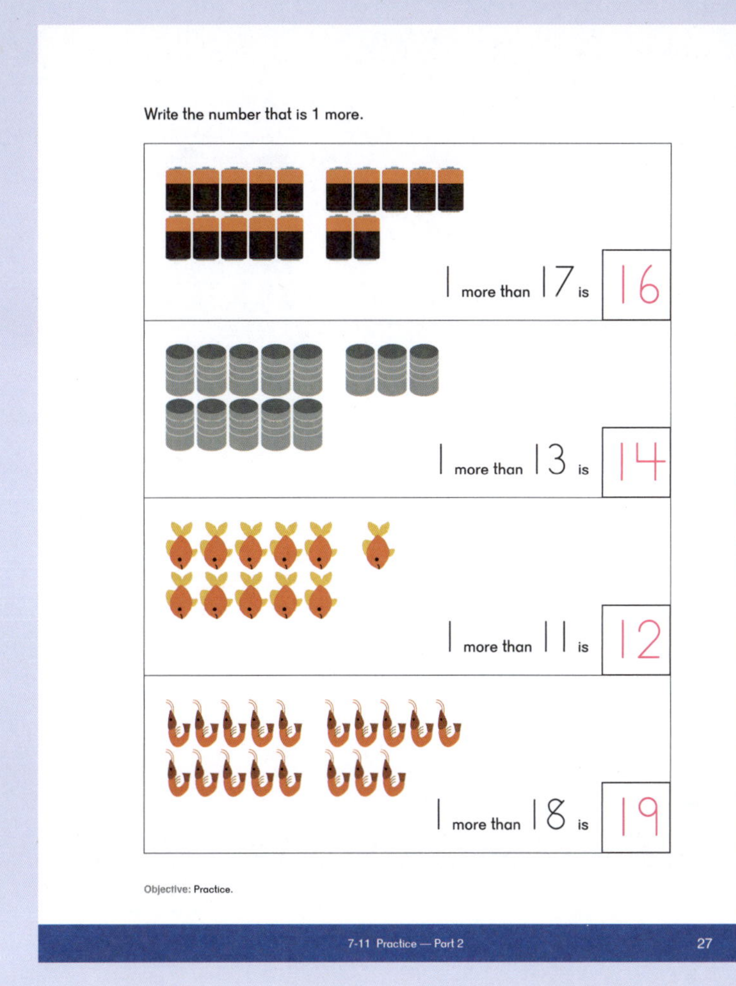
Write the number that is 1 more.

1 more than 17 is 16

1 more than 13 is 14

1 more than 11 is 12

1 more than 18 is 19

Objective: Practice.

7-11 Practice — Part 2 27

28

Cross 1 off.
Write the number that is 1 less.

1 less than 20 is 19

1 less than 15 is 14

1 less than 17 is 16

1 less than 12 is 11

Objective: Practice.

Exercise 11 • page 21

28 7-11 Practice — Part 2

Exercise 1 • pages 1–2

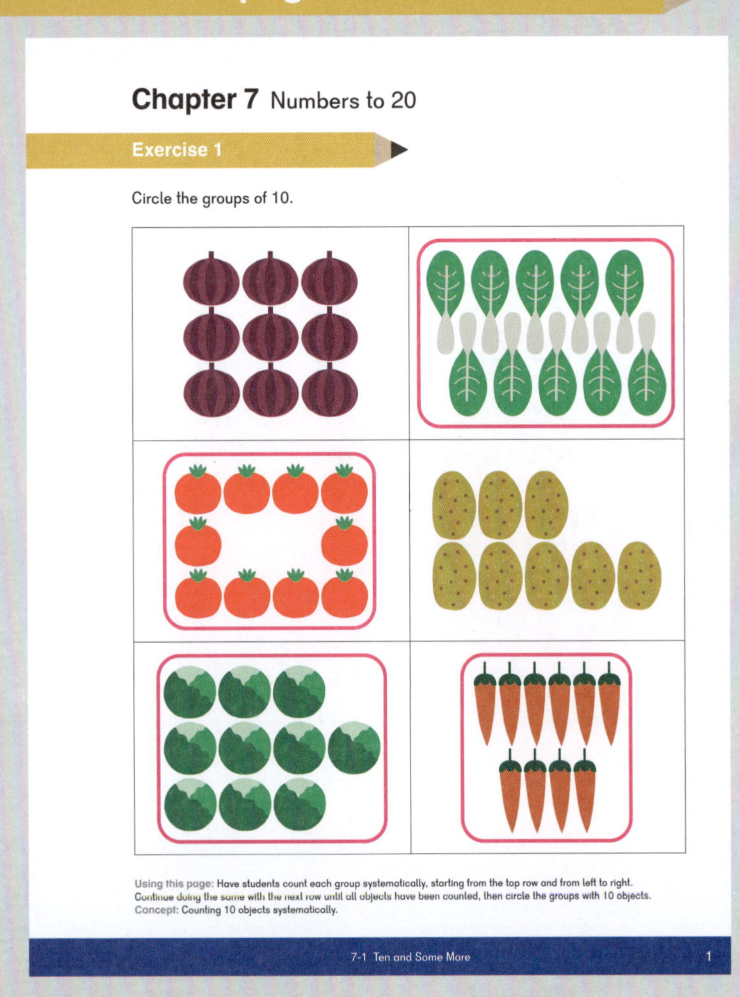

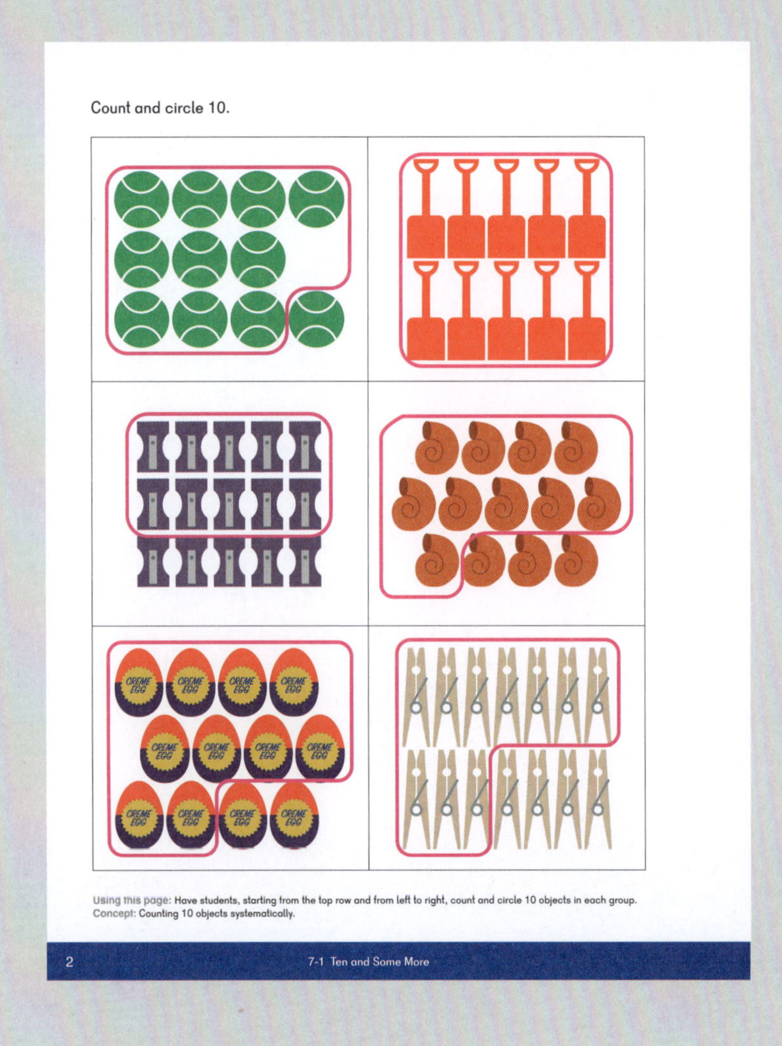

Exercise 2 • pages 3–4

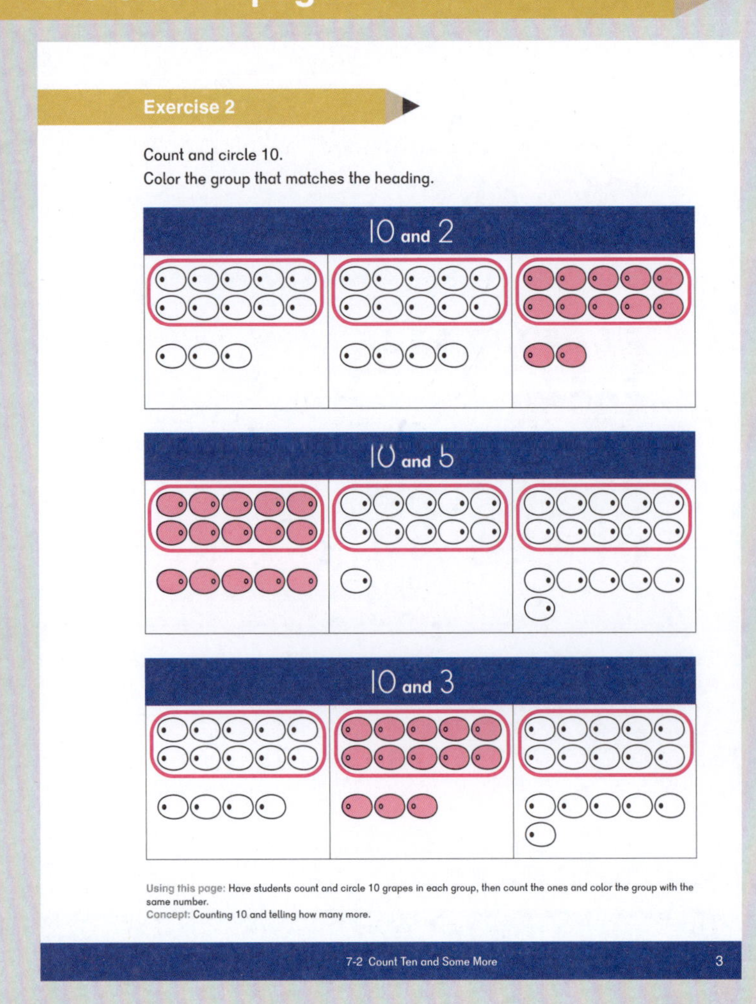

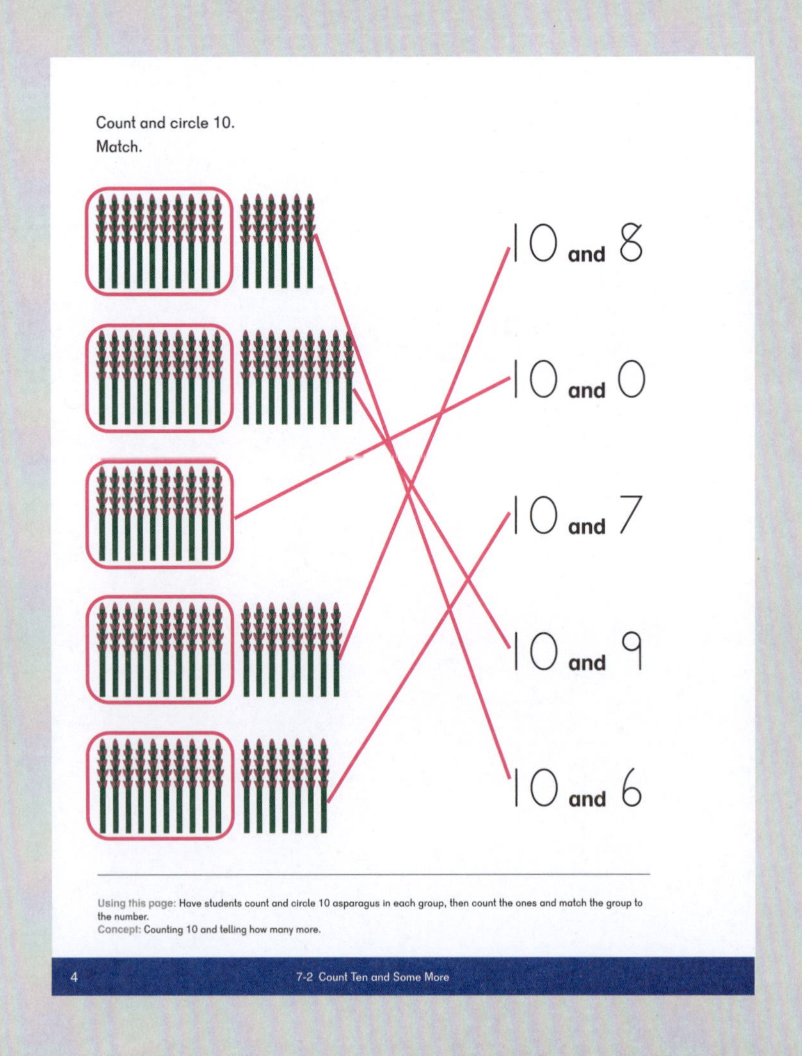

Exercise 3 • pages 5–6

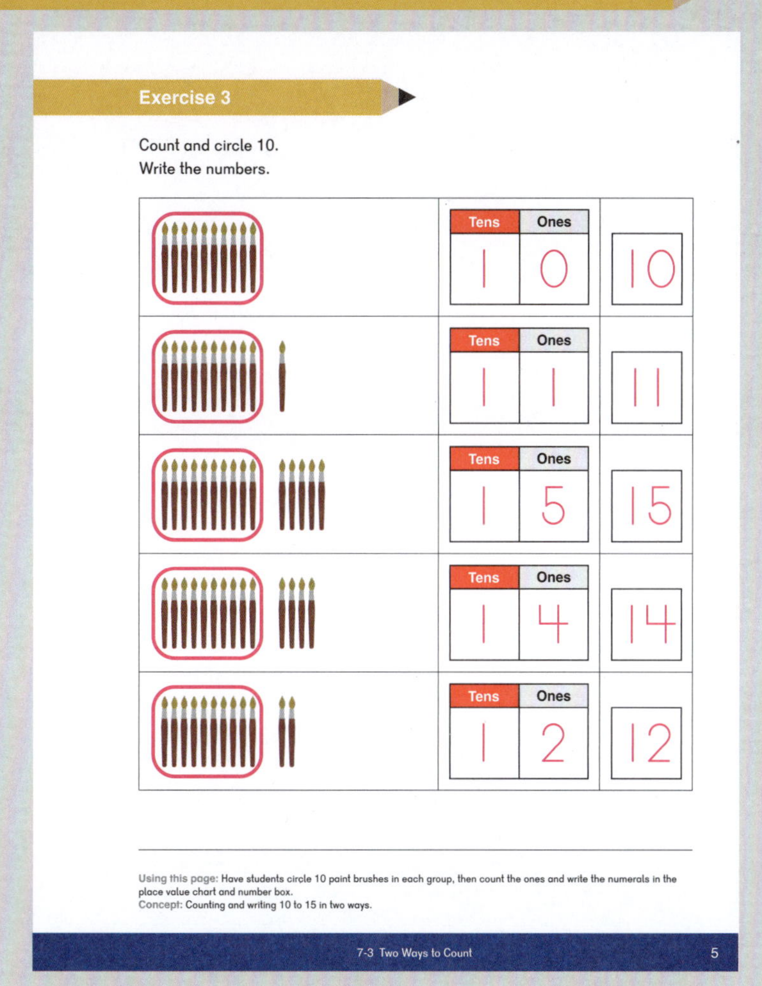

Exercise 3

Count and circle 10.
Write the numbers.

	Tens	Ones	
	1	0	10
	1	1	11
	1	5	15
	1	4	14
	1	2	12

Using this page: Have students circle 10 paint brushes in each group, then count the ones and write the numerals in the place value chart and number box.
Concept: Counting and writing 10 to 15 in two ways.

7-3 Two Ways to Count 5

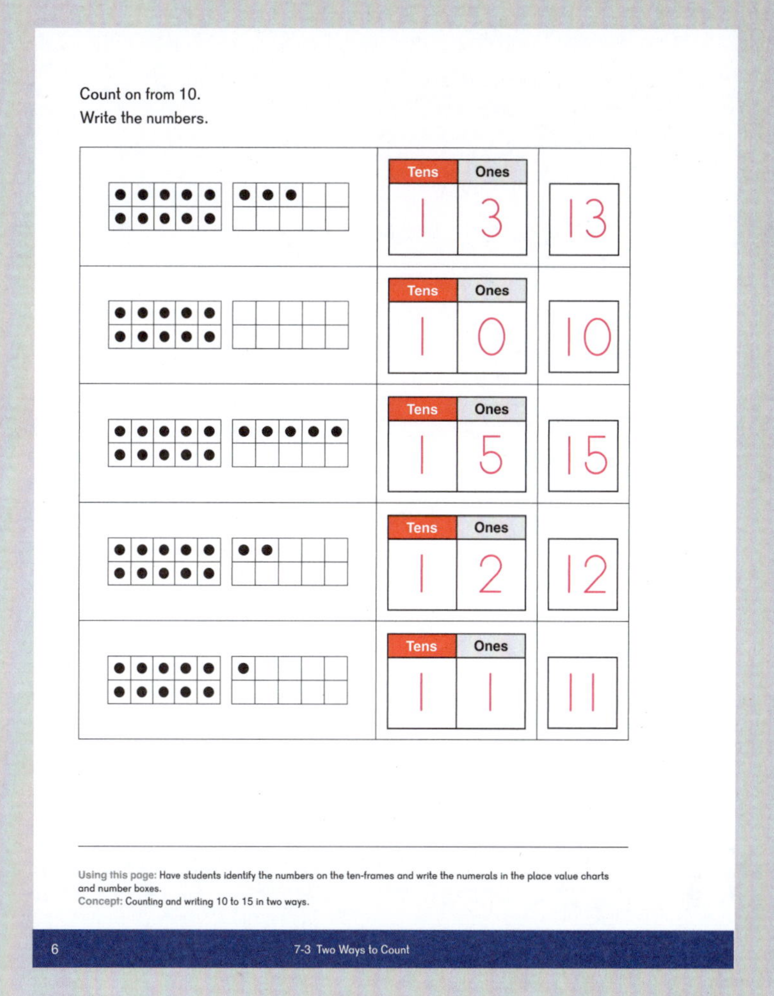

Count on from 10.
Write the numbers.

	Tens	Ones	
	1	3	13
	1	0	10
	1	5	15
	1	2	12
	1	1	11

Using this page: Have students identify the numbers on the ten-frames and write the numerals in the place value charts and number boxes.
Concept: Counting and writing 10 to 15 in two ways.

6 7-3 Two Ways to Count

Exercise 4 • pages 7–8

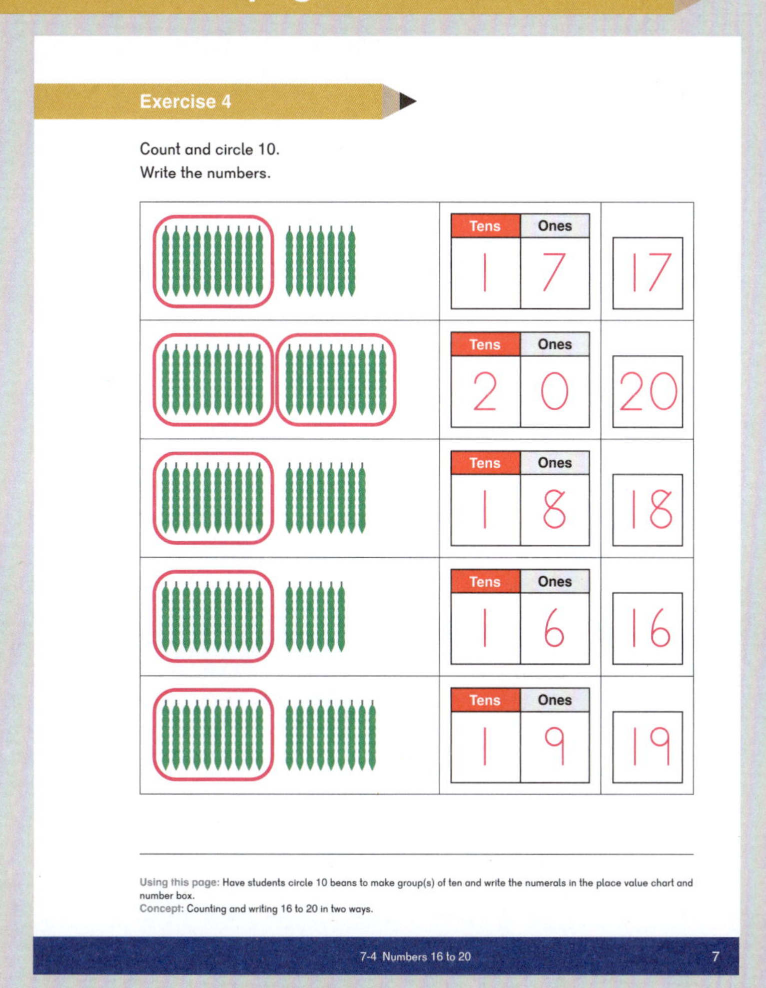

Exercise 4

Count and circle 10.
Write the numbers.

	Tens	Ones	
	1	7	17
	2	0	20
	1	8	18
	1	6	16
	1	9	19

Using this page: Have students circle 10 beans to make group(s) of ten and write the numerals in the place value chart and number box.
Concept: Counting and writing 16 to 20 in two ways.

7-4 Numbers 16 to 20 7

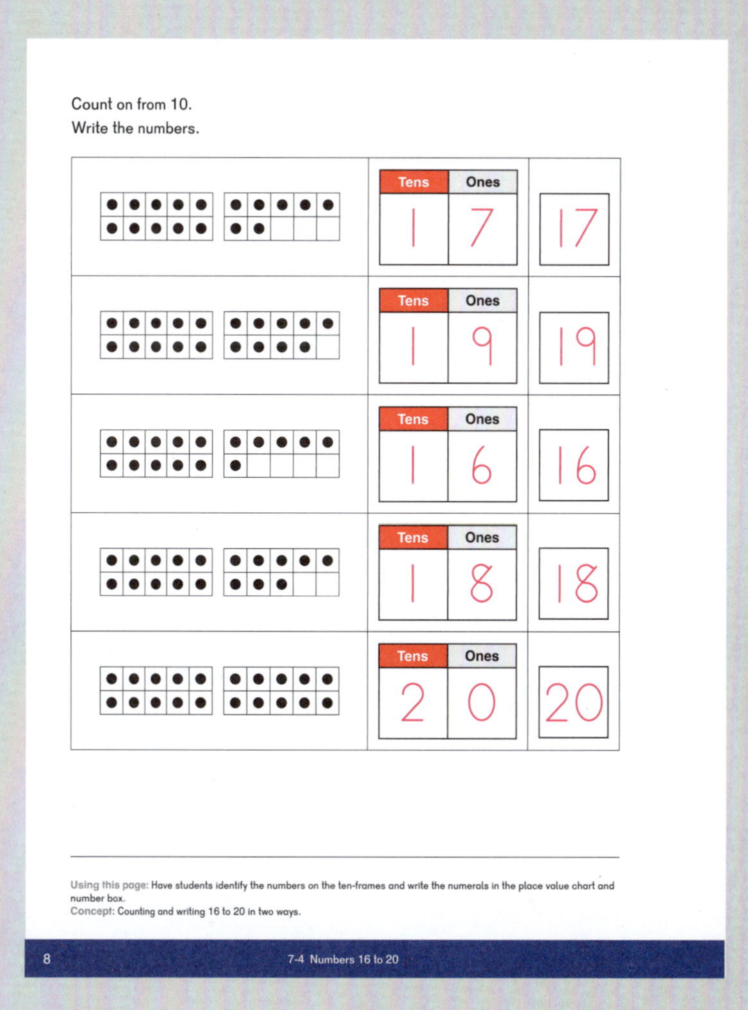

Count on from 10.
Write the numbers.

	Tens	Ones	
	1	7	17
	1	9	19
	1	6	16
	1	8	18
	2	0	20

Using this page: Have students identify the numbers on the ten-frames and write the numerals in the place value chart and number box.
Concept: Counting and writing 16 to 20 in two ways.

8 7-4 Numbers 16 to 20

Exercise 5 • pages 9–10

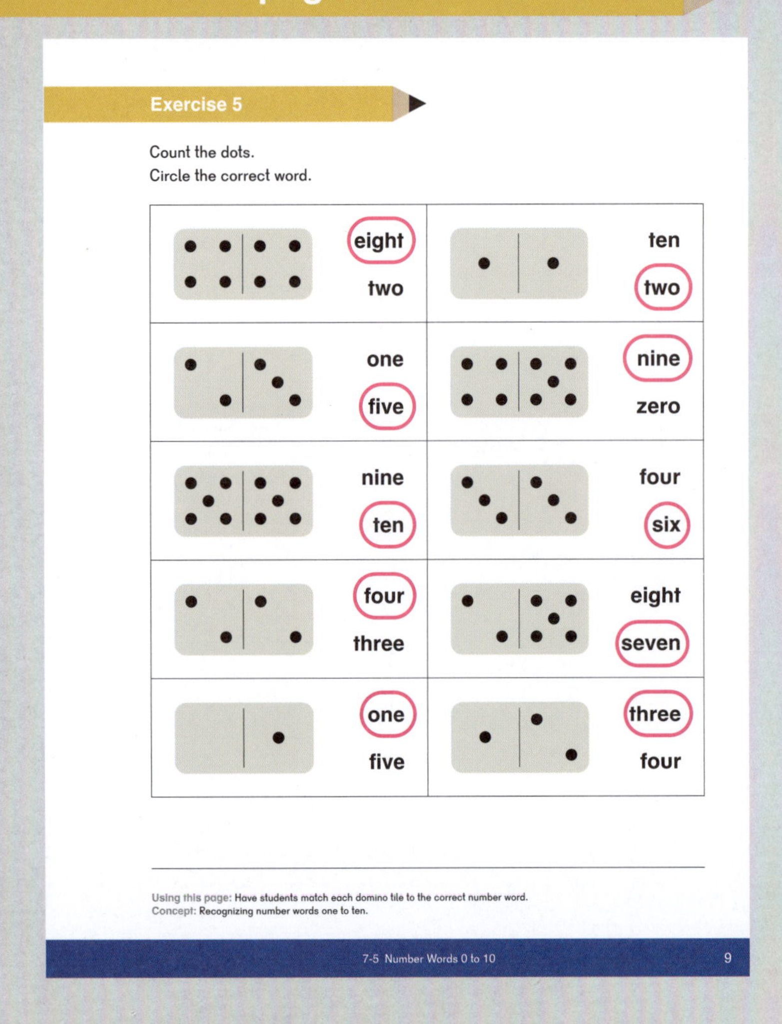

Exercise 5

Count the dots.
Circle the correct word.

	eight two		ten two
	one five		nine zero
	nine ten		four six
	four three		eight seven
	one five		three four

Using this page: Have students match each domino tile to the correct number word.
Concept: Recognizing number words one to ten.

7-5 Number Words 0 to 10 9

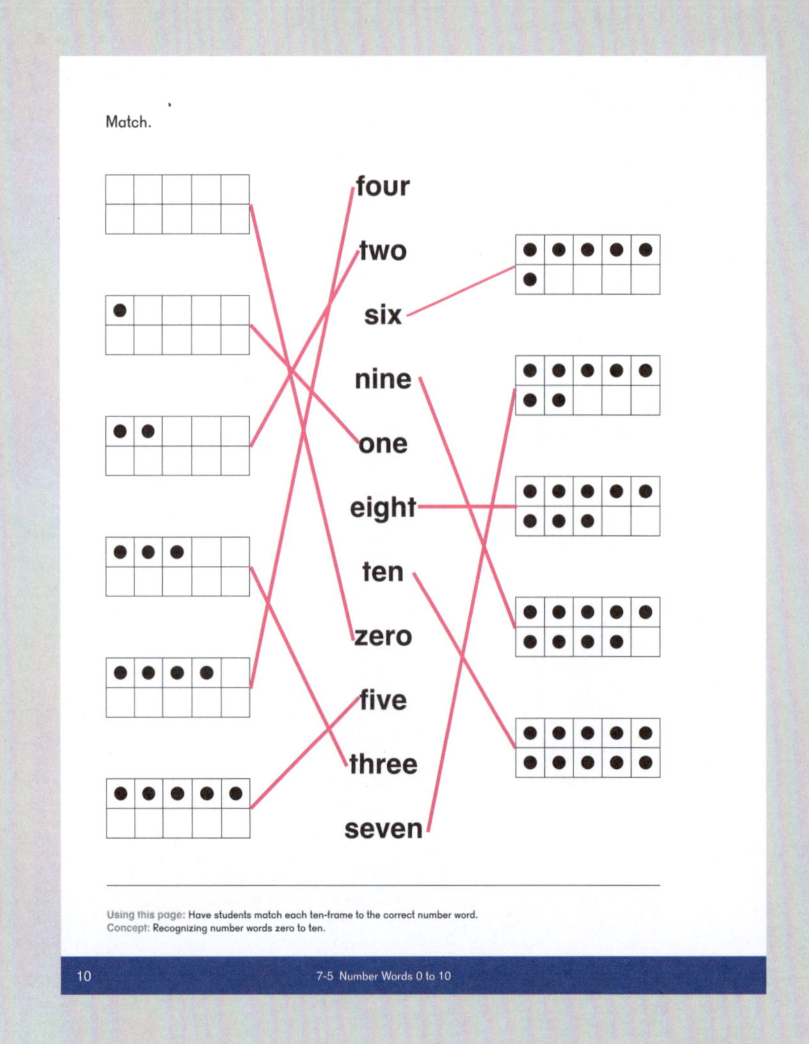

Match.

four
two
six
nine
one
eight
ten
zero
five
three
seven

Using this page: Have students match each ten-frame to the correct number word.
Concept: Recognizing number words zero to ten.

10 7-5 Number Words 0 to 10

Exercise 6 • pages 11–12

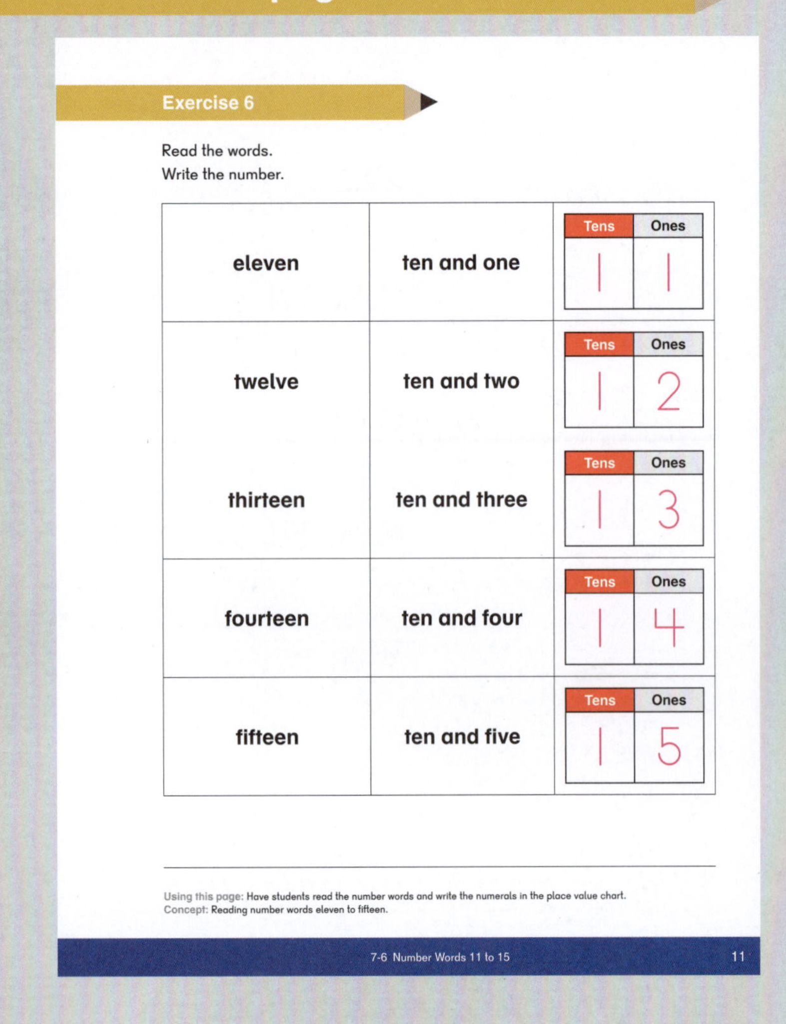

Exercise 6

Read the words.
Write the number.

		Tens	Ones
eleven	ten and one	1	1
twelve	ten and two	1	2
thirteen	ten and three	1	3
fourteen	ten and four	1	4
fifteen	ten and five	1	5

Using this page: Have students read the number words and write the numerals in the place value chart.
Concept: Reading number words eleven to fifteen.

7-6 Number Words 11 to 15 11

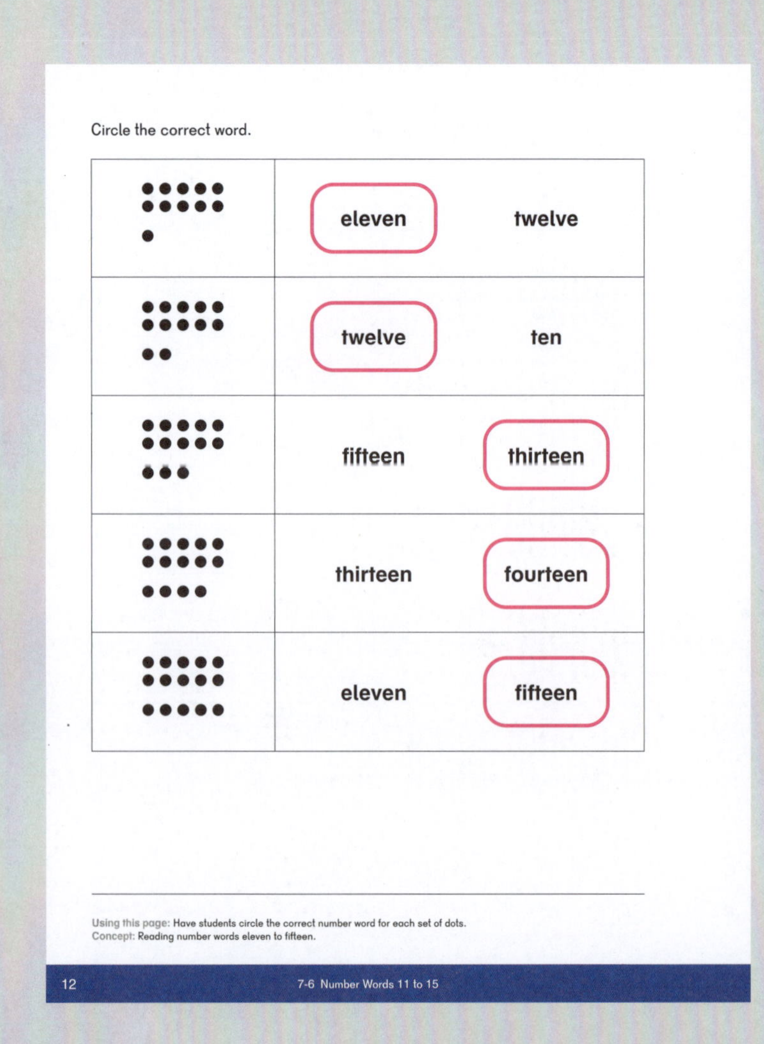

Circle the correct word.

	eleven	twelve
	twelve	ten
	fifteen	thirteen
	thirteen	fourteen
	eleven	fifteen

Using this page: Have students circle the correct number word for each set of dots.
Concept: Reading number words eleven to fifteen.

12 7-6 Number Words 11 to 15

Exercise 7 • pages 13–14

Exercise 7

Read the words.
Write the number.

		Tens	Ones
sixteen	ten and six	1	6
seventeen	ten and seven	1	7
eighteen	ten and eight	1	8
nineteen	ten and nine	1	9
twenty	two tens	2	0

Using this page: Have students circle 10 beans to make group(s) of ten and write the numerals in the place value chart.
Concept: Reading number words sixteen to twenty.

7-7 Number Words 16 to 20 13

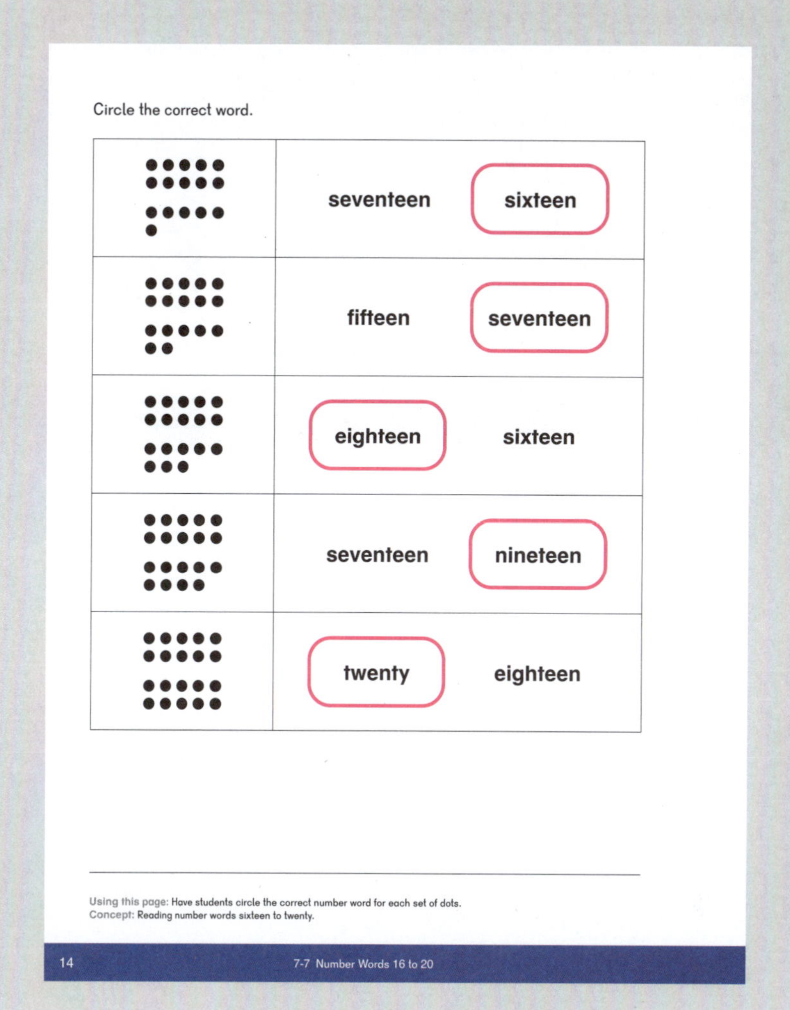

Circle the correct word.

Dots		
16 dots	seventeen	sixteen (circled)
17 dots	fifteen	seventeen (circled)
18 dots	eighteen (circled)	sixteen
19 dots	seventeen	nineteen (circled)
20 dots	twenty (circled)	eighteen

Using this page: Have students circle the correct number word for each set of dots.
Concept: Reading number words sixteen to twenty.

14 7-7 Number Words 16 to 20

Exercise 8 • pages 15–16

Exercise 8

Paste the puzzle by number order.

Before using this page: Precut the puzzle strips at the back of the book.
Using this page: Have students order the puzzle strips from 11 to 20, then paste them accordingly in the outline.
Concept: Ordering numbers 11 to 20.

7-8 Number Order 15

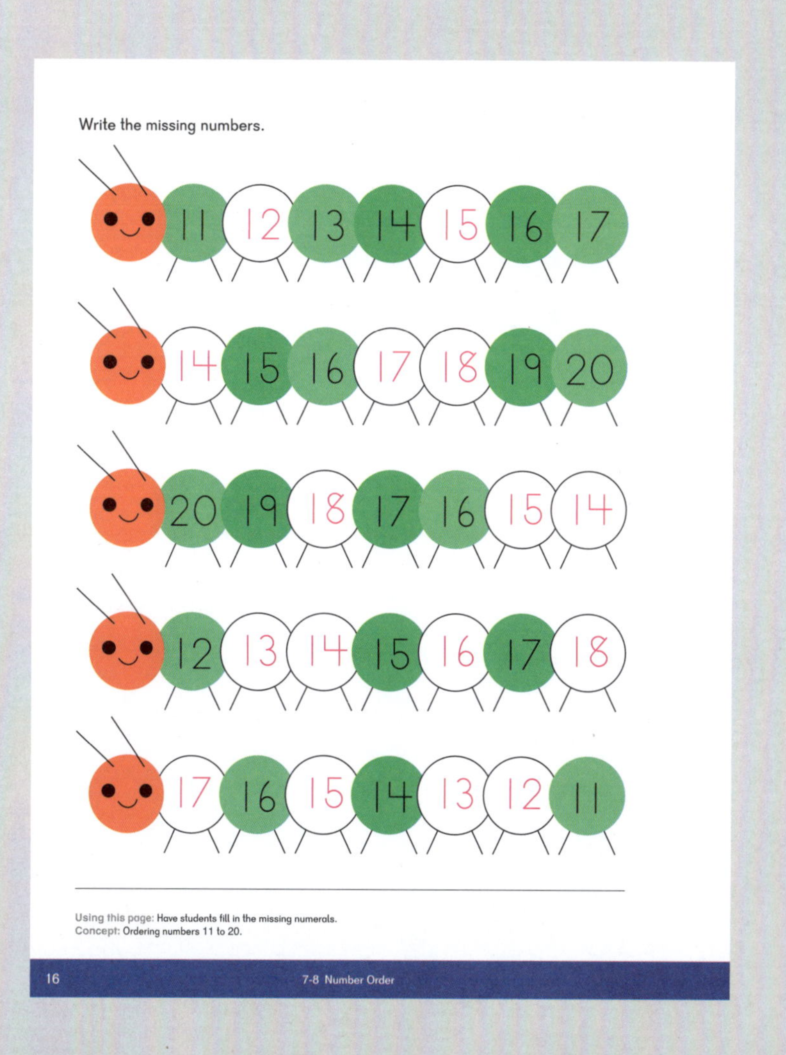

Write the missing numbers.

Using this page: Have students fill in the missing numerals.
Concept: Ordering numbers 11 to 20.

16 7-8 Number Order

Exercise 9 • pages 17–18

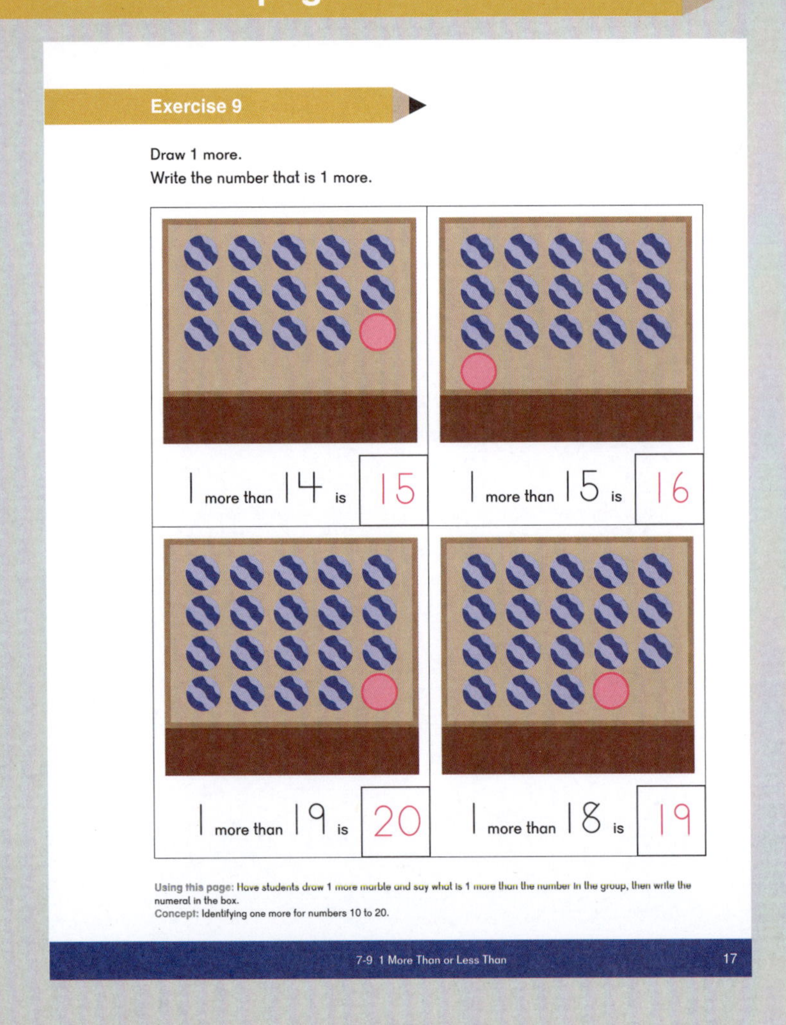

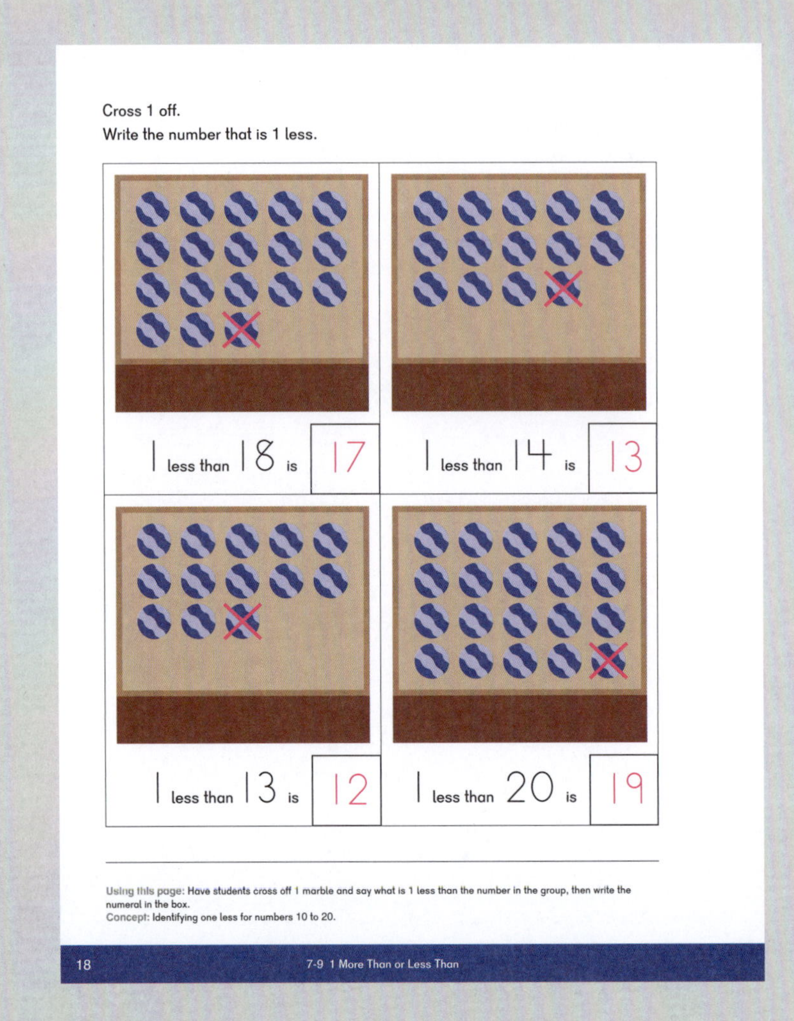

Exercise 10 • pages 19–20

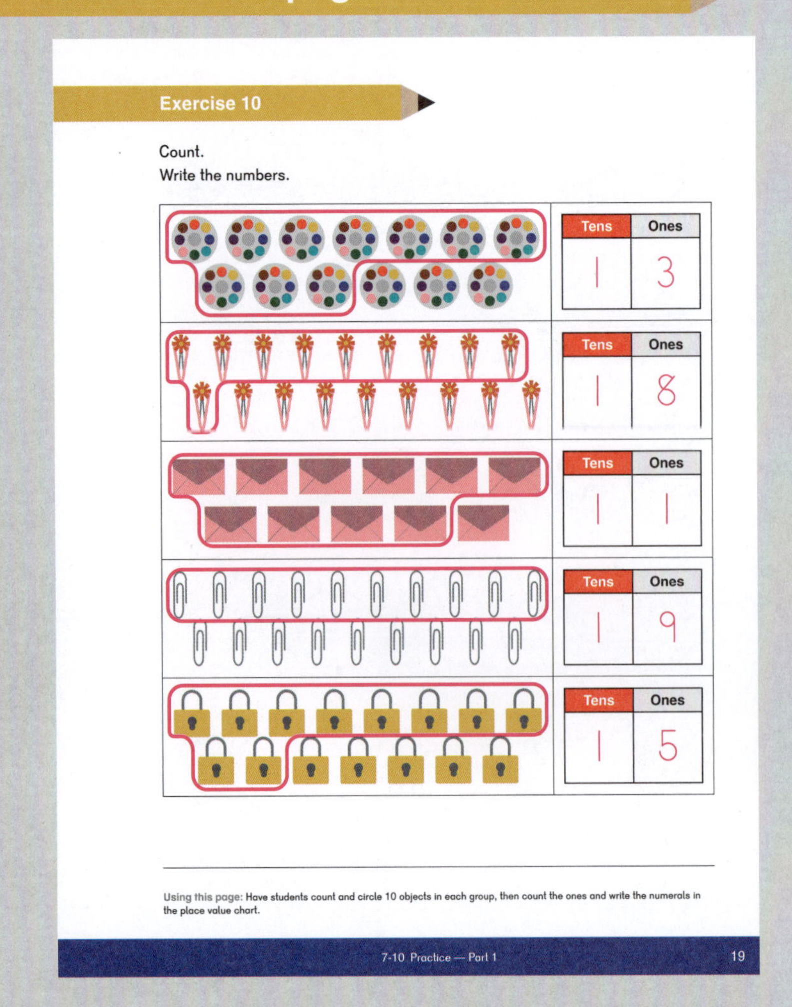

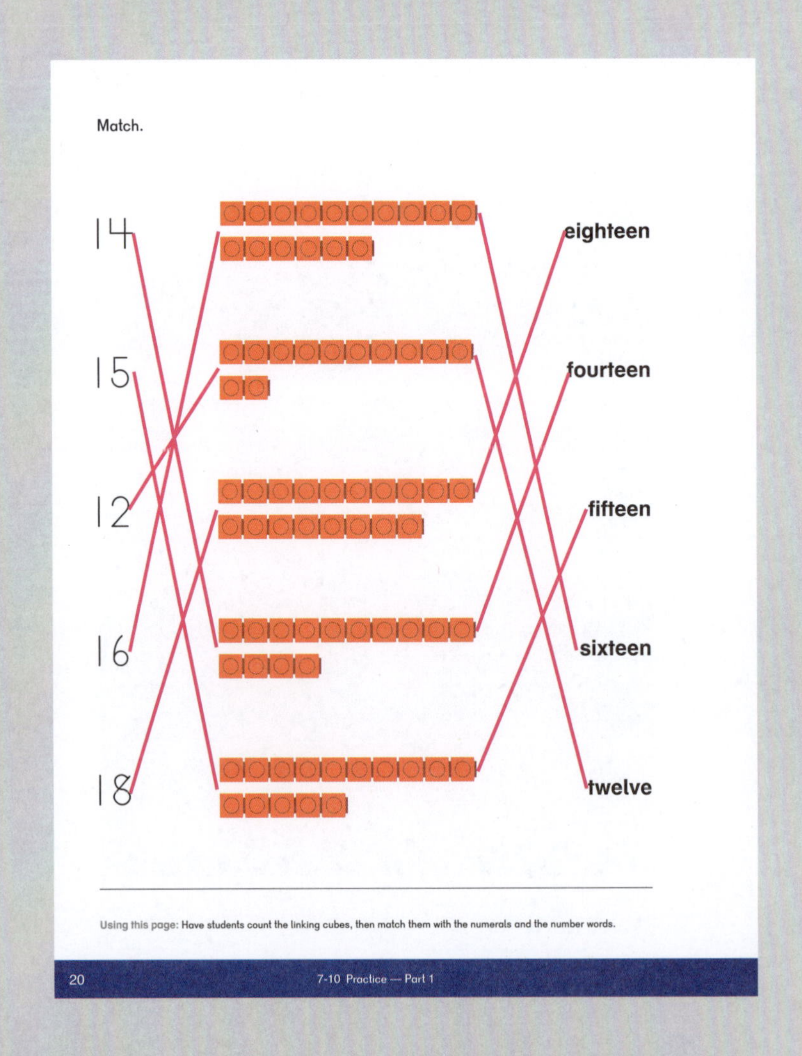

Exercise 11 • pages 21–22

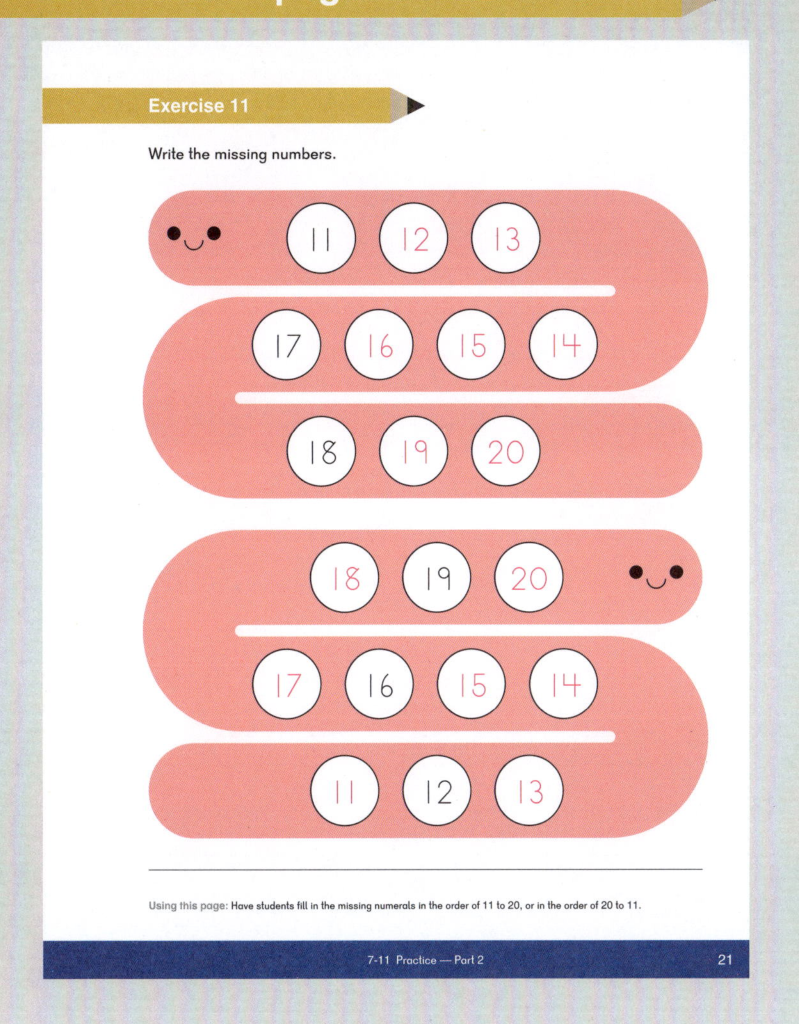

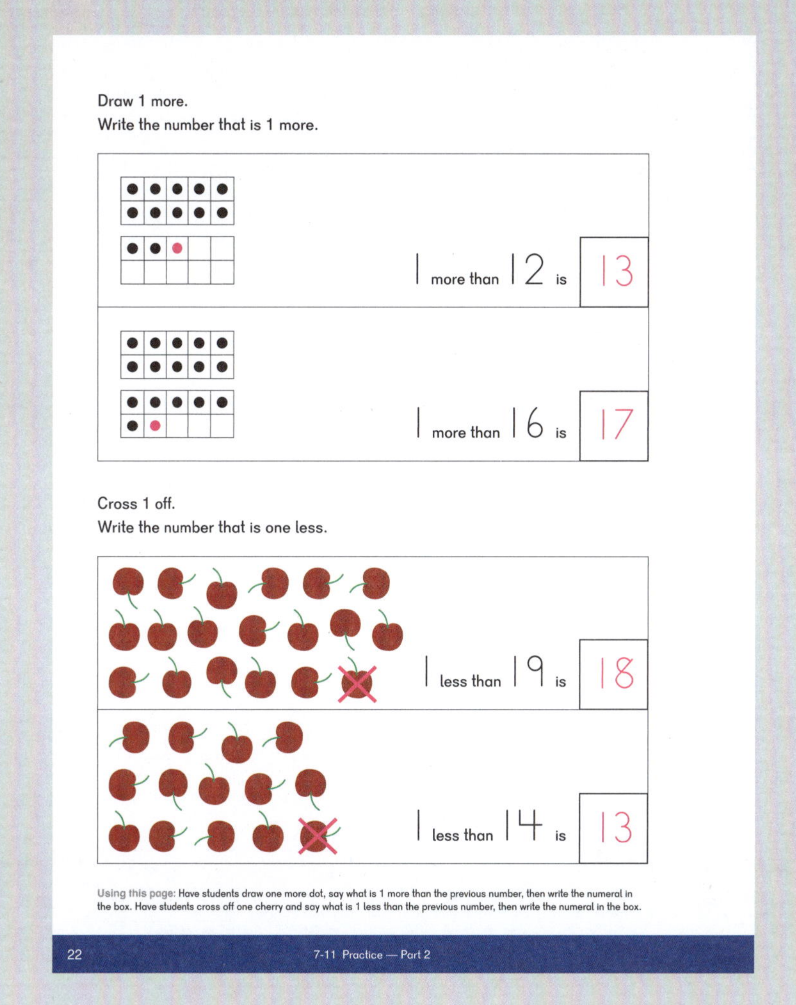

Notes

Suggested number of class periods: 14–15

	Lesson	Page	Resources	Objectives
	Chapter Opener	p. 39	TB: p. 29	
1	Putting Numbers Together — Part 1	p. 40	TB: p. 30 WB: p. 23	Count to find how many objects there are in all.
2	Putting Numbers Together — Part 2	p. 41	TB: p. 31 WB: p. 25	Count and model two sets to find how many altogether.
3	Parts Making a Whole	p. 43	TB: p. 33 WB: p. 27	Understand that parts can be put together to make a whole.
4	Look for a Part	p. 45	TB: p. 36 WB: p. 29	Find parts in a number bond.
5	Number Bonds for 2, 3, and 4	p. 47	TB: p. 38 WB: p. 31	Find two parts that make 2, 3, and 4.
6	Number Bonds for 5	p. 50	TB: p. 41 WB: p. 33	Find 2 parts that make 5.
7	Number Bonds for 6	p. 53	TB: p. 45 WB: p. 35	Find 2 parts that make 6.
8	Number Bonds for 7	p. 56	TB: p. 49 WB: p. 37	Find 2 parts that make 7.
9	Number Bonds for 8	p. 59	TB: p. 53 WB: p. 39	Find 2 parts that make 8.
10	Number Bonds for 9	p. 61	TB: p. 55 WB: p. 41	Find 2 parts that make 9.
11	Number Bonds for 10	p. 64	TB: p. 59 WB: p. 45	Find 2 parts that make 10.
12	Practice — Part 1	p. 68	TB: p. 64 WB: p. 49	Practice number bonds to 6 and 7.
13	Practice — Part 2	p. 70	TB: p. 66 WB: p. 51	Practice number bonds to 8 and 9.
14	Practice — Part 3	p. 71	TB: p. 67 WB: p. 53	Practice finding the missing part in a number bond to 10.
	Workbook Solutions	p. 73		

Notes

Lessons in this chapter reinforce basic number skills and familiarize students with number bonds to 10.

A number bond is a pictorial representation of a number and the parts that make it. It can be represented on paper simply by writing numbers representing the whole and the parts and connecting the parts to the whole with a line. At early levels, a circle or other shape is drawn around the numbers to help set them apart visually.

The order of the parts of the number bonds does not matter. Both "5 is 2 and 3 " and "5 is 3 and 2" are acceptable and reinforce the commutative property of addition.

In the pictorial representation of a number bond, the orientation or the shapes used to enclose the numbers is not significant.

Students will see the number bonds in different orientations because they must learn it is the relationship between the numbers that is important, not their visual representation.

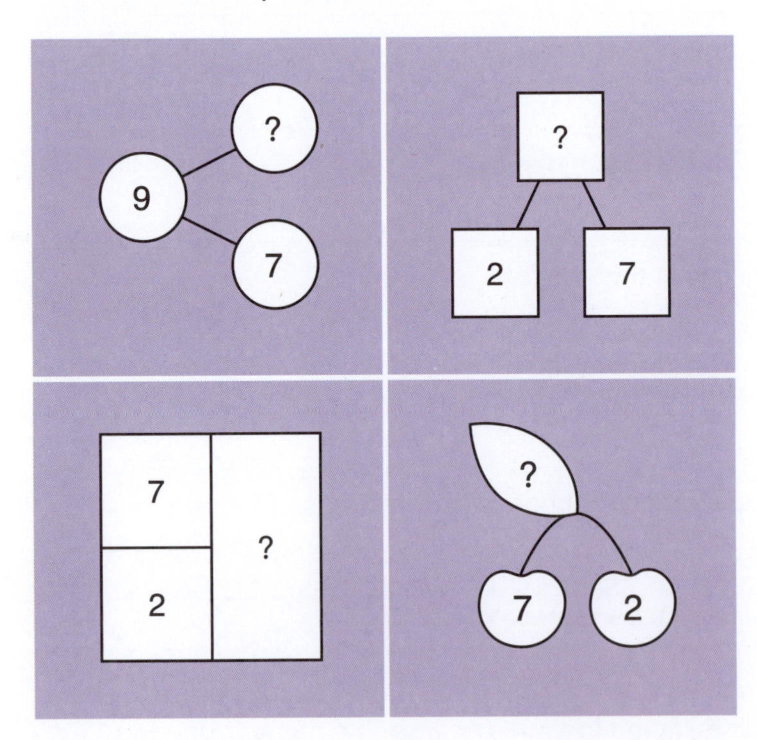

Number bonds generally refer to a whole and only two parts in **Dimensions Math® Kindergarten** textbooks. Students may be curious about drawing more parts. For example, "3 and 1 and 2 make 6" would be represented by 3 parts attached to the whole. However, these are not really helpful later when students use number bonds (two parts and a whole) to learn or to explain strategies for mental calculation. Theoretically, numbers can be split into infinitely many parts. Thus multiple parts and a whole are not really "number bonds" as the term is generally used in this series.

In subsequent chapters, students will be using number bonds to show the strategies they use to add and subtract within ten. Number bonds will also be used when adding and subtracting two-digit numbers in **Dimensions Math® 1A** and **1B**.

This chapter uses "___ and ___ make___" instead of the formal addition and equal symbols (+ and =) which will be introduced in **Chapter 9: Addition**. Students who are familiar with the symbols should not be discouraged from using them.

By the end of **Dimensions Math® Kindergarten B**, students should know their number bonds to 10 with automaticity. Meaning, that when given two parts, the students know the whole without counting.

There are many practice games and activities to help students achieve this goal. Games and activities are included in the lessons and they should be used throughout the year to help students move from counting to knowing their number bonds to 10. **Number Bond Bracelets** is an activity in many of the lessons. Struggling students can use these bracelets in other games, so you may want to introduce that activity first in those lessons as needed.

It is assumed that all students will have access to recording tools. When a lesson refers to a whiteboard, any writing materials can be used.

Materials

- 10 empty water bottles
- 10-sided die
- Attribute bears
- Bags or baskets
- Beads
- Blank index cards
- Construction paper
- Counters
- Crayons, markers, or paint daubers
- Cups
- Dice
- Dominoes
- Dry erase markers
- Hexagon shape, such as a pattern block, that can be traced
- Items with a round face students can trace
- Large hula hoops
- Linking cubes
- Painter's tape
- Pipe cleaners (chenille stems)
- Pool noodle
- Small ball
- Stickers
- Strips of construction paper in a variety of colors
- Traffic cones
- Two-color counters
- Whiteboards

Blackline Masters

- 7 Days of the Week Template
- Blank Number Bond Template
- Blank Ten-frame
- Domino Parking Lot
- Hexagon Template
- Number Bond Book
- Number Bond Cards
- Number Bond Recording Sheet
- Number Bond Story Template
- Number Cards
- Number Cards — Large
- Number Path
- Number Word Cards
- Octopus Template with Bond
- Part-Whole Recording Worksheet
- Picture Cards
- See Saw Template
- Ten-frame Cards

Storybooks

- Any picture book provides opportunities to create a number story from images on the page
- *Quack and Count* by Keith Baker
- *Ten Friends* by Bruce Goldstone
- *Domino Addition* by Lynette Long
- *Ten for Me* by Barbara Mariconda
- *Ten Monkey Jamboree* by Dianne Ochiltree
- *Mice Mischief: Math Facts in Action* by Caroline Stills
- *One More Bunny: Adding from One to Ten* by Rick Walton

Letters Home

- Chapter 8 Letter

Notes

Lesson Materials

- Two-color counters, 5 per student

29

Have students discuss what they see in the picture on page 29. Ask them which items can be put into groups and how many there would be in each group altogether. Provide students up to 5 two-color counters to represent the sets they see on the page.

Students may see:

- There are 2 orange flowers and 3 blue flowers. There are 5 flowers in all.
- There is 1 big tree and 2 small trees. There are 3 trees altogether.
- Mei has 1 baseball and Sofia has 1 soccer ball. There are 2 balls.
- There are 3 gray birds and 1 red bird. There are 4 birds in all.

Have students act out additional number stories using items in the classroom. For example:

- There are 2 girls and 1 boy reading books. There are 3 students reading books in all.
- There are 4 students. 2 students are wearing red shirts and 2 students are wearing blue shirts.
- There are 5 students. 4 students have black shoes and 1 student has white shoes.

For each story, ask students questions about the different groups. For example:

- How many girls are reading ? How many boys are reading? How many students are reading in all?
- How many students are there in all? How many have blue shirts? How many have red shirts?

Textbook page 30 can be used as a continuation of this lesson, or as a second lesson as needed.

Extend

★ Number Story Pictures

Materials: Number Cards (BLM) 1 to 10, stickers or Picture Cards (BLM)

Have students draw their own number story pictures similar to the **Chapter Opener**. Have them share their pictures and ask the other students to tell numbers stories based on the pictures.

Objective

- Count to find how many objects there are in all.

Lesson Materials

- Linking cubes, 5 each of 2 colors per student or pair of students

30

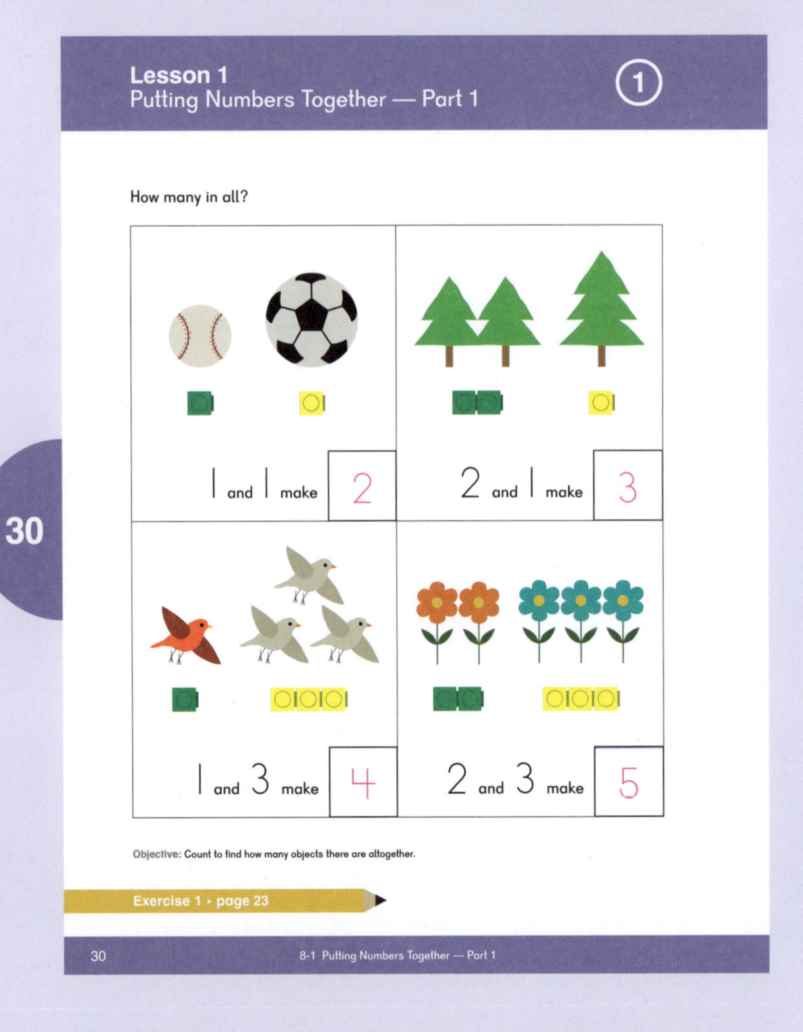
Lesson 1
Putting Numbers Together — Part 1
1

How many in all?

1 and 1 make 2

2 and 1 make 3

1 and 3 make 4

2 and 3 make 5

Objective: Count to find how many objects there are altogether.

Exercise 1 • page 23

30 8-1 Putting Numbers Together — Part 1

Explore

Provide each student or pair of students with 10 linking cubes, 5 each of 2 different colors. Using the **Chapter Opener** as a guideline, have students tell their own number stories based on objects that are in the classroom. Keep the whole or total items within 10.

As students tell stories, have them act out the stories with the 2 colors of linking cubes. Example:

- At the table, there are 4 student chairs and 1 teacher chair. There are 5 chairs in all.
- Students build a tower of 5 cubes with 4 of one color and 1 of another color.

Learn

Using the examples from **Explore** or the **Chapter Opener**, have students say how many of each item there are in all. Write their stories on the board using the terms "and" and "make." For example, "4 and 1 make 5," or, "2 and 2 make 4."

Small Group Activities

Textbook Page 30

▲ Telling Stories

Materials: 5 linking cubes in one color, 5 linking cubes in another color, Part-Whole Recording Worksheet (BLM), bags or baskets

Place the cubes in a bag or basket. Partners take turns grabbing a handful of cubes and telling a story to their partner. For example, if 4 green and 2 yellow cubes are grabbed, the student may tell a story such as, "There are 4 green apples and 2 yellow apples in the basket. How many apples are there altogether?" Have each student complete a Part-Whole Recording Worksheet (BLM) with the numbers used in each story.

Exercise 1 • page 23

Objective

- Count and model two sets to find how many altogether.

Lesson Materials

- Two-color counters, 5 per student

Lesson 2
Putting Numbers Together — Part 2
2

Look and talk.
How many altogether?

Objective: Count and model two sets for each item to find how many altogether.

8-2 Putting Numbers Together — Part 2 31

31

Explore

Call 5 students to the front of the classroom. Ask other students to tell stories about the 5 students. As they share stories, have students build a model with the two-color counters. For example, "2 students have brown hair and 3 students have blond hair." Students model with the counters.

Learn

Have students discuss the different items on page 31. Discuss how the animals are the same and different. Students should identify ways the animals can be sorted and say how many there are altogether. Students may see:

Snails:

- 1 shell is orange and 1 shell is blue. 1 and 1 make 2.
- There are 2 shells on the sand and 0 shells in the water. 2 and 0 make 2.

Turtles:

- There are 2 turtles in the water and 2 turtles on land. 2 and 2 make 4.
- There are 3 green turtles and 1 brown turtle. 3 and 1 make 4.

Fish:

- There are 2 big fish and 1 small fish. 2 and 1 make 3.
- There is 1 fish swimming to the left and 2 fish swimming to the right. 1 and 2 make 3.

Seagulls:

- There are 4 seagulls on the beach and 1 seagull is flying. 4 and 1 make 5.
- There is 1 seagull with feet showing and 4 with no feet showing. 1 and 4 make 5.
- There are 2 seagulls facing right and 3 seagulls facing left. 2 and 3 make 5.

Whole Group Activity

▲ Show Me

Ask students to show you multiple ways to make a specific number on their fingers.

For example, ask students to show you 5. Most will likely show one hand with 5 fingers. Ask them to show 5 another way, and students may show 2 fingers on one hand and 3 on the other, or 4 fingers on one hand and 1 on the other.

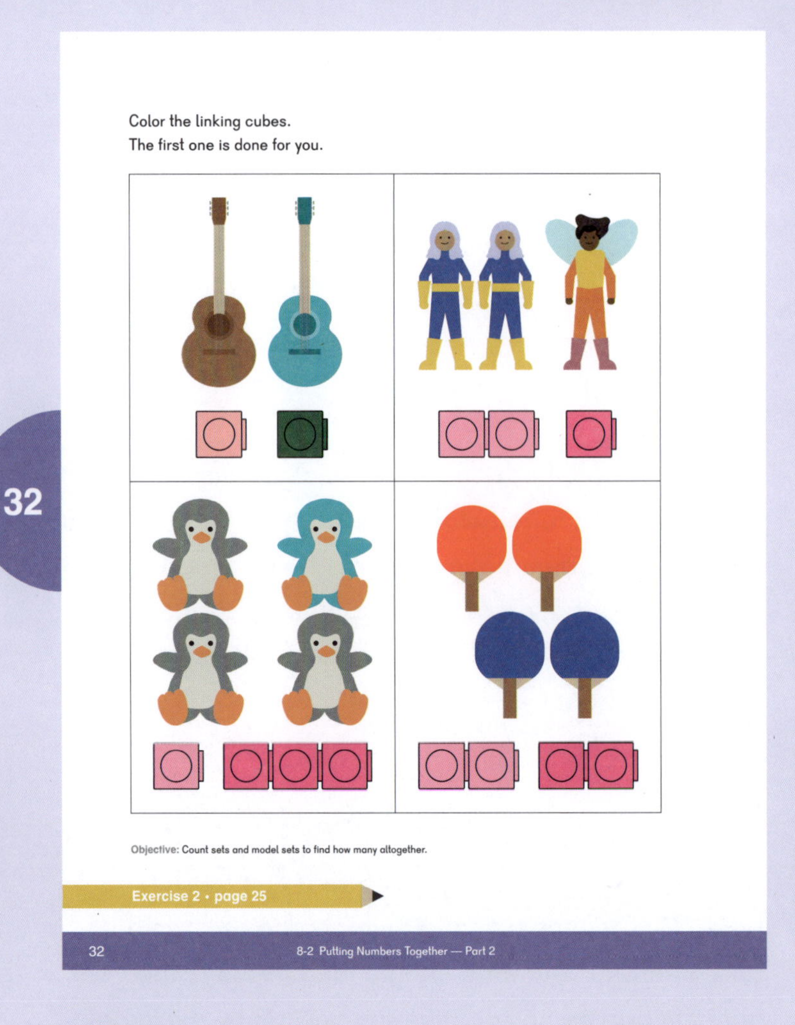

32

Small Group Activities

Textbook Pages 31–32. Ask students, "How are the guitars the same? How are they different?" Ask students to tell numbers stories about each group of objects. Students may say, "1 brown guitar and 1 blue guitar makes 2 guitars altogether." Repeat for other objects on the page.

▲ Toss-Up

Materials: 5 two-color counters per pair of students

Pairs of students take turns tossing the specified number of two-color counters. Count how many of each color lands upright and say how many they make in all. For example, toss:

And say, "4 red counters and 1 yellow counter make 5 counters in all." Partners can check each other's counting. Students will discuss the game further in the next lesson.

Extend

★ Under the Cup: Number Words

Materials: 5 counters and 1 cup per pair of students, Number Word Cards (BLM) 1 to 5

Students work in pairs. Player 1 hides some of the counters under the cup and leaves the rest to be seen. Player 2 counts the amount showing and finds the Number Word Card (BLM) for how many are under the cup. Players switch roles and play continues.

Exercise 2 • page 25

Objective

- Understand that parts can be put together to make a whole.

Lesson Materials

- Two-color counters, 5 per pair of students
- Number Bond Recording Sheet (BLM)
- 3 large hula hoops
- Painter's tape

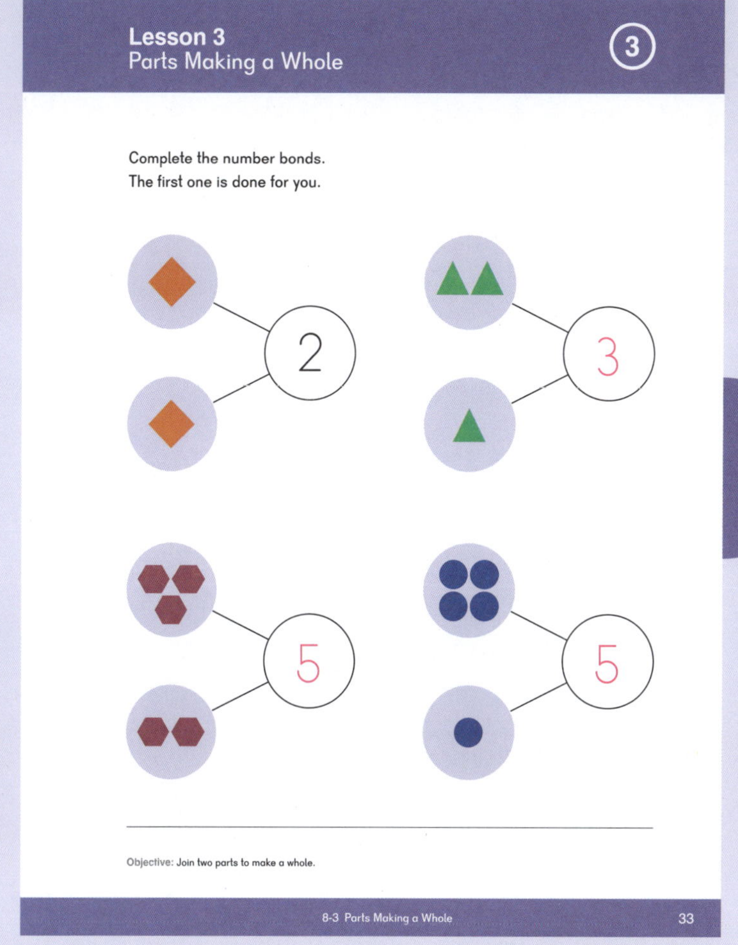

33

Explore

Ask students about the combinations they remember from playing **Toss-Up** with 5 counters in the previous lesson. Tell students you are going to use a way of organizing the counters called a "number bond."

Using a large number bond made of hula hoops or painter's tape, invite groups of students to stand in the parts and then move to the whole. For example, you might invite 2 girls and 1 boy to stand in the parts. When they move to the whole, students will see that 2 and 1 makes 3. Repeat with other combinations of numbers up to 5.

Learn

Provide students with a Number Bond Recording Sheet (BLM). Have them play the **Toss-Up** game using the number bond template to organize their counters. For example: While playing the game, students should slide the two parts together into the "whole."

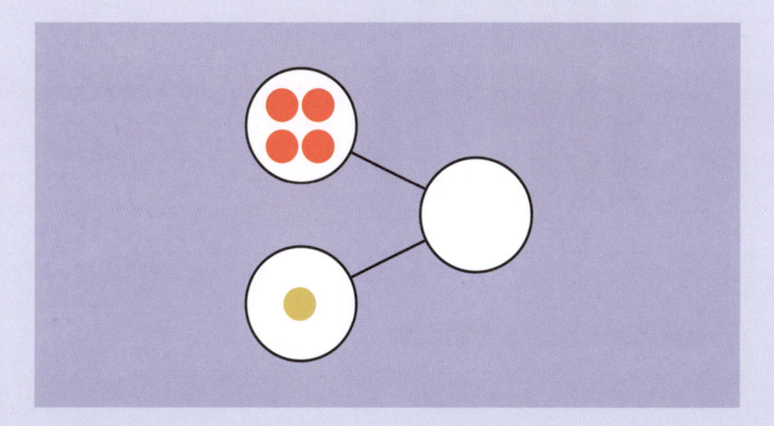

Encourage students to use the terms for "parts" and "whole:"

- Part of the counters are red and part of the counters are yellow. How many counters in all?
- 4 red counters and 1 yellow counter make 5 counters altogether.

After playing the game, ask students how they can write the parts in the corresponding number bond. Show some examples.

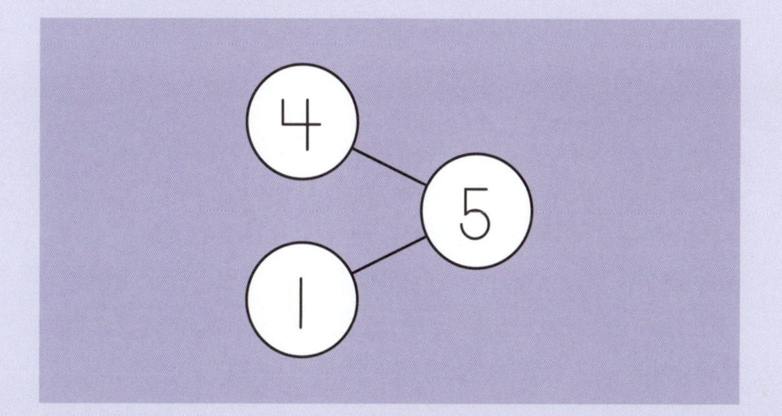

Whole Group Activity

▲ Show Me the Bond

On the board, write either the two parts, or one part and the whole in a number bond up to 5. Students then write the number that would complete the number bond.

Small Group Activities

Textbook Pages 33–35. Page 33 does not differentiate by categories. The emphasis is on putting the shapes together to see how many in all. On pages 34 and 35, encourage students to find the parts based on the color of the crayons. For example, "Part of the crayons are pink and part of them are brown."

▲ How Many Ways?

Materials: 5 two-color counters per student, Number Bond Story Template (BLM)

Have students place counters in the Number Bond Story Template (BLM) to create as many different combinations of parts to make 5 as possible.

Exercise 3 • page 27

Extend

★ Ten-frame to Number Bond

Materials: Classroom objects (such as paper clips and erasers that can be sorted) or two-color counters, Blank Number Bond Template (BLM), Blank Ten-frame (BLM)

Have students work in pairs and take turns. One student puts two sets of objects together with a whole up to 10 (they can also use two-color counters). Their partner sorts the objects into sets and represents the sets on a Blank Ten-frame (BLM), then fills in a Blank Number Bond Template (BLM).

34

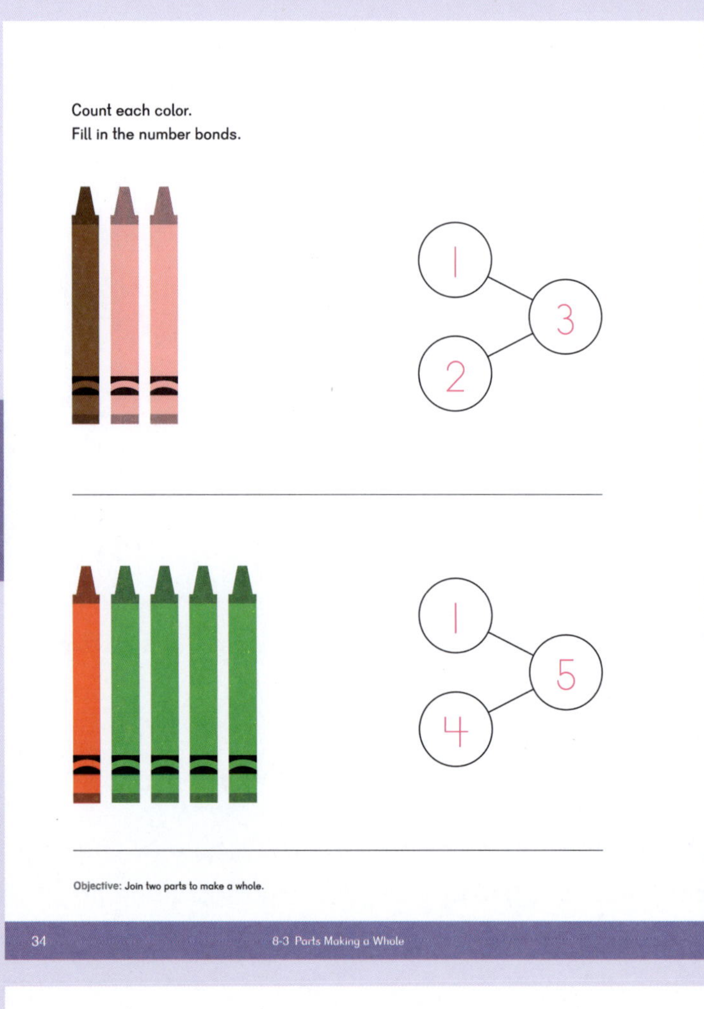

35

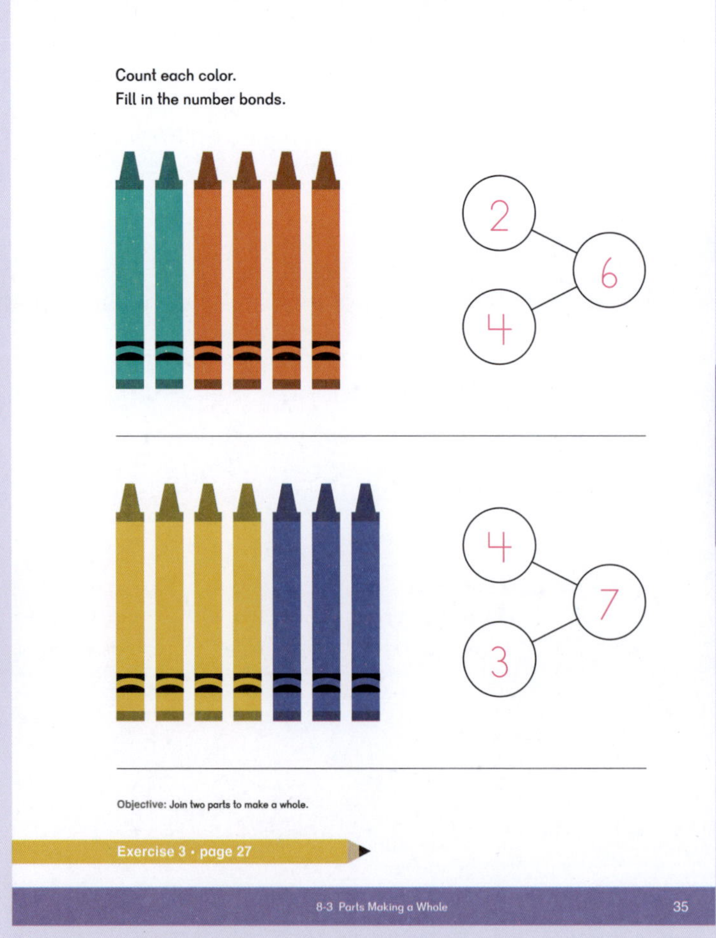

Objective

- Find parts in a number bond.

Lesson Materials

- 3 large hula hoops
- Painter's tape
- Attribute bears
- Blank Number Bond Template (BLM)

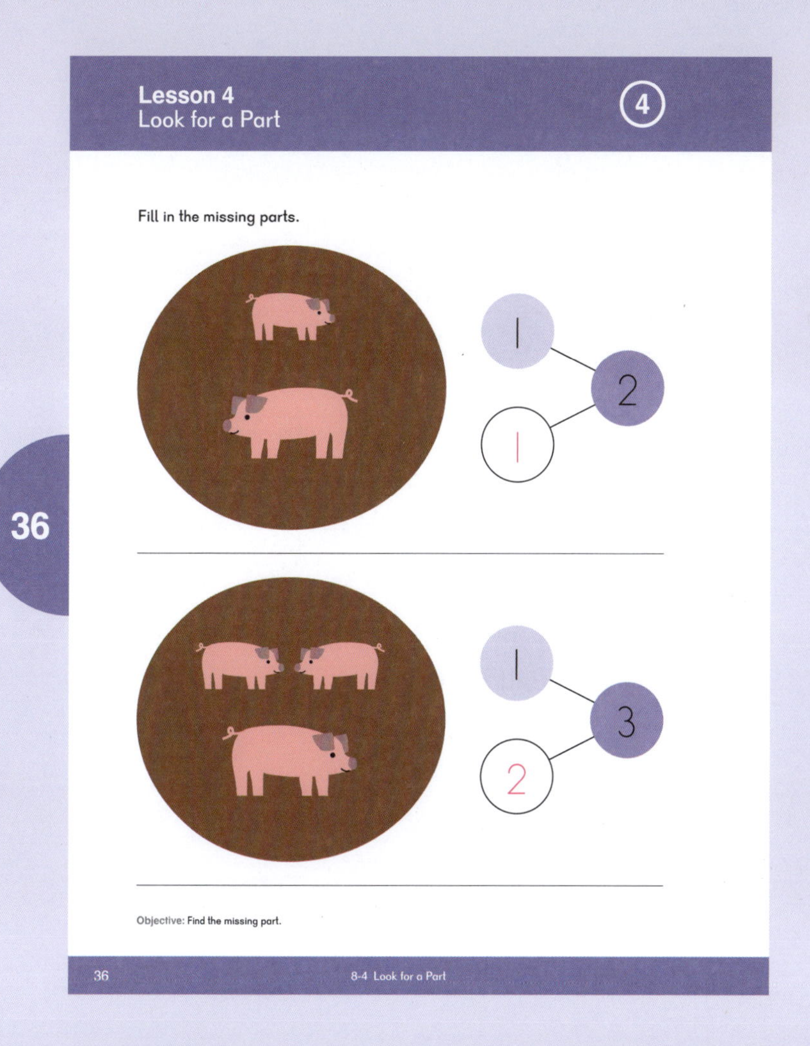

Explore

Have students act out a situation similar to the **Explore** activity in Lesson 3. Choose two students to stand in the "whole" part of the number bonds based on shirt color or shoe type. Ask the students to find something that is different between them and have them step to the two parts. Guide students to see, for example, that there were 2 students, 1 is wearing sneakers and 1 is wearing boots.

Repeat with up to 5 students sorting themselves into 2 groups.

Learn

Give pairs of students a handful of up to 5 attribute bears. Ask the partners to find 2 bears that are similar, but not exactly the same. Have students explain how they are different: color, size, laying down vs. sitting, etc. Using the Blank Number Bond Template (BLM), have students share how many there are in all, then how many of one part and how many of the second part.

Whole Group Activity

▲ Complete the Bond

Materials: Number Cards — Large (BLM) 2 to 5

Hold up a Number Card — Large (BLM) showing 2, 3, 4, or 5. Tell students that they will be finding the missing part to the whole on the number cards.

For example, if you hold up Number Cards — Large (BLM) 4, show one part with your fingers (2 fingers), and the students will show the missing part with their fingers (2 fingers). 2 and 2 make 4.

Small Group Activities

Textbook Pages 36–37

▲ Family Bonds

Materials: Items with a round face students can trace

Have students create a number bond representing their family. They might choose to show adults in one part, children in the other part, and the whole family in the whole. To aid in drawing the circles, give students items to trace (bowls, cups, plates, etc.).

Exercise 4 • page 29

Extend

★ Make 5

Materials: 4 sets of Number Cards (BLM) 0 to 5

Provide small groups of students with 4 sets of Number Cards (BLM) 0 to 5. Students shuffle the deck and arrange the cards facedown in a grid.

Players take turns flipping 2 cards. If they make 5, the student keeps the cards. The player with the most cards at the end wins.

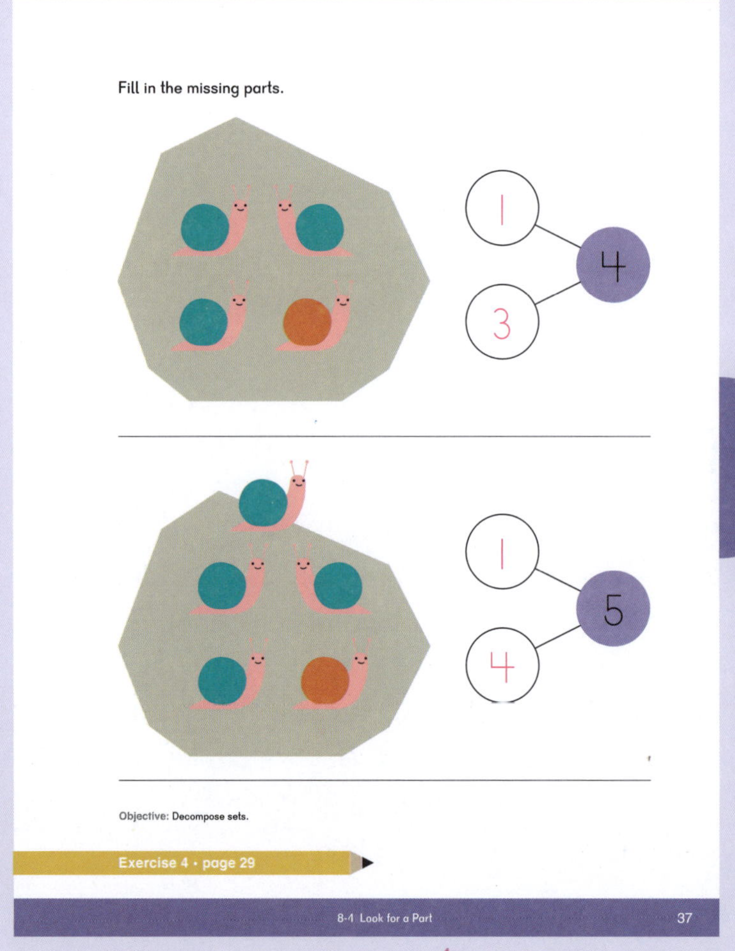

Lesson 5 Number Bonds for 2, 3, and 4

Objective

- Find two parts that make 2, 3, and 4.

Lesson Materials

- Blank Number Bond Template (BLM)
- Two-color counters, 4 per student

Explore

Provide pairs of students with two-color counters and the Blank Number Bond Template (BLM). Have students show and record number bonds that make 2, then 3, then 4.

Learn

Looking at page 38, tell students that Emma has a puzzle. She wants to make each number (2, 3, and 4) in as many ways as she can with 2 parts.

Discuss that it doesn't matter in what order we put the counters in the number bond. The total is the same.

Students should see the following:

- 1 and 1 make 2.
- 2 and 1 make 3 **and** 1 and 2 make 3.
- 3 and 1 make 4 **and** 1 and 3 make 4.
- 2 and 2 make 4.

Students may also extend to include:

- 2 and 0 make 2 **and** 0 and 2 make 2.
- 3 and 0 make 3 **and** 0 and 3 make 3.
- 4 and 0 make 4 **and** 0 and 4 make 4.

Whole Group Activity

▲ Show Me the Bond

On the board, write either the two parts, or one part and the whole in a number bond up to 4. Students then write the number that would complete the number bond.

Small Group Activities

Textbook Pages 39–40

▲ Number Bond Bracelets

Materials: 4 pipe cleaners and 9 beads per student, painter's tape, 4 hula hoops and cut-up pool noodles (optional)

Create bracelets for the numbers 2–4:

- One with 2 beads of the same color
- One with 3 beads of the same color
- One with 4 beads of the same color

Twist the ends together to make a bracelet, or "ring." Fold a piece of painter's tape over and write the number 2, 3, or 4, on each respectively, with the student's name on the back.

Students will be making a set of the bracelets from the number bonds from 5 to 10 in future lessons.

To use the bracelet for 4, for example, have students begin with the 4 beads together, then slide 1 to a side and note that 1 and 3 make 4. Move all the beads together again to show 4, then slide 2 to a side and see that 2 and 2 also make 4.

A hula-hoop and cut up pool noodles make a nice oversized display number bracelet.

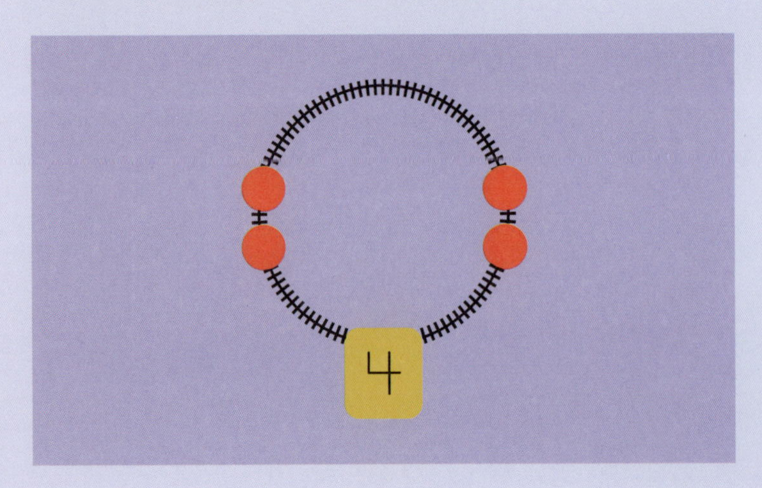

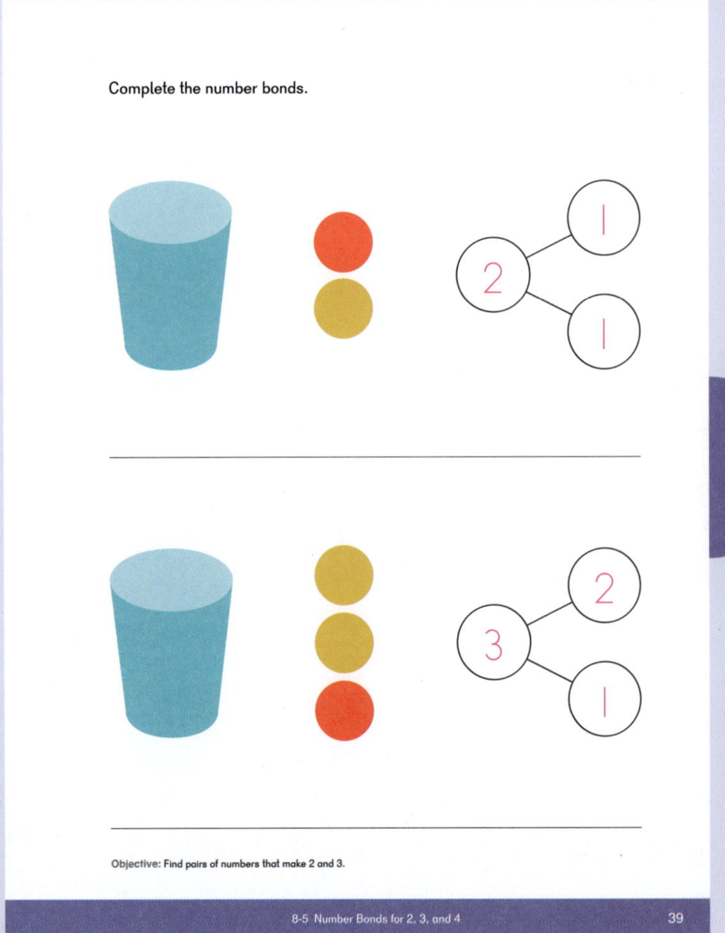
Complete the number bonds.

2 — 1, 1

3 — 2, 1

Objective: Find pairs of numbers that make 2 and 3.

8-5 Number Bonds for 2, 3, and 4 39

39

Exercise 5 • page 31

Extend

★ Super Number Bond

Provide students with a bond similar to the one shown below. Ask students to complete the bond using the numbers 0 through 9. Challenge students to make as many variations as possible.

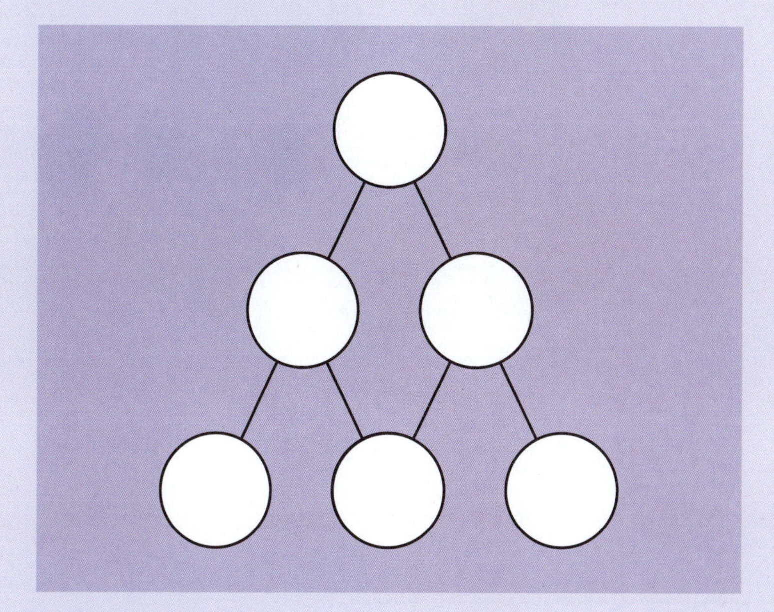

One solution:

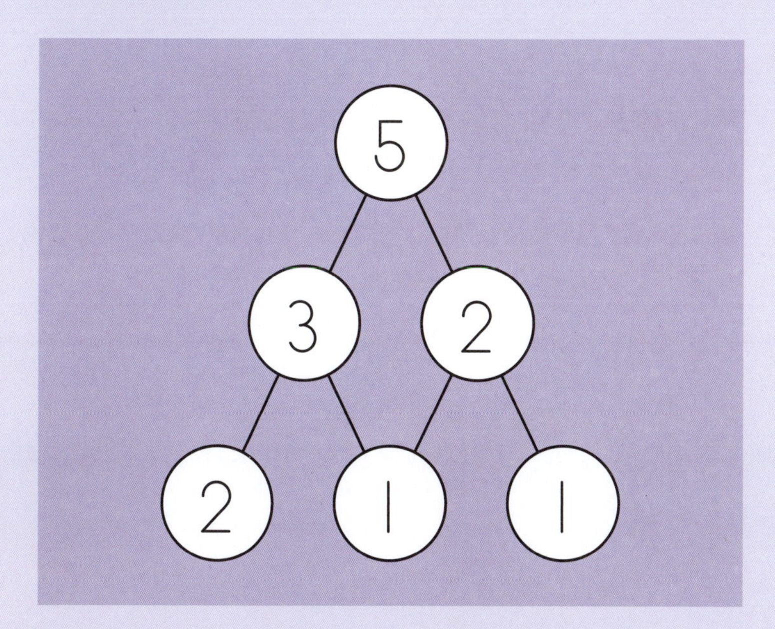

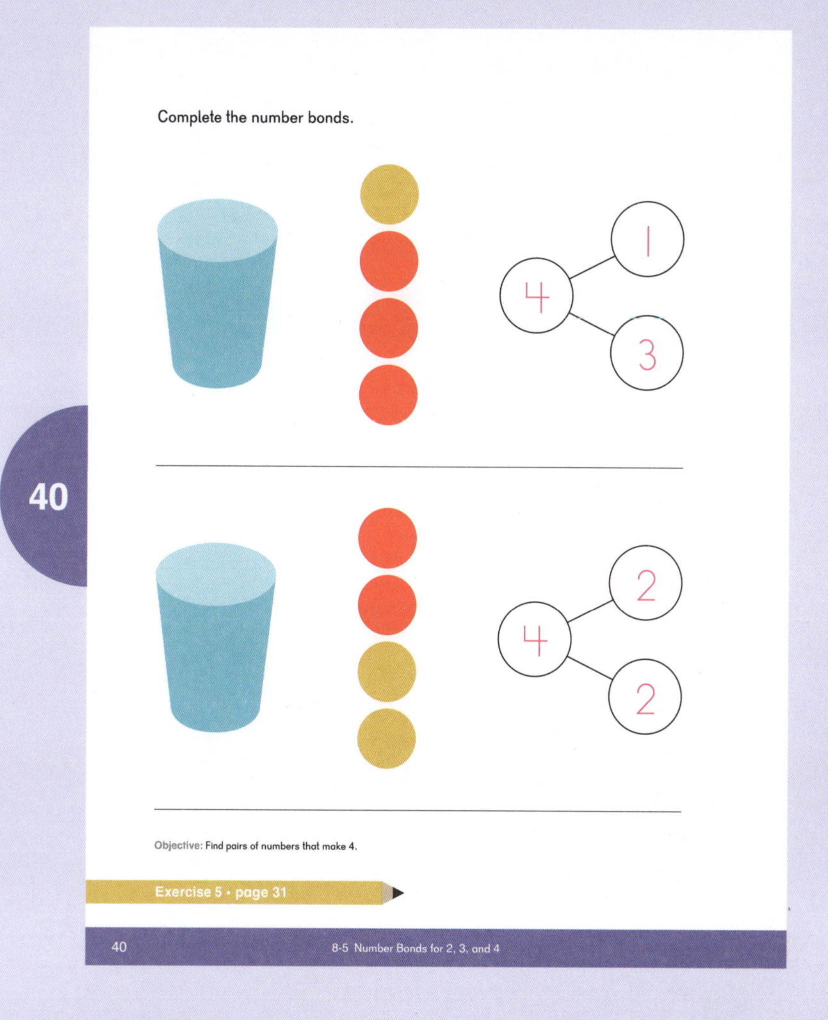

★ Building Bonds

Materials: Number Cards (BLM) 0 to 9

Have students make as many correct number bonds as they can using Number Cards (BLM). Use the Number Cards (BLM) 0 to 9 only once in each bond.

Lesson 6 Number Bonds for 5

Objective

- Find 2 parts that make 5.

Lesson Materials

- 20 linking cubes, 10 each of 2 different colors, per pair of students

41

Explore

Provide each pair of students with 20 linking cubes, 10 of each of 2 colors. Have the partners make 5 as many ways as they can with the 2 colors of cubes. Have students share their answers.

Learn

Look at textbook page 41. Discuss Emma's cube towers. Ask questions like, "How are each teams' towers like Emma's? How are they different?"

Have students organize their cubes in a step pattern similar to Emma's. Have students identify all of the number bonds for 5 from the block pattern.

Whole Group Activities

▲ Show Me the Bond

On the board, write either the two parts, or one part and the whole in a number bond up to 5. Students then write the number that would complete the number bond.

▲ Show Me

Ask students to show you 5 on their fingers. Ask students to show you multiple ways to make the number using their fingers.

Small Group Activities

Textbook Pages 42–44

▲ Number Bond Bracelets

Materials: 1 pipe cleaner and 5 beads (all the same color) per student

Have each student create a bracelet for the number 5 as described in the previous lesson.

▲ Go Fish

Materials: 4 sets of either Ten-frame Cards (BLM) or Number Cards (BLM) per small group

Shuffle 4 sets of either Ten-frame Cards (BLM) or Number Cards (BLM), using cards from 0 to 5. Deal 4 cards to each player. Players play **Go Fish** with the cards, where a pair is a set of cards that together makes 5.

Use Ten-frame Cards (BLM) for students who are still counting. Number Cards (BLM) can be used for students who understand the abstract number without counting.

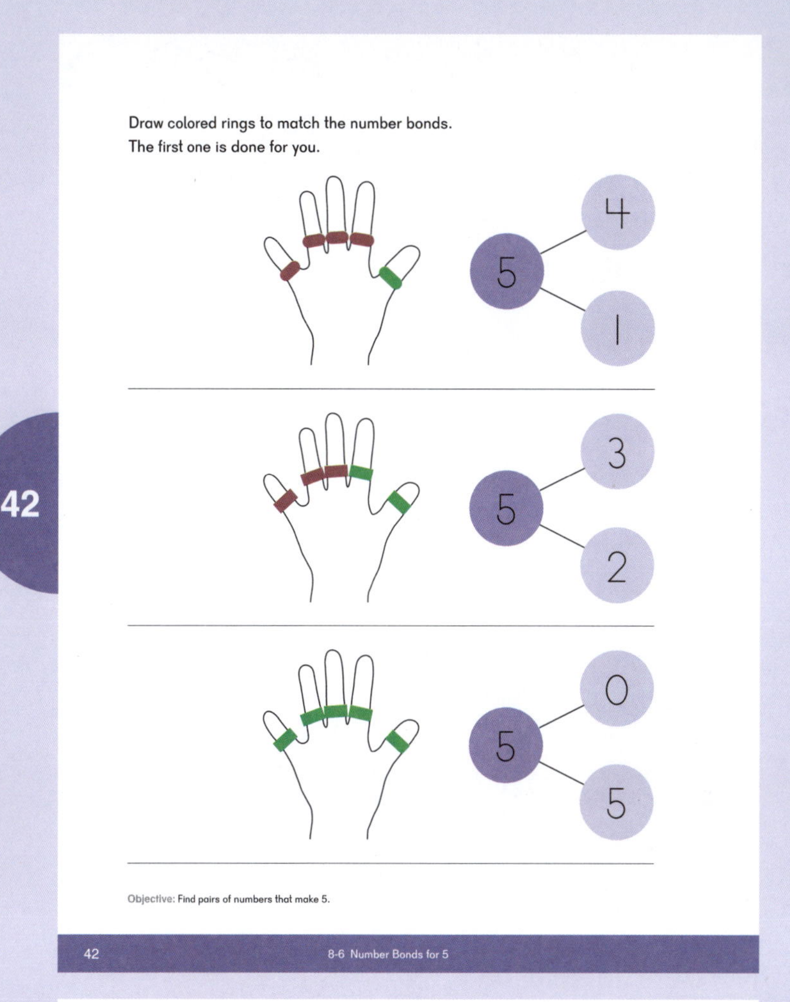

▲ Domino Parking Lot

Materials: Dominoes, Domino Parking Lot (BLM)

Use only dominoes with pips that add up to 5, and only columns 1–5 on the Domino Parking Lot (BLM). Students sort the dominoes based on how many pips there are on the two halves.

▲ Under the Cup

Materials: 5 counters and 1 cup or 5 linking cubes, number bond bracelets (optional)

Give students 5 counters and a cup. One student will hide some of the counters under the cup, then the other student will figure out how many are hiding based on how many are visible. Number bond bracelets can be used to help determine how many are hiding.

Alternatively, students can play with 5 linking cubes assembled into a tower. Player 1 breaks the tower behind his back. He shows one part of the tower to Player 2. Player 2 says how many cubes are hidden behind Player 1's back.

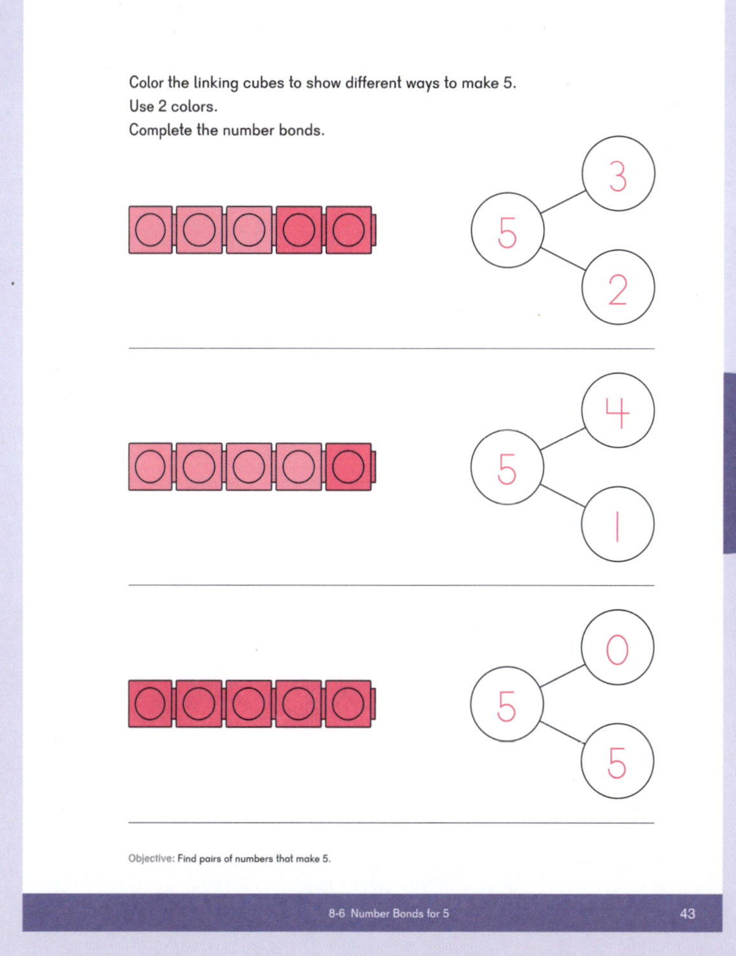

Exercise 6 • page 33

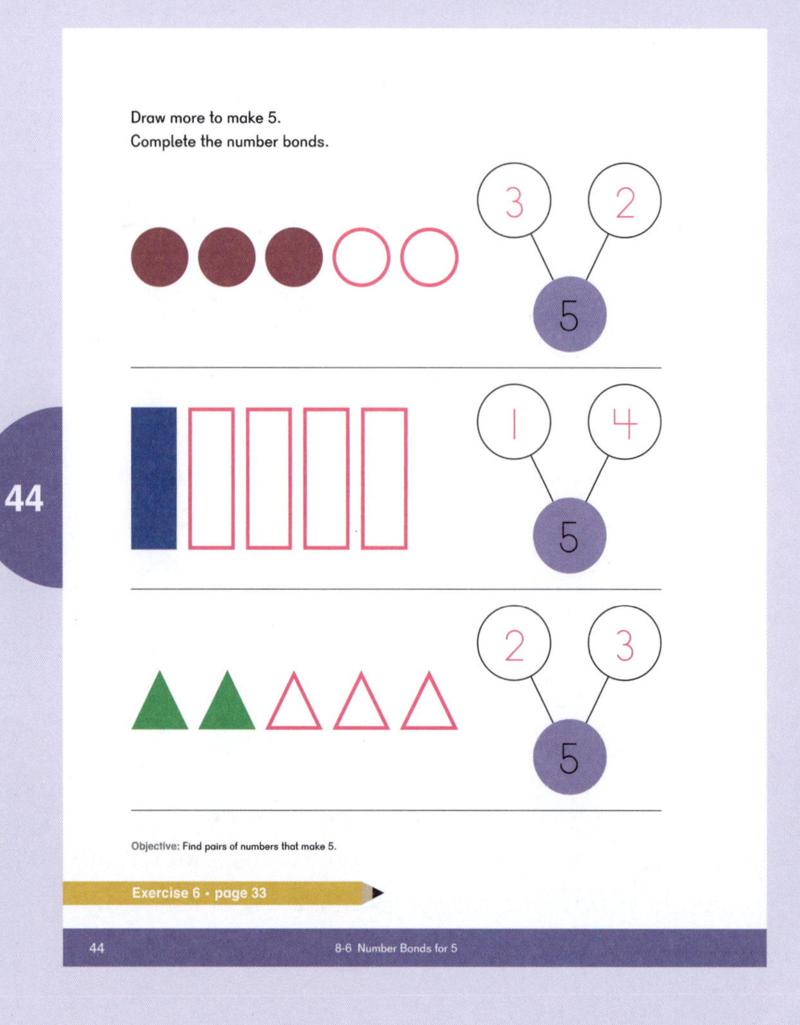

Extend

★ **Number Bond Puzzler**

Read the following statements to students:

- We know that Mei has 3 as a whole.
- Emma's number bond has parts less than 5.
- Mei's number bond has two parts that are 1 number apart from each other.
- Dion's number bond whole is the same as the smaller part of Mei's.
- Emma's parts don't match Dion's or Mei's parts.
- If Mei's whole is 3, what are the other numbers?

Then ask:

- What do we know?
- Can we find Mei's parts?

Reread the statements as many times as necessary. Statements could be recorded for students to listen to on their own.

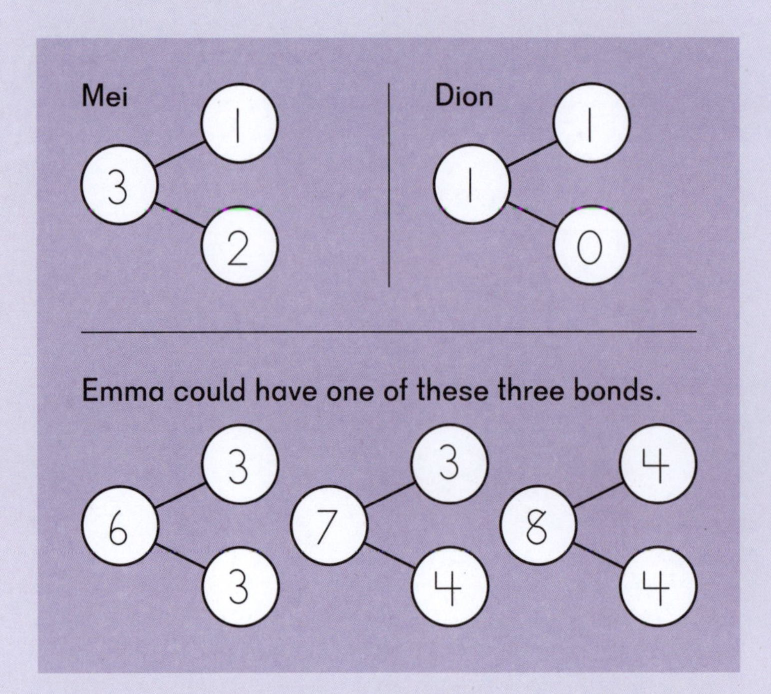

Lesson 7 Number Bonds for 6

Objective

- Find 2 parts that make 6.

Lesson Materials

- Two-color counters, 6 per student
- Blank Number Bond Template (BLM)

45

Explore

Provide each student with 6 two-colored counters and a Blank Number Bond Template (BLM). Ask them to explore ways to make 6 using the counters.

Learn

Have students discuss the different ways to make 6 using the illustration of dogs on page 45. Student responses may include:

- There is 1 gray dog and 5 brown dogs. There are 6 dogs in all. (There are 6 dogs. 1 is gray and 5 are brown.)
- There are 2 dogs with spots and 4 dogs with no spots. There are 6 dogs in all. (There are 6 dogs in all. 2 dogs have spots. 4 dogs have no spots.)
- 3 dogs are standing and 3 dogs are running. There are 6 dogs altogether. (There are 6 dogs in all. 3 dogs are running. 3 dogs are not running.)
- 3 dogs are on the yellow rug and 3 dogs are on the green carpet. There are 6 dogs altogether.

Have students model the number stories about the dogs using their counters on a number bonds template.

▲ Show Me

Ask students to show you 6 on their fingers. Ask students to show you multiple ways to make the number using their fingers.

Whole Group Activities

▲ Show Me the Bond

On the board, write either the two parts, or one part and the whole in a number bond up to 6. Students then write the number that would complete the number bond.

Small Group Activities

Textbook Pages 46–48

▲ Number Bond Bracelets

Materials: 1 pipe cleaner and 6 beads per student, painter's tape

Create bracelets as described in Lesson 5 for the number 6.

▲ Go Fish

Materials: 4 sets of either Ten-frame Cards (BLM) or Number Cards (BLM) per small group

Play the game as described in Lesson 6, but using cards from 0 to 6.

Use Ten-frame Cards (BLM) for students who are still counting. Number Cards (BLM) can be used for students who understand the abstract number without counting.

▲ Toss-Up

Materials: 6 two-color counters per pair of students

Play as directed in Lesson 2, using 6 two-color counters.

▲ Domino Parking Lot

Materials: Dominoes, Domino Parking Lot (BLM)

Use only dominoes with pips that add up to 6, and only columns 1–6 on the Domino Parking Lot (BLM). Students sort the dominoes based on how many pips there are on the two halves.

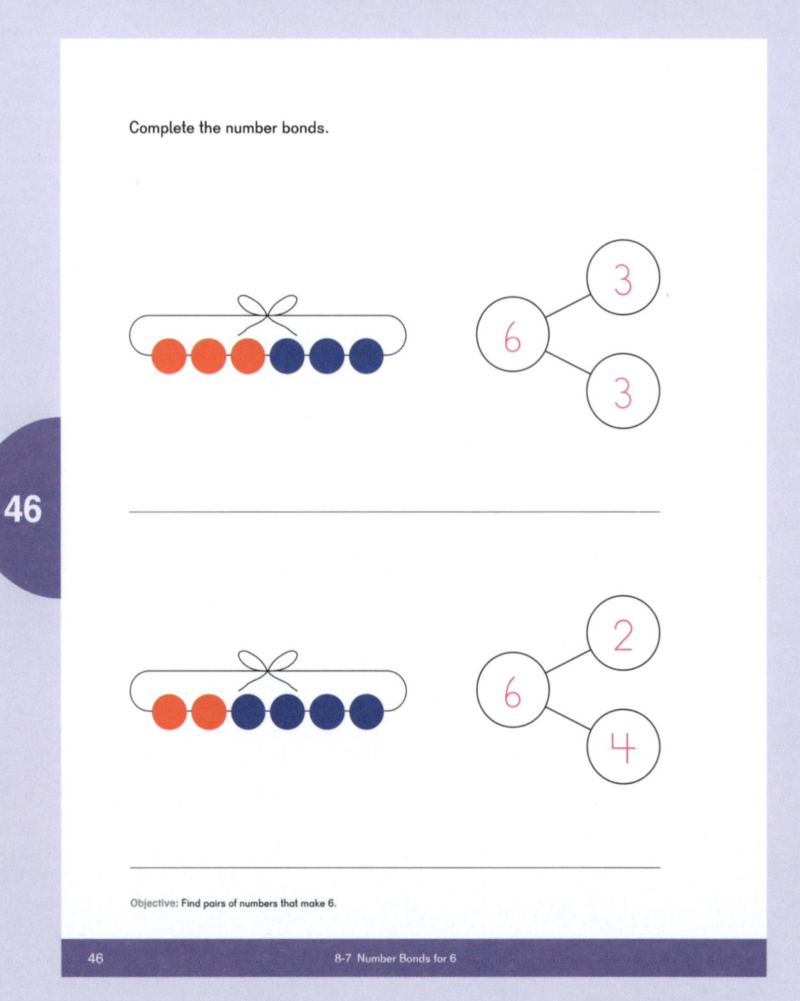

Exercise 7 • page 35

Extend

★ What Comes in Groups of 6?

Materials: Hexagon shape from pattern blocks, or Hexagon Template (BLM)

Have students think of things that come in 6. Then have them draw the objects in two sets.

For example, there are six sides on a hexagon. Have them trace around a hexagon or a cut-out Hexagon Template (BLM), then color the sides using two different colors. Have them do that as many ways as possible, filling in a number bond for each way shown. Repeat with other objects.

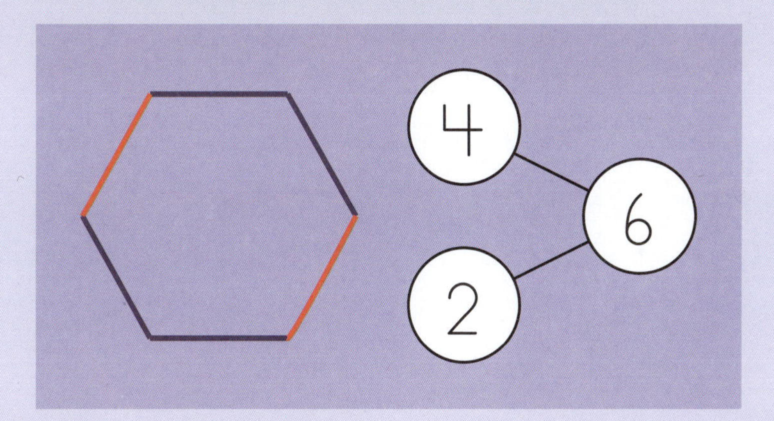

Things that come in 6:

- Legs on an ant
- Cans of soda
- Sides on a die
- Players on a hockey team
- A carton of eggs

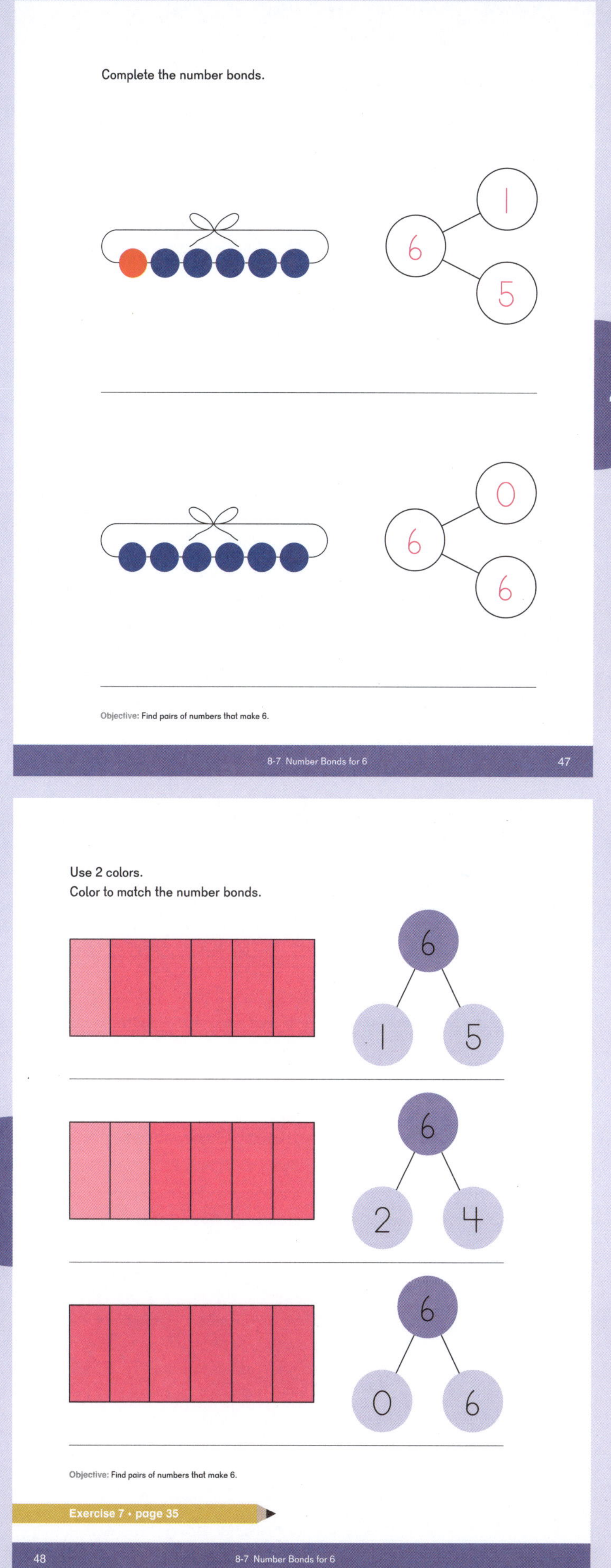

Lesson 8 Number Bonds for 7

Objective

- Find 2 parts that make 7.

Lesson Materials

- Two-color counters, 7 per student
- Blank Number Bond Template (BLM)

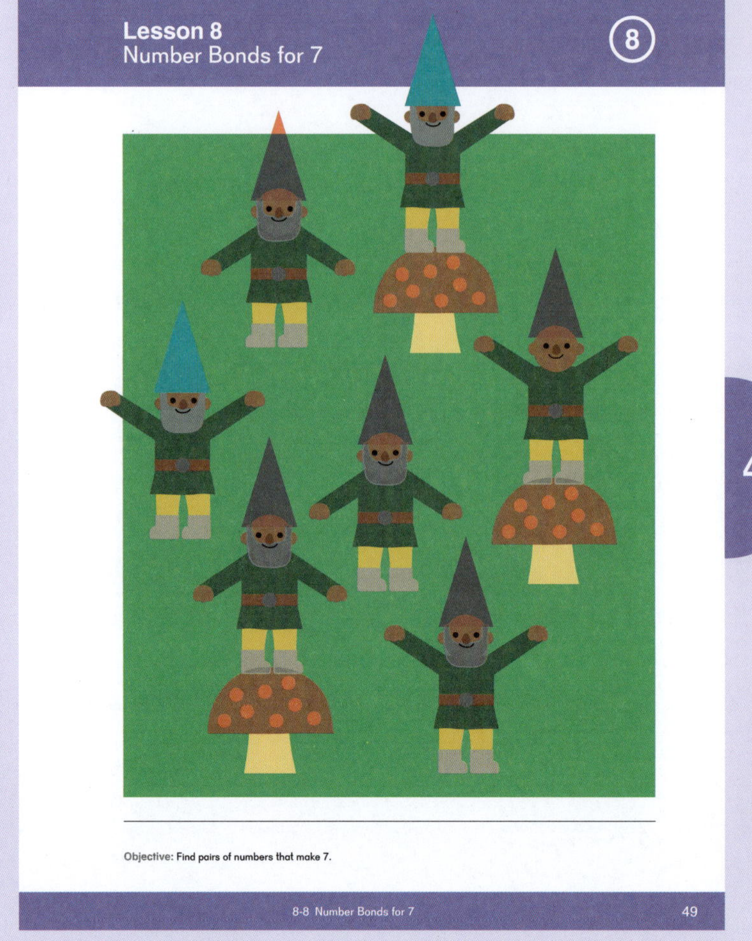

49

Explore

Provide each student with 7 two-colored counters and a Blank Number Bond Template (BLM). Ask students to explore ways to make 7 using the counters.

Learn

Look at the illustration on page 49. Have students make up and share different number stories for the 7 gnomes. Have students model number stories about the gnomes using their counters on a number bonds template. For example:

- There is 1 gnome without a beard and 6 gnomes with beards. There are 7 gnomes in all. (There are 7 gnomes, 1 has a beard and 6 do not have beards.)
- There are 2 gnomes with blue hats and 5 gnomes with orange hats.
- There are 3 gnomes on toadstools and 4 gnomes on the ground.
- There are 4 gnomes with their hands in the air and 3 gnomes with their hands at their sides.

Whole Group Activities

▲ Show Me the Bond

On the board, write either the two parts or one part and the whole in a number bond up to 7. Students then write the number on their personal boards that would go in either the whole or part.

▲ Show Me

Ask students to show you 7 on their fingers. Ask students to show you multiple ways to make the number using their fingers.

Small Group Activities

Textbook Pages 50–52

▲ Number Bond Bracelets

Materials: 1 pipe cleaner and 7 beads per student, painter's tape, Number Bond Book (BLM)

Create bracelets as described in Lesson 5 for the number 7. Students use the bracelet made in this lesson and prior lessons to complete number bonds in their Number Bond Book (BLM) showing ways to make 2 through 7.

▲ Under the Cup

Materials: 7 counters, 1 cup, number bond bracelets (optional)

Play the game as described in Lesson 6, but for number bonds for 7.

▲ Toss-Up Stories

Materials: 8 two-color counters per pair of students

Play as directed in Lesson 2, using 8 two-color counters. After a throw, students write or illustrate a story to match their tosses.

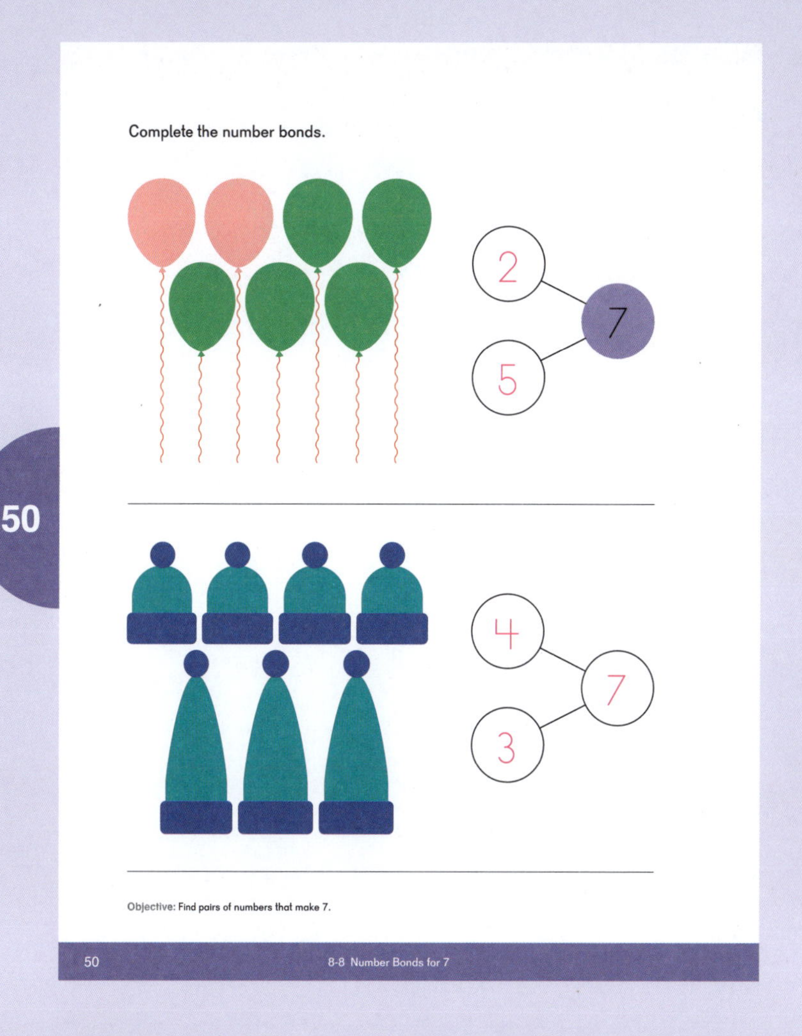

▲ Match

Materials: 2 sets of Ten-frame Cards (BLM) 0 to 7

Students arrange the cards faceup in a grid. Students take turns finding two cards that go together to make 7.

★ Memory

Materials: 2 sets of Ten-frame Cards (BLM) 0 to 7

Students arrange the cards facedown in a grid. Students take turns finding two cards that go together to make 7.

▲ Number Bond Chains

Materials: Index cards, strips of paper in a variety of colors

Provide students with different color strips of paper and an index card. Have them choose and write a number bond for 7 on the index card, then make a chain with two colors to show the parts. Attach the card to the chain and display in the classroom.

Exercise 8 • page 37

Extend

★ Days of the Week

Materials: 7-day calendar strip or 7 Days of the Week Template (BLM)

Have students draw two things they do on the days of a typical week. For example:

- Go to school vs. not go to school
- Go to sports practice vs. no sports practice
- Go home on the bus vs. walk home from school
- Do a chore vs. do a different chore

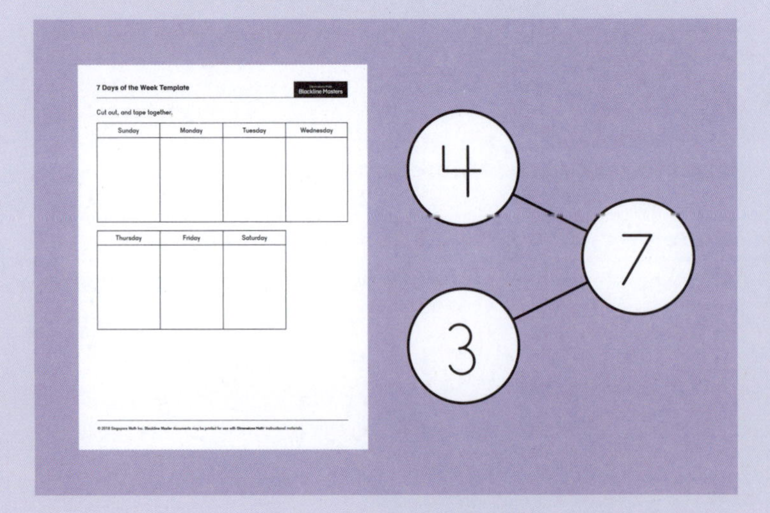

Objective

- Find 2 parts that make 8.

Lesson Materials

- Two-color counters, 8 per student
- Blank Number Bond Template (BLM)

Explore

Provide each student with 8 two-color counters and a Blank Number Bond Template (BLM). Ask students to explore ways to make 8 using the counters.

Learn

Look at page 53. Have students make up and share number stories for the 8 flowers.

Have students model stories about the flowers using their counters on a Blank Number Bond Template (BLM).

For example:

- There is 1 pink flower and 7 red flowers. 1 and 7 make 8.
- There are 2 flowers with no leaves and 6 flowers with leaves. 2 and 6 make 8.
- There are 3 flowers with blue centers and 5 flowers with yellow centers. 3 and 5 make 8.
- 4 flowers are short and 4 flowers are tall. (4 flowers are in one box and 4 flowers in the other box.) 4 and 4 make 8.

Whole Group Activity

▲ Show Me the Bond

On the board, write either the two parts or one part and the whole in a number bond up to 8. Students then write the number on their personal boards that would go in either the whole or part.

Small Group Activities

Textbook Page 54

▲ Number Bond Bracelets

Materials: 1 pipe cleaner and 8 beads per student, painter's tape, Number Bond Book (BLM)

Create bracelets as described in Lesson 5 for the number 8. Using the Number Bond Book (BLM), have students record all the ways to make 8.

▲ Under the Cup

Materials: 8 counters, 1 cup, number bond bracelets (optional)

Play the game as described in Lesson 6, but for number bonds for 8.

▲ Toss-Up Stories

Materials: 8 two-color counters per pair of students

Play as directed in Lesson 2, using 8 two-color counters. After a throw, students write or illustrate a story to match their tosses.

▲ Number Bond Chains

Materials: Index cards, strips of paper in a variety of colors

Play as directed in the previous lesson, creating number bond chains for the number 8.

Exercise 9 • page 39

54

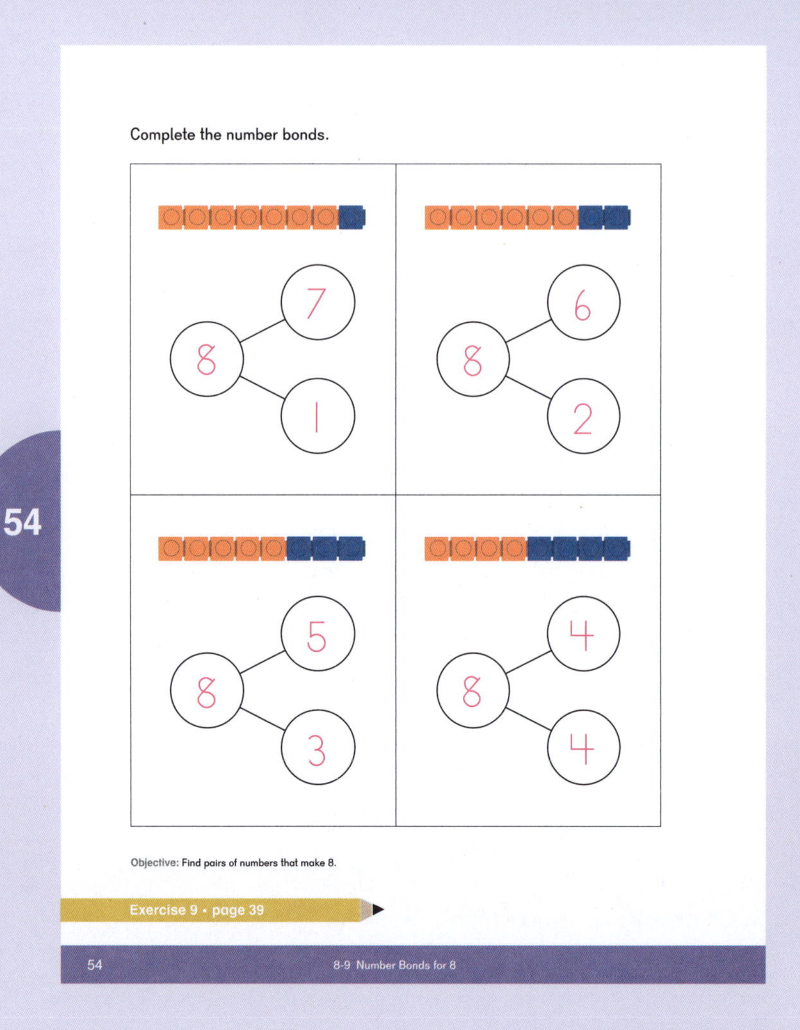

Complete the number bonds.

8 → 7, 1	8 → 6, 2
8 → 5, 3	8 → 4, 4

Objective: Find pairs of numbers that make 8.

Exercise 9 • page 39

54 8-9 Number Bonds for 8

Extend

★ What Comes in Groups of 8?

Materials: Octopus Template with Bond (BLM), crayons or markers

Have students think of things that come in 8. Then have them draw the objects in two or more colors.
For example, there are eight legs on an octopus. Have them use Octopus Template with Bond (BLM) to color octopuses using two or more different colors for the legs. Have them do that as many ways as possible, filling in the number bond for each way shown.

Things that come in groups of 8:

- Legs on a spider or octopus
- Sides on a stop sign
- Notes in a musical scale
- Party hats in a package

Lesson 10 Number Bonds for 9

Objective

- Find 2 parts that make 9.

Lesson Materials

- Two-color counters, 9 per student
- Blank Number Bond Template (BLM)

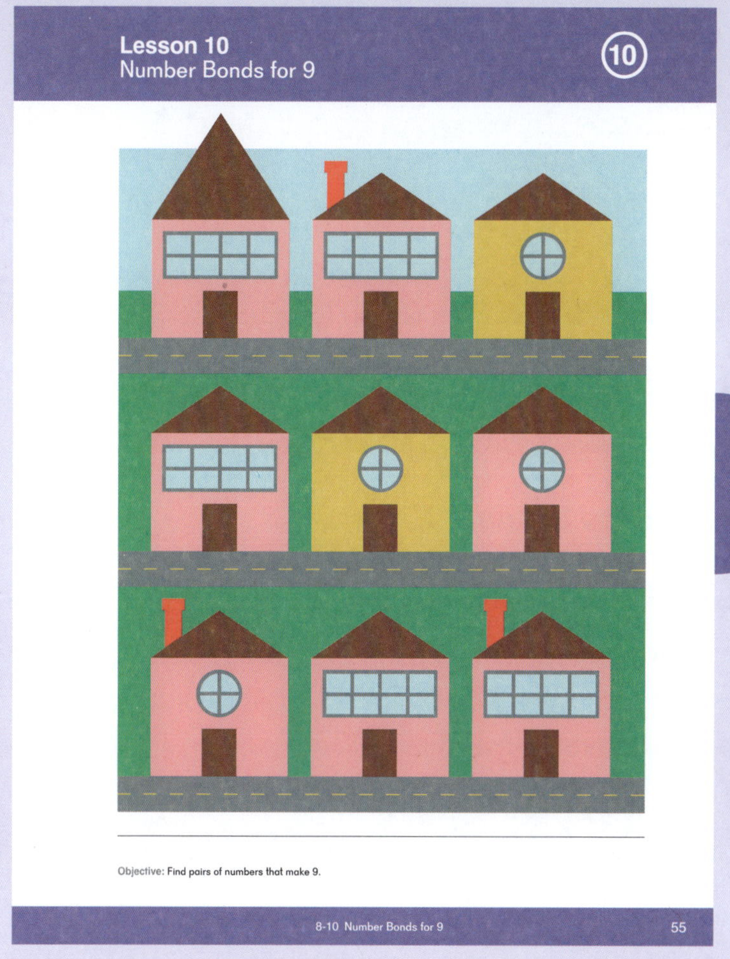

55

Explore

Provide students 9 two-colored counters and a Blank Number Bond Template (BLM). Ask students to explore ways to make 9 using the counters.

Learn

Look at page 55. By this lesson, students should be comfortable discussing and sharing the number bonds they see in the picture. Challenge students to find all of the combinations that make 9.

Have students model the number stories about the houses using their counters and a Blank Number Bond Template (BLM). For example, on page 55, students may see:

- There is 1 house with a tall roof and 8 with shorter roofs. 1 and 8 make 9.
- There are 2 yellow houses and 7 pink houses. 2 and 7 makes 9.
- There are 3 houses with chimneys and 6 houses with no chimneys. 3 and 6 make 9.
- 4 houses have windows that are circles and 5 houses have windows that are rectangles. 4 and 5 make 9.
- 9 houses have doors (or triangle roofs) and 0 houses don't have doors (flat roofs). 9 and 0 make 9.

Whole Group Activity

▲ Show Me the Bond

On the board, write either the two parts, or one part and the whole in a number bond up to 9. Students then write the number on their personal boards that would go in either the whole or part.

Small Group Activities

Textbook Pages 56–58

▲ Number Bond Bracelets

Materials: 1 pipe cleaner and 9 beads per student, painter's tape, Number Bond Book (BLM)

Create bracelets as described in Lesson 5 for the number 9. Using the Number Bond Book (BLM), have students record all the ways to make 9.

▲ Go Fish

Materials: 4 sets of either Ten-frame Cards (BLM) or Number Cards (BLM) per small group

Play the game as described in Lesson 6, but using cards from 0 to 9.

Use Ten-frame Cards (BLM) for students who are still counting. Number Cards (BLM) can be used for students who understand the abstract number without counting.

▲ Toss-Up Stories

Materials: 9 two-color counters per pair of students

Play as directed in Lesson 2, using 9 two-color counters. After a throw, students write or illustrate a story to match their tosses.

▲ Number Bond Chains

Materials: Index cards, strips of paper in a variety of colors

Play as directed in the previous lesson, creating number bond chains for the number 9.

▲ Number Bond Stories

Materials: Art paper, crayons or markers

Illustrate and share stories for number bonds for 9.

▲ Match

Materials: 2 sets of Ten-frame Cards (BLM) 0 to 9

Students arrange the cards faceup in a grid. Students take turns finding two cards that go together to make 9.

★ Memory

Materials: 2 sets of Ten-frame Cards (BLM) 0 to 9

Students arrange the cards facedown in a grid. Students take turns finding two cards that go together to make 9.

56

Complete the number bonds.

9 — 1, 8

9 — 2, 7

9 — 3, 6

9 — 4, 5

Objective: Find pairs of numbers that make 9.

56 8-10 Number Bonds for 9

Exercise 10 • page 41

Extend

★ Match Me

Materials: Markers, crayons, or paint daubers, blank index cards — 2 per student

Have each player use 2 colors to create a dot card showing 9 dots.

After each player has created a dot card, give each player another blank card. Player 1 then describes the dot card they made to Player 2, who tries to recreate Player 1's card on their blank index card.

Players trade roles and repeat. When both players have tried to recreate their partner's card, have them compare their artwork to the original and discuss what is similar or different.

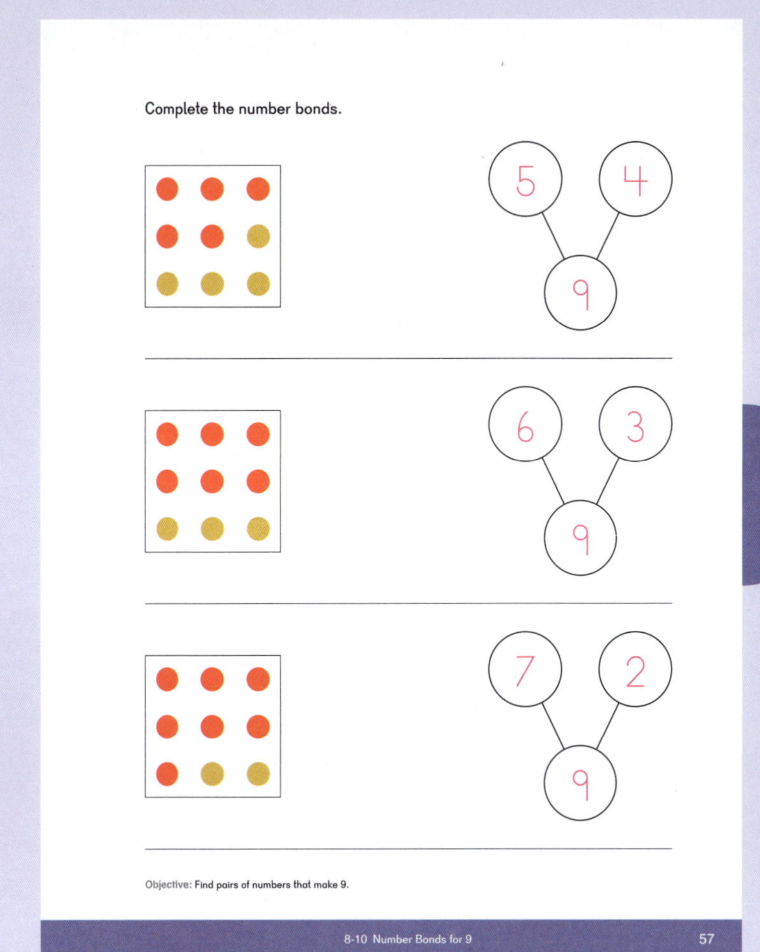

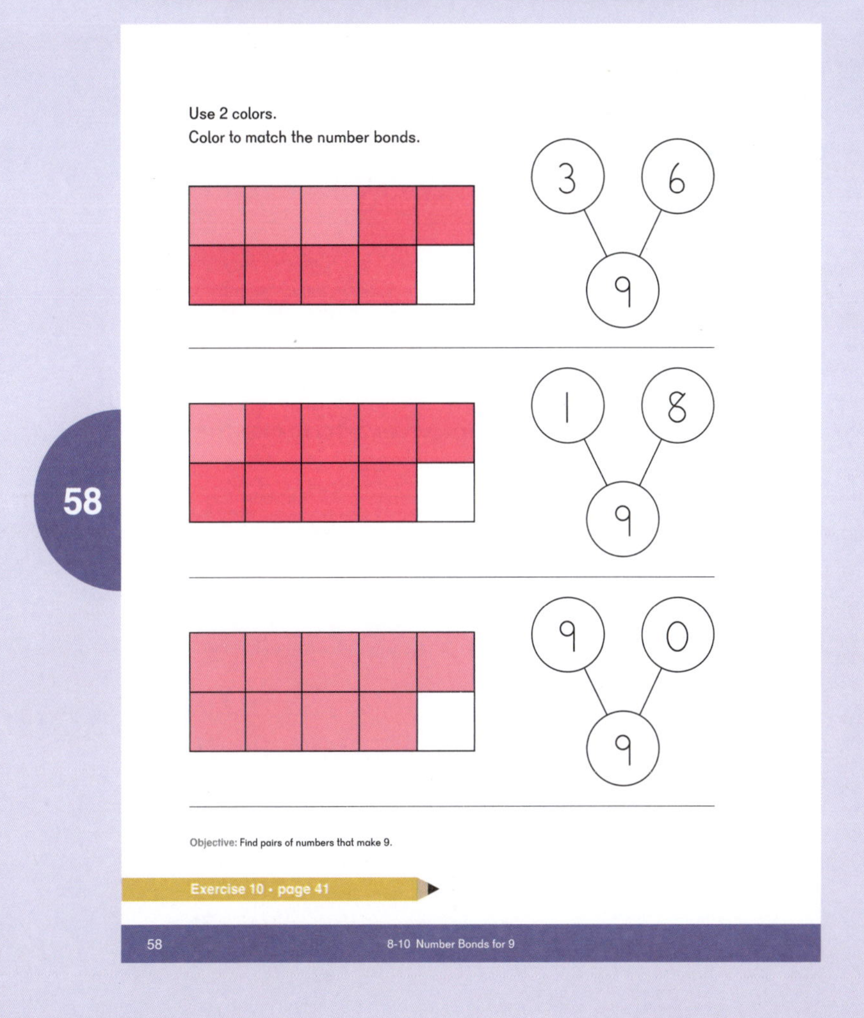

Lesson 11 Number Bonds for 10

Objective

- Find 2 parts that make 10.

Lesson Materials

- Two-color counters, 10 per student
- Blank Number Bond Template (BLM)
- Number Bond Recording Sheet (BLM)
- Blank Ten-frames (BLM)

Explore

Provide students with 10 two-color counters, a Blank Number Bond Template (BLM), and a Number Bond Recording Sheet (BLM). Ask students to explore ways to make 10 using the counters on their number bond template while recording them on their recording sheet.

Learn

Pass out Blank Ten-frames (BLM) to students. Ask them to place counters on the ten-frames, as you discuss ways to make 10 from **Explore**. Have students discuss the patterns they see on the ten-frames.

Whole Group Activity

▲ Show Me

Ask students to show you 10 on their fingers. Ask students to show you multiple ways to make the number using their fingers.

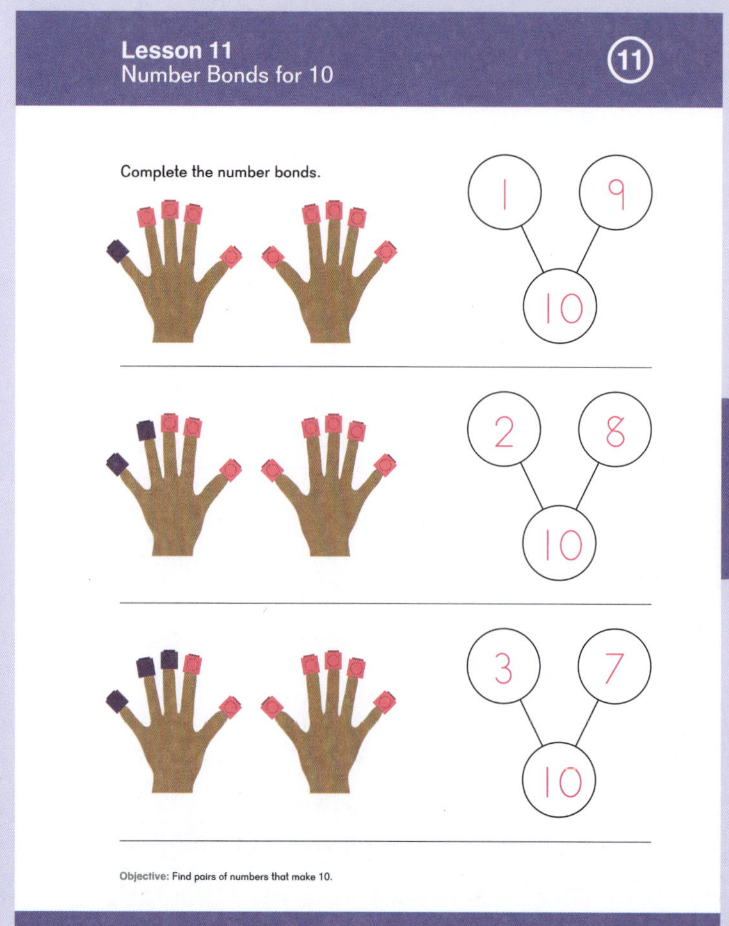

Lesson 11
Number Bonds for 10 (11)

Complete the number bonds.

1, 9 → 10

2, 8 → 10

3, 7 → 10

Objective: Find pairs of numbers that make 10.

8-11 Number Bonds for 10 59

59

60

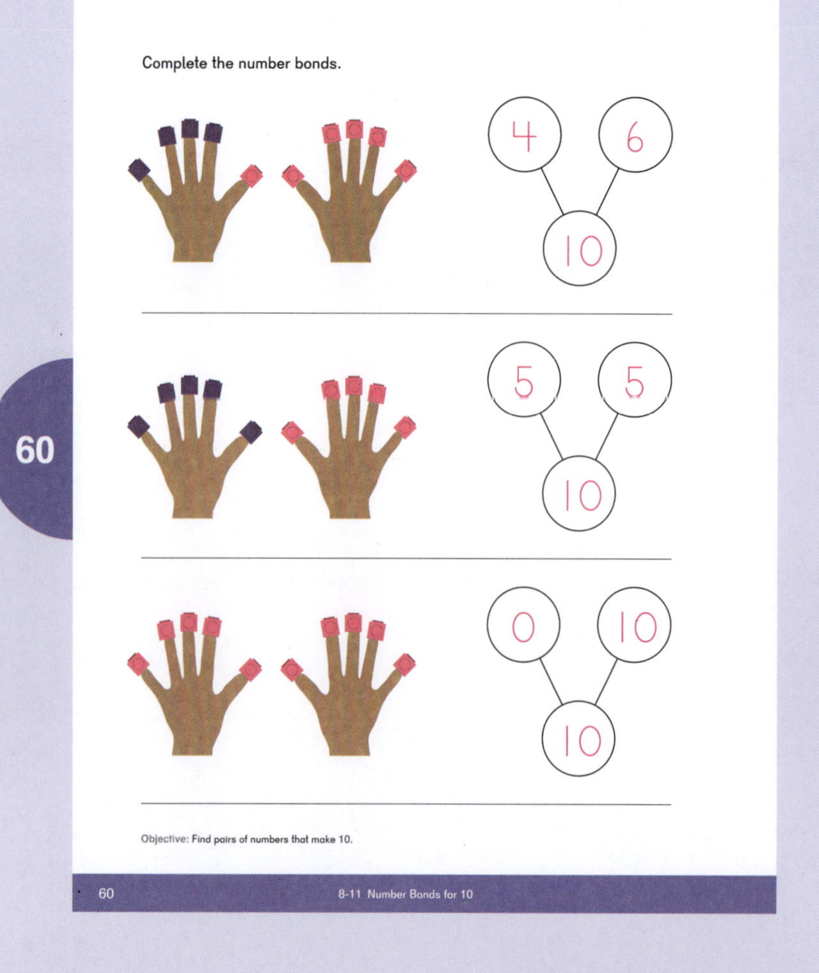

Complete the number bonds.

4, 6 → 10

5, 5 → 10

0, 10 → 10

Objective: Find pairs of numbers that make 10.

60 8-11 Number Bonds for 10

Small Group Activities

Textbook Pages 59–63

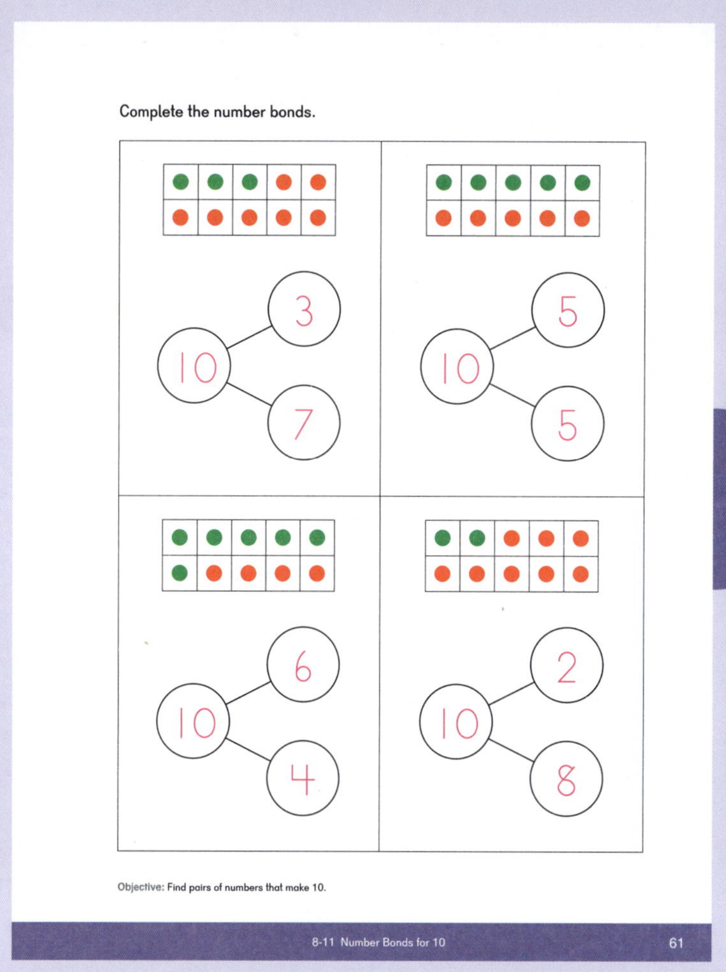
Complete the number bonds.

10, 3, 7

10, 5, 5

10, 6, 4

10, 2, 8

Objective: Find pairs of numbers that make 10.

8-11 Number Bonds for 10 61

61

▲ Number Bond Bracelets

Materials: 1 pipe cleaner and 10 beads per student, painter's tape, Number Bond Book (BLM)

Create bracelets as described in Lesson 5 for the number 10. Using the Number Bond Book (BLM), have students record all the ways to make 10.

▲ Go Fish

Materials: 4 sets of either Ten-frame Cards (BLM) or Number Cards (BLM) per small group

Play the game as described in Lesson 6, but using cards from 0 to 10.

Use Ten-frame Cards (BLM) for students who are still counting. Number Cards (BLM) can be used for students who understand the abstract number without counting.

▲ Number Bond Chains

Materials: Index cards, strips of paper in a variety of colors

Play as directed in the previous lesson, creating number bond chains for the number 10.

▲ Domino Trains

Materials: Dominoes

Using a set of dominoes, have students find and connect two numbers that make 10.

To play, students turn all dominoes facedown, then each player draws 7 dominoes. One domino from the pile is turned faceup to begin. Players try to make a 10 with a domino from their "hands." If they can't make that number, they draw from the pile until they can play.

The first player to get rid of all of his dominoes is the winner.

▲ Match

Materials: 2 sets of Ten-frame Cards (BLM) 0 to 10

Students arrange the cards faceup in a grid. Students take turns finding two cards that go together to make 10.

★ Memory

Materials: 2 sets of Ten-frame Cards (BLM) 0 to 10

Students arrange the cards facedown in a grid. Students take turns finding two cards that go together to make 10.

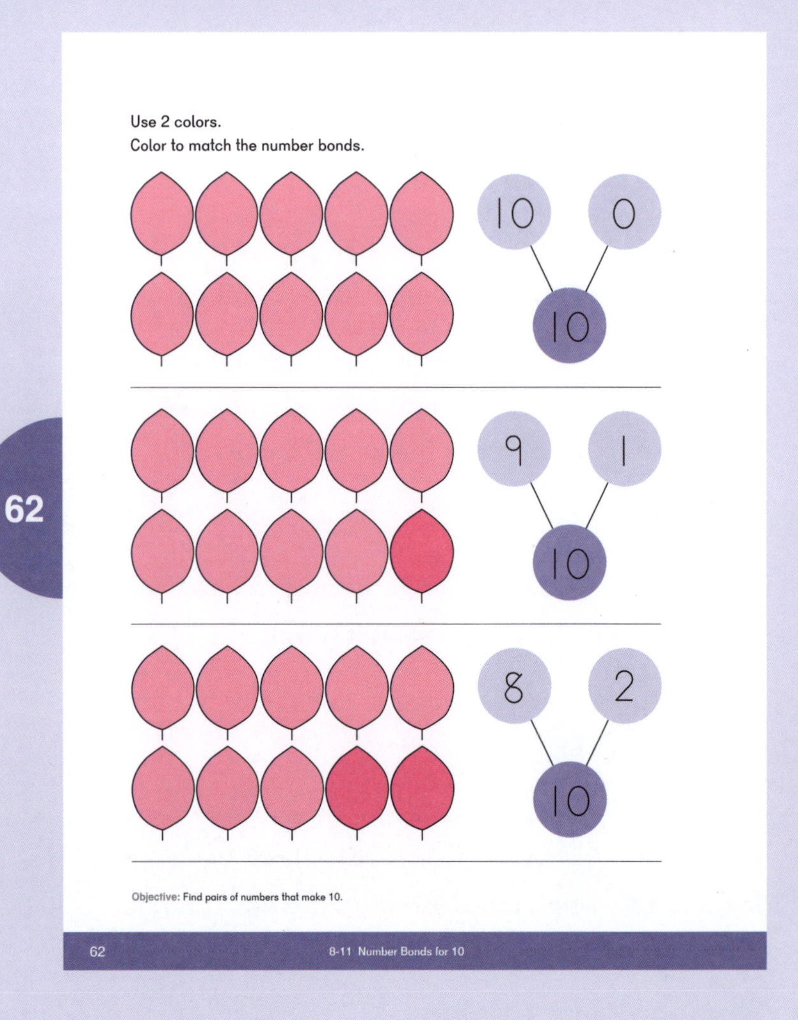

Take it Outside

★ Number Bond Ninjas

Materials: 10 traffic cones, construction paper cones (2 different colors)

Create paper cones by loosely rolling up construction paper and taping it into the shape.

Set up the gym or playground by putting 10 traffic cones on the halfway line.

Divide students into 2 teams. Each team picks 2 or 3 players to be Guards, depending on the class size, and the rest play as Ninjas. Each Ninja gets a paper cone, with each team getting a different color.

The Ninjas run to the traffic cones and place their paper cones on one without being tagged by a Guard. If they are tagged, they go to Jail (a designated area, as in the game Capture the Flag). Tagged Ninjas can then be released from jail if tagged by a teammate.

At any time during the game, you can whistle for a break and announce, "6 cones are capped! How many are not?" Students reply, "4! 6 and 4 make 10!"

Once all 10 of the traffic cones are capped, the game ends. The team with more of their color on the cones wins.

Exercise 11 • page 45

Extend

★ **See Saw**

Materials: Number Bonds Cards (BLM) 0 to 10, See Saw Template (BLM) – 1 per player

Use Number Bond Cards (BLM) to 5 or 10. Each player draws a card and finds the total (whole) of the two parts given on the card. They put the cards on one side of the See Saw Template (BLM) and trade their Templates and cards with a partner. Partners look for a number bond that has the same total (whole) to make the See Saw balance.

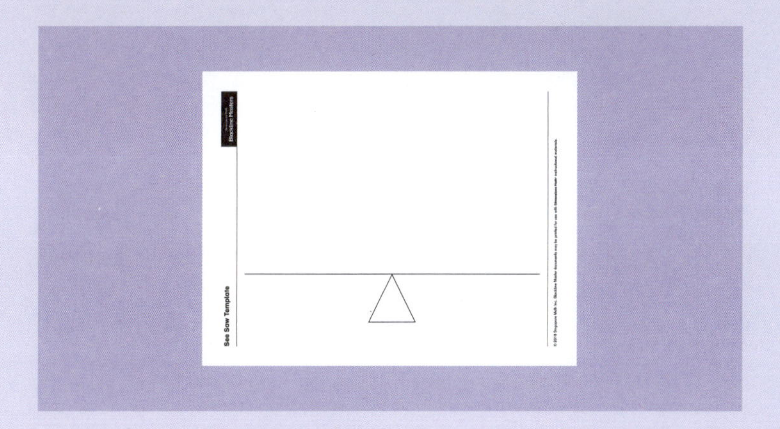

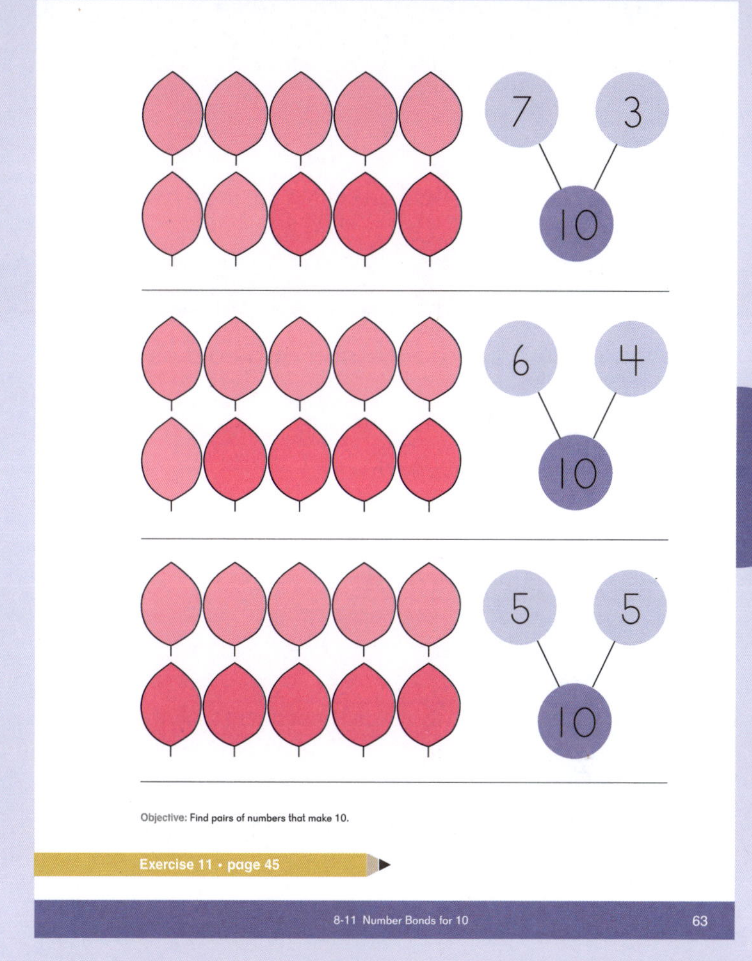

63

Objective

- Practice number bonds to 6 and 7.

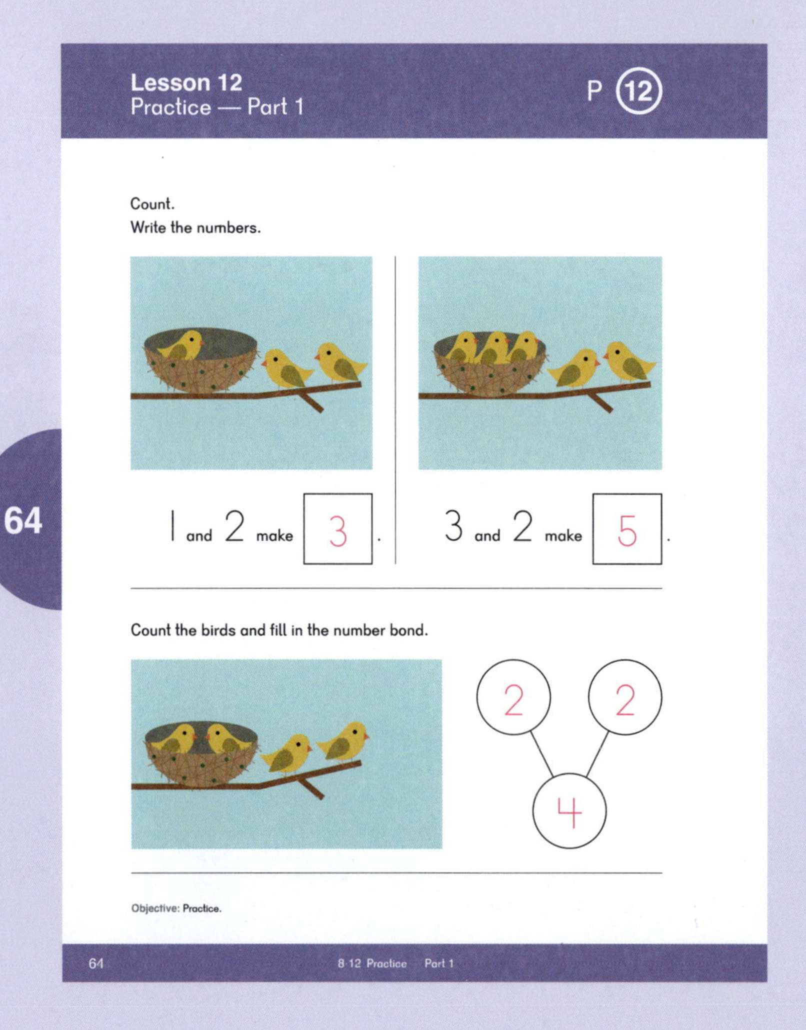

64

Lesson 12
Practice — Part 1

P 12

Count.
Write the numbers.

1 and 2 make 3.

3 and 2 make 5.

Count the birds and fill in the number bond.

2 2
4

Objective: Practice.

64 8.12 Practice Part 1

Practice lessons are designed for further practice and assessment as needed.

Students can complete the textbook pages and workbook pages as practice and/or as assessment.

Use activities and extensions from the chapter for students not being assessed.

Small Group Activities

▲ Rainbow Bonds

Materials: Number Path (BLM), crayons or markers

Have students connect each number bond to 10 with different colored crayons or markers. This should result in a rainbow.

▲ Cover All

Materials: Number Path (BLM), counters, 10-sided die or Number Cards (BLM) 0 to 10

A player rolls the die. She chooses two numbers that together match the roll and places counters on those numbers on her number path. The goal is to cover all the numbers on the number path. When playing with a group, the first player to cover all numbers on her number path is the winner.

Solitaire version: The player rolls the die. She puts counters on two numbers that match the roll. For example, if a player rolls a 7, she could place her counters on any of the pairs that make 7:

- 7 and 0
- 6 and 1
- 5 and 2
- 4 and 3

Once all the numbers are covered, the game is over.

Exercise 12 • page 49

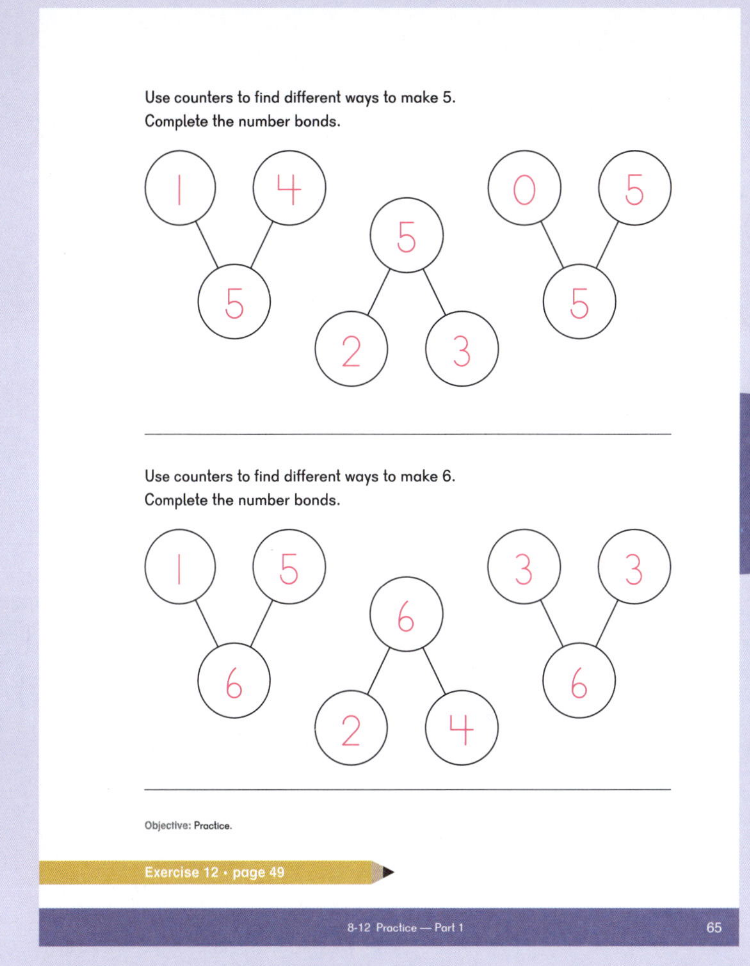
Use counters to find different ways to make 5.
Complete the number bonds.

1, 4 → 5; 5 → 2, 3; 0, 5 → 5

Use counters to find different ways to make 6.
Complete the number bonds.

1, 5 → 6; 6 → 2, 4; 3, 3 → 6

Objective: Practice.

Exercise 12 • page 49

8-12 Practice — Part 1 65

65

Lesson 13 Practice — Part 2

Objective

- Practice number bonds to 8 and 9.

Exercise 13 • page 51

66

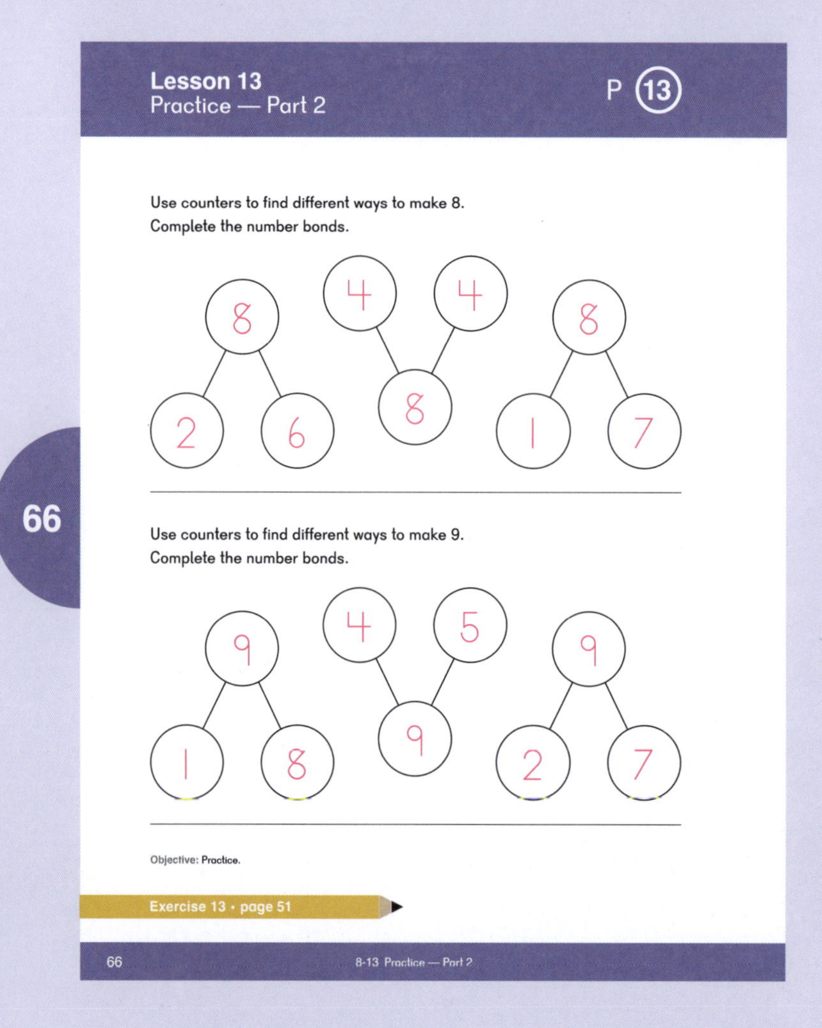

Objective

- Practice finding the missing part in a number bond to 10.

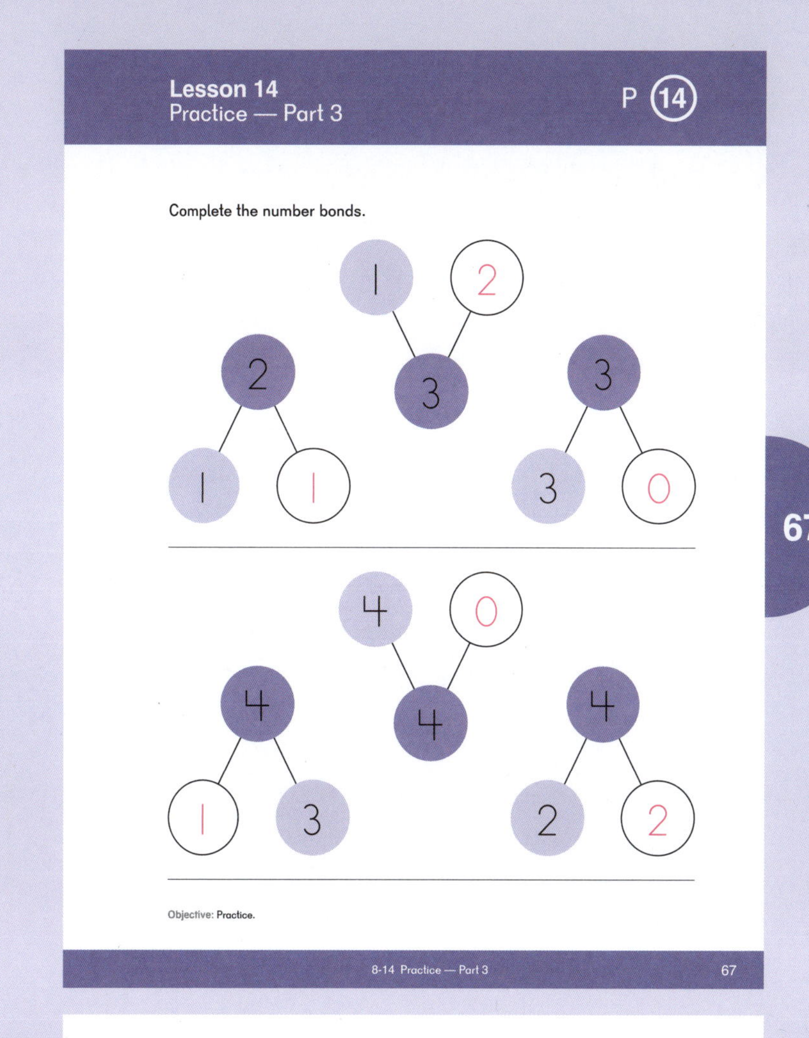

67

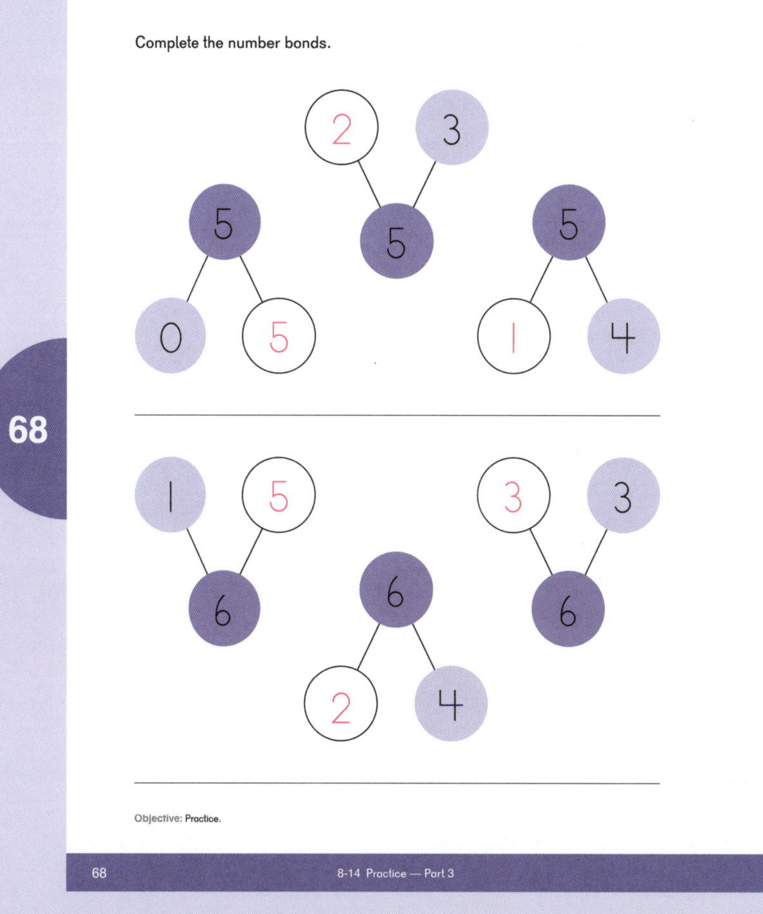

68

Exercise 14 • page 53

Extend

★ Bowling Bonds

Materials: 10 “Bowling Pins” made from water bottles, small ball, whiteboard and dry erase markers for recording score

Have players set up “bowling pins” in a standard format or by looking at **Dimensions Mathematics® Kindergarten A** textbook page 57.

Player 1 rolls the ball and tries to knock down as many pins as he can. After the first roll, that student records the number of pins still standing on a number bond with a whole of 10.

Once a player has calculated the number left standing, he rolls again and completes the second part of the number bond. The player with the most pins knocked down with both rolls is the winner.

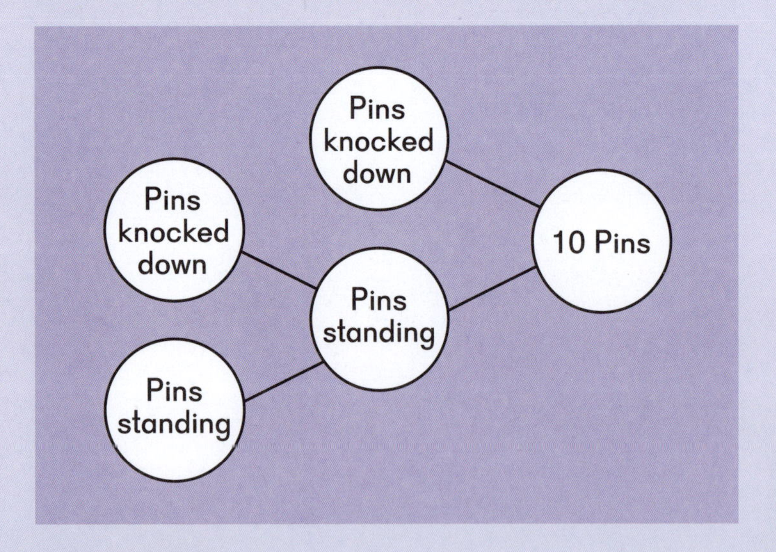

For example:

- Roll #1: There were 10 pins standing, and I knocked down 3 pins. There are 7 pins still standing.
- Roll #2: There were 7 pins standing, and I knocked down 4 pins. There are 3 pins still standing.

69

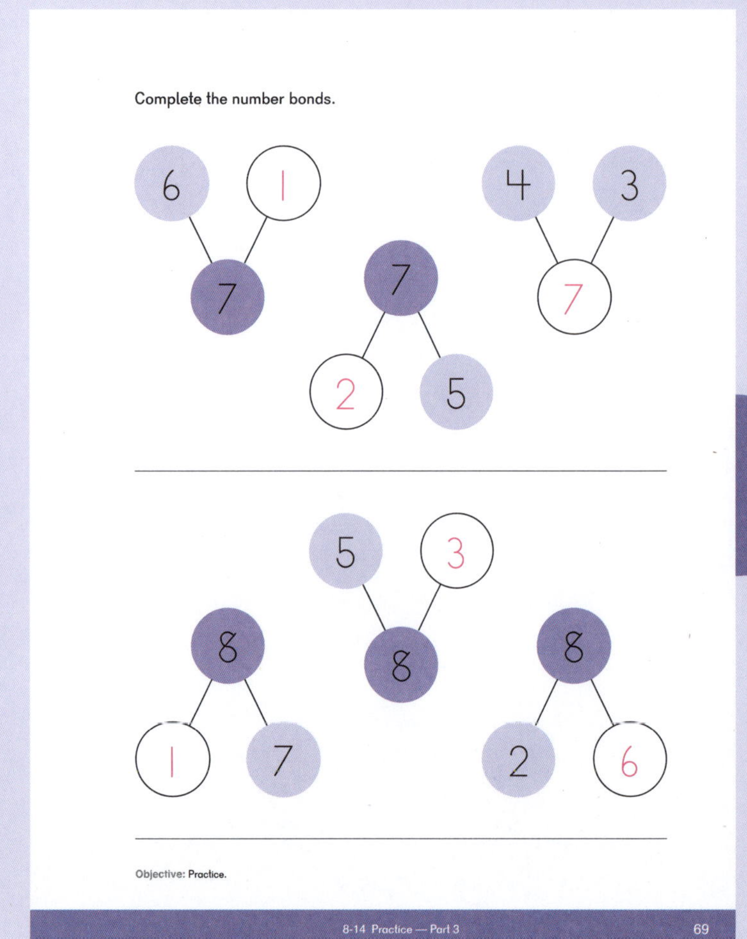
Complete the number bonds.

Objective: Practice.

8-14 Practice — Part 3 69

70

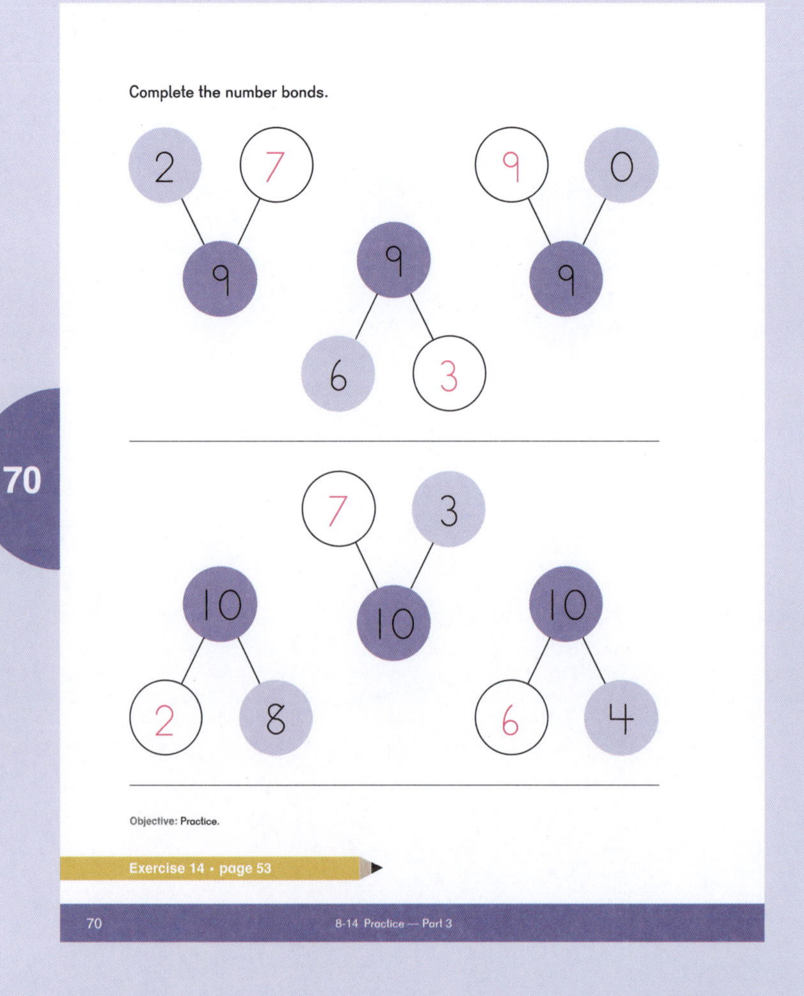
Complete the number bonds.

Objective: Practice.

Exercise 14 • page 53

70 8-14 Practice — Part 3

Exercise 1 • pages 23–24

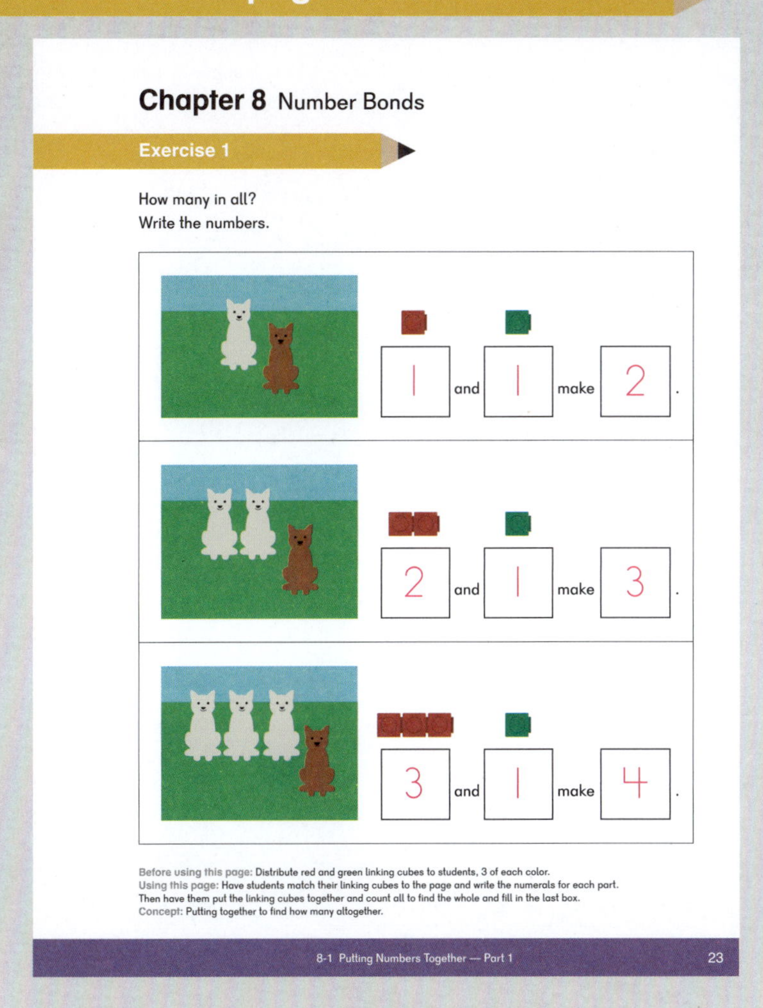

Chapter 8 Number Bonds

Exercise 1

How many in all?
Write the numbers.

1 and 1 make 2.

2 and 1 make 3.

3 and 1 make 4.

Before using this page: Distribute red and green linking cubes to students, 3 of each color.
Using this page: Have students match their linking cubes to the page and write the numerals for each part. Then have them put the linking cubes together and count all to find the whole and fill in the last box.
Concept: Putting together to find how many altogether.

8-1 Putting Numbers Together — Part 1 23

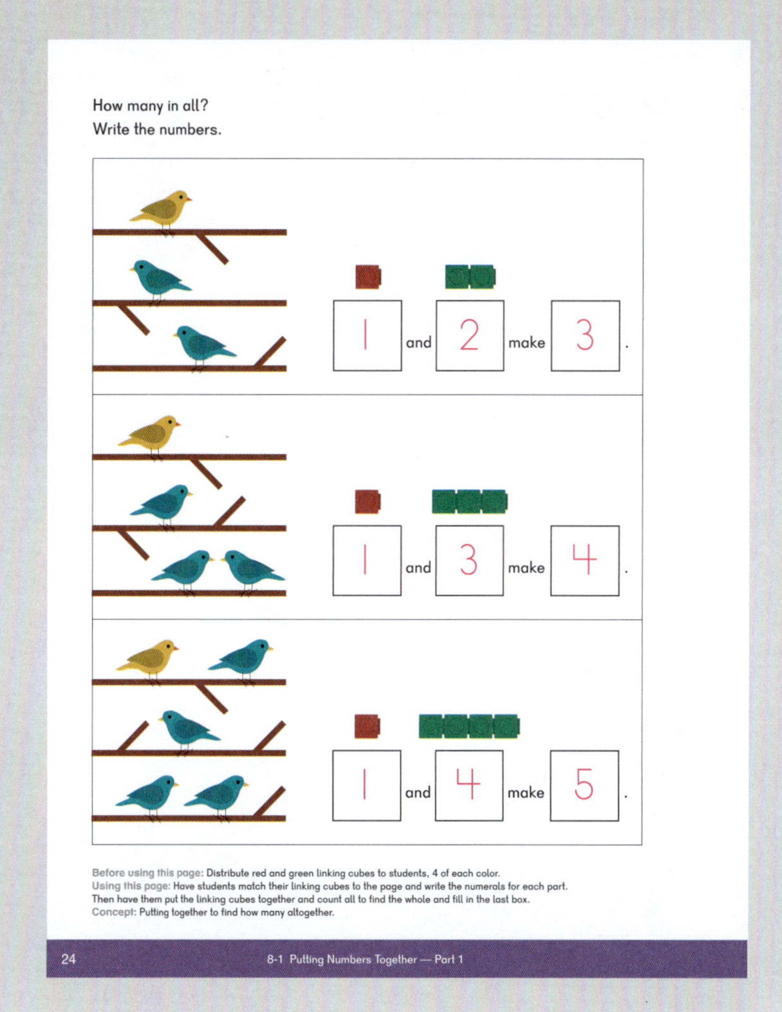

How many in all?
Write the numbers.

1 and 2 make 3.

1 and 3 make 4.

1 and 4 make 5.

Before using this page: Distribute red and green linking cubes to students, 4 of each color.
Using this page: Have students match their linking cubes to the page and write the numerals for each part. Then have them put the linking cubes together and count all to find the whole and fill in the last box.
Concept: Putting together to find how many altogether.

24 8-1 Putting Numbers Together — Part 1

Exercise 2 • pages 25–26

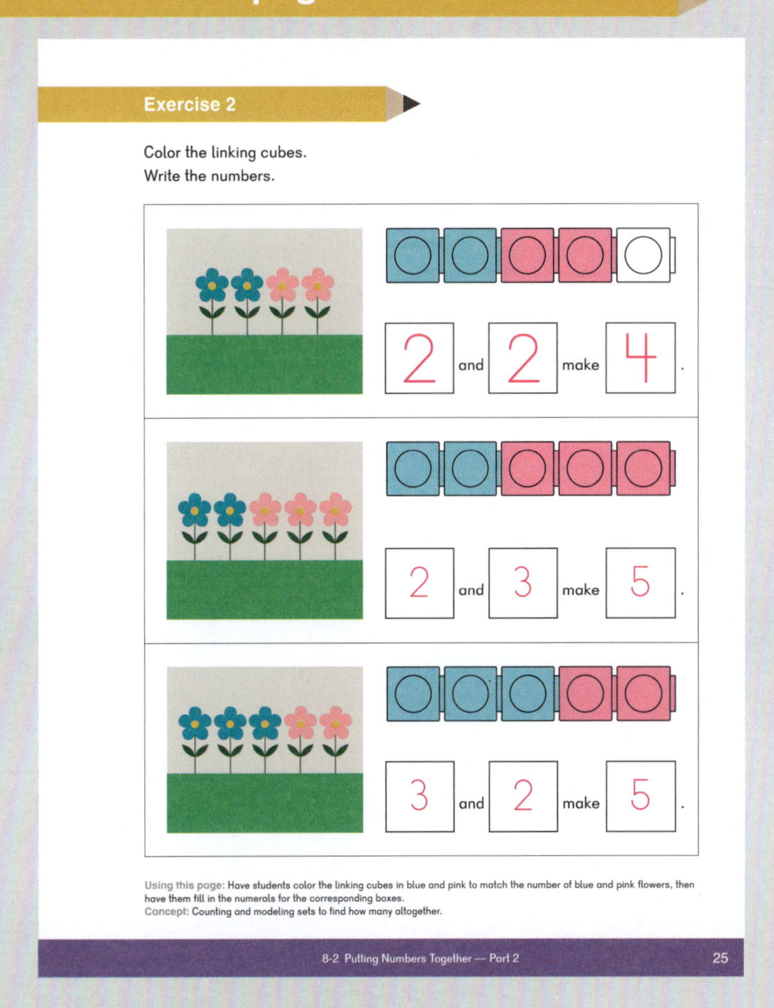

Exercise 2

Color the linking cubes.
Write the numbers.

2 and 2 make 4.

2 and 3 make 5.

3 and 2 make 5.

Using this page: Have students color the linking cubes in blue and pink to match the number of blue and pink flowers, then have them fill in the numerals for the corresponding boxes.
Concept: Counting and modeling sets to find how many altogether.

8-2 Putting Numbers Together — Part 2 25

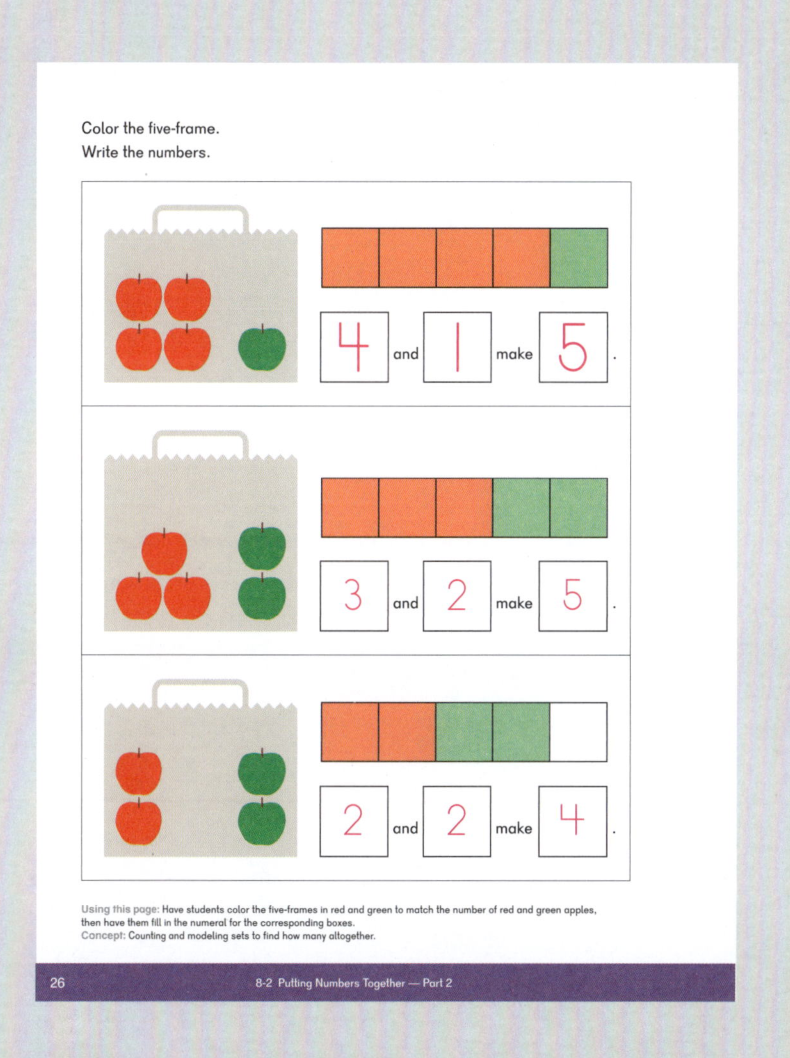

Color the five-frame.
Write the numbers.

4 and 1 make 5.

3 and 2 make 5.

2 and 2 make 4.

Using this page: Have students color the five-frames in red and green to match the number of red and green apples, then have them fill in the numeral for the corresponding boxes.
Concept: Counting and modeling sets to find how many altogether.

26 8-2 Putting Numbers Together — Part 2

Exercise 3 • pages 27–28

Exercise 3

Complete the number bonds.

3

5

4

4

Using this page: Have students put the two parts of buttons together to find the whole, then write the numeral to complete the number bond.
Concept: Joining two parts to make a whole.

8-3 Parts Making a Whole 27

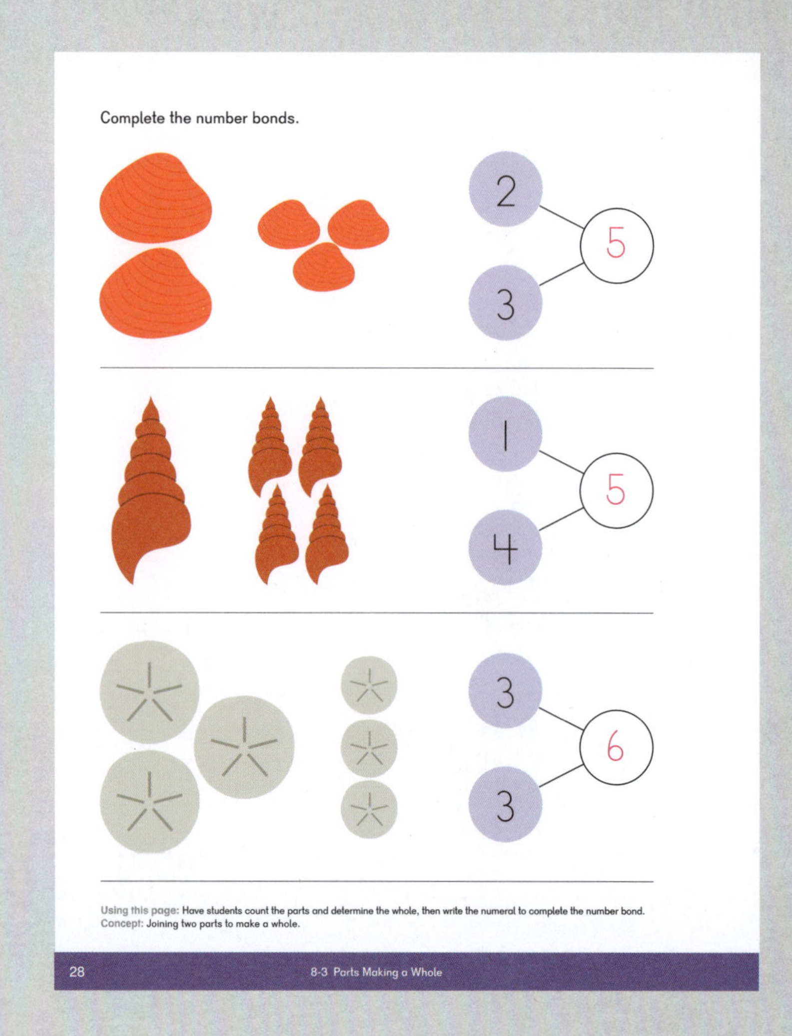

Complete the number bonds.

2, 3, 5

1, 4, 5

3, 3, 6

Using this page: Have students count the parts and determine the whole, then write the numeral to complete the number bond.
Concept: Joining two parts to make a whole.

28 8-3 Parts Making a Whole

Exercise 4 • pages 29–30

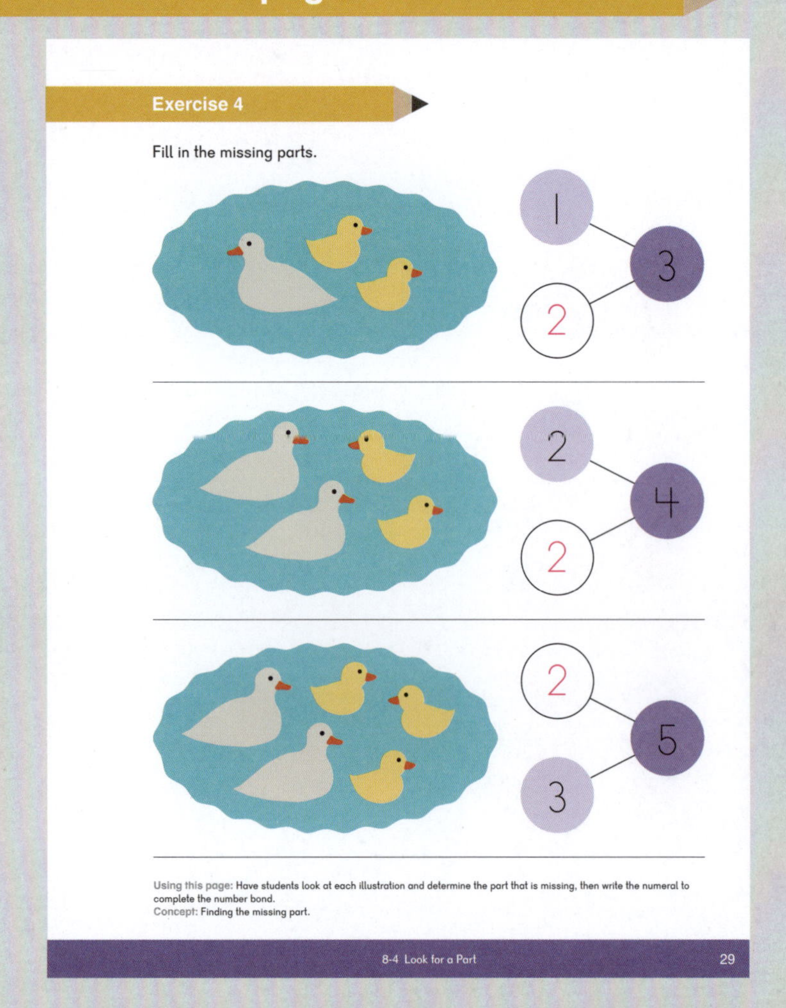

Exercise 4

Fill in the missing parts.

1, 2, 3

2, 2, 4

2, 3, 5

Using this page: Have students look at each illustration and determine the part that is missing, then write the numeral to complete the number bond.
Concept: Finding the missing part.

8-4 Look for a Part 29

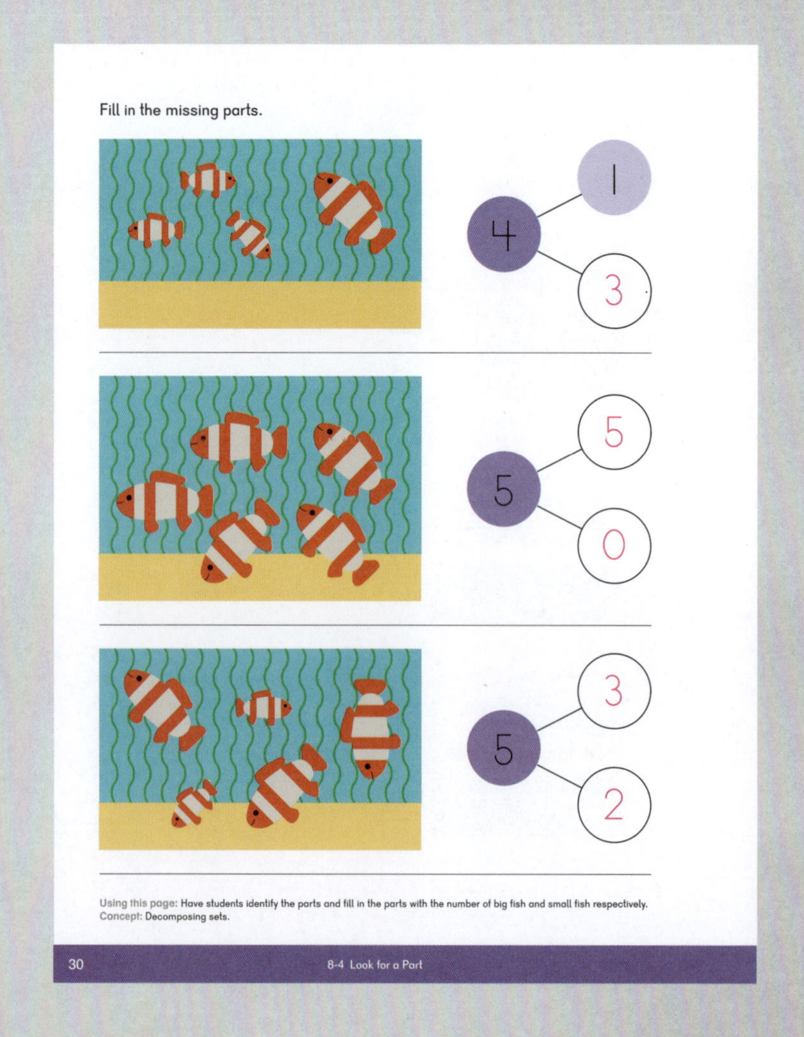

Fill in the missing parts.

4, 1, 3

5, 5, 0

5, 3, 2

Using this page: Have students identify the parts and fill in the parts with the number of big fish and small fish respectively.
Concept: Decomposing sets.

30 8-4 Look for a Part

Exercise 5 • pages 31–32

Exercise 5

Complete the number bonds.

Using this page: Have students identify the number of pink and purple linking cubes and complete the number bond.
Concept: Finding pairs of numbers that make 2 and 3.

8-5 Number Bonds for 2, 3, and 4 31

Complete the number bonds.

Using this page: Have students identify the number of pink and purple linking cubes and complete the number bond.
Concept: Finding pairs of numbers that make 4.

32 8-5 Number Bonds 2, 3, and 4

Exercise 6 • pages 33–34

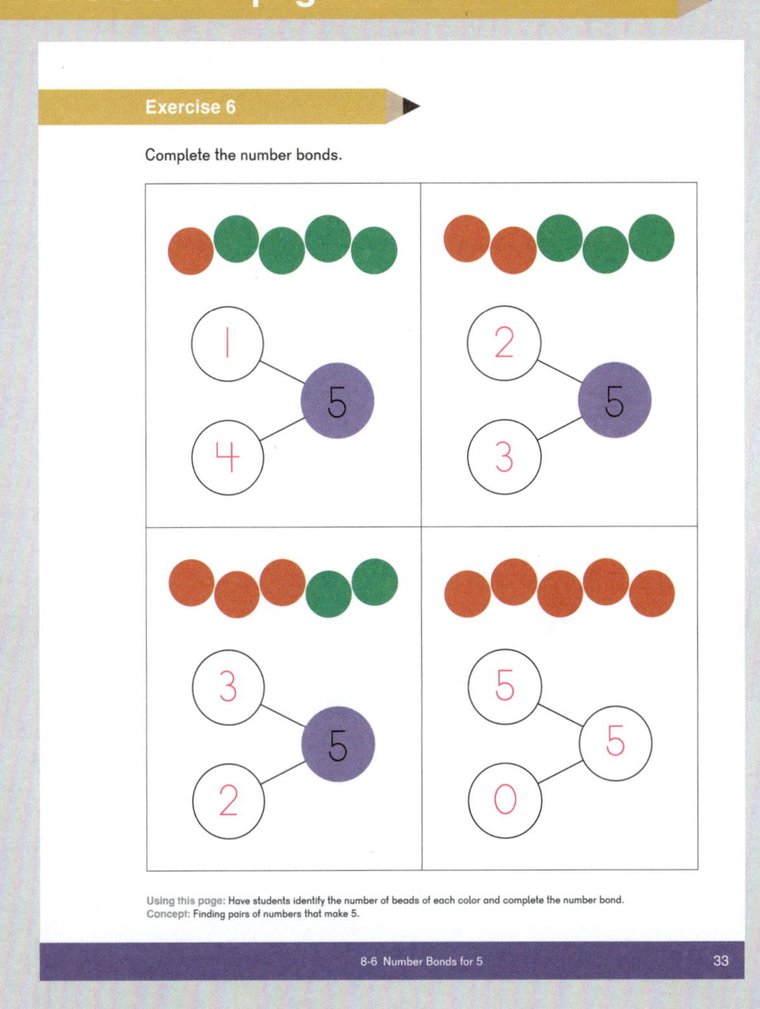

Exercise 6

Complete the number bonds.

Using this page: Have students identify the number of beads of each color and complete the number bond.
Concept: Finding pairs of numbers that make 5.

8-6 Number Bonds for 5 33

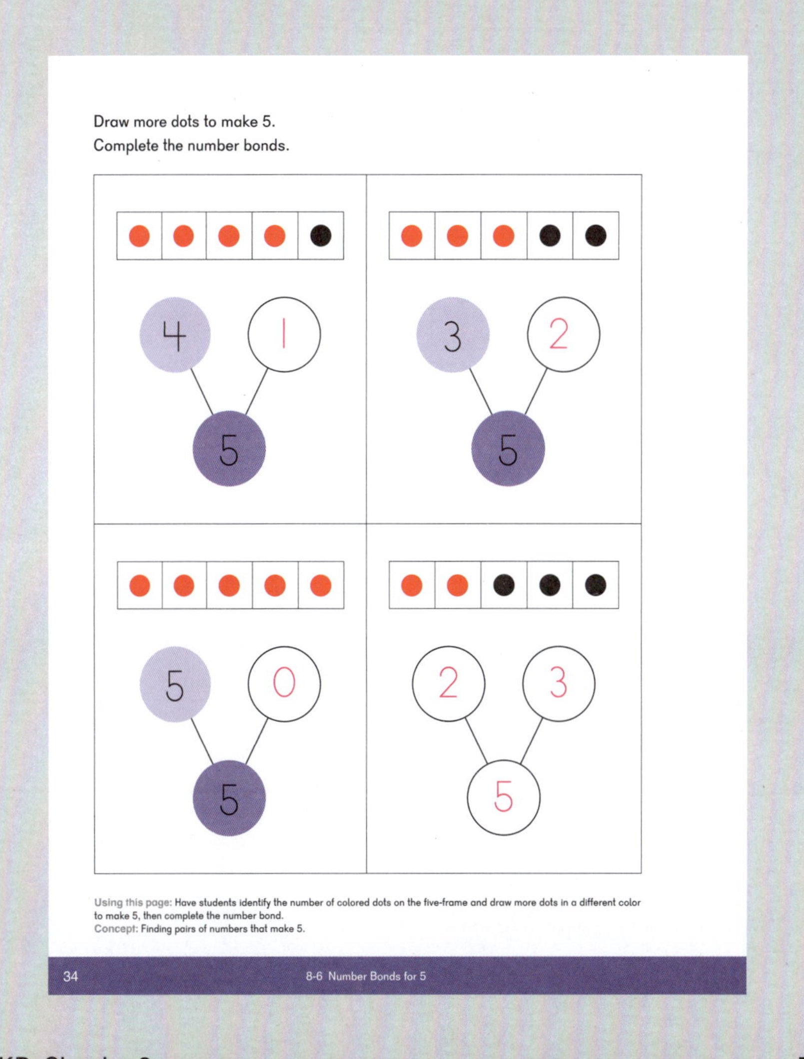

Draw more dots to make 5.
Complete the number bonds.

Using this page: Have students identify the number of colored dots on the five-frame and draw more dots in a different color to make 5, then complete the number bond.
Concept: Finding pairs of numbers that make 5.

34 8-6 Number Bonds for 5

Exercise 7 • pages 35–36

Exercise 7

Complete the number bonds.

1, 5 → 6	2, 4 → 6
3, 3 → 6	6, 0 → 6

Using this page: Have students identify the number of beads of each color and complete the number bond.
Concept: Finding pairs of numbers that make 6.

8-7 Number Bonds for 6 35

Draw more to make 6.
Complete the number bonds.

5, 1 → 6	3, 3 → 6
2, 4 → 6	4, 2 → 6

Using this page: Have students identify the number of each shape and draw more to make 6, then complete the number bond.
Concept: Finding pairs of numbers that make 6.

36 8-7 Number Bonds for 6

Exercise 8 • pages 37–38

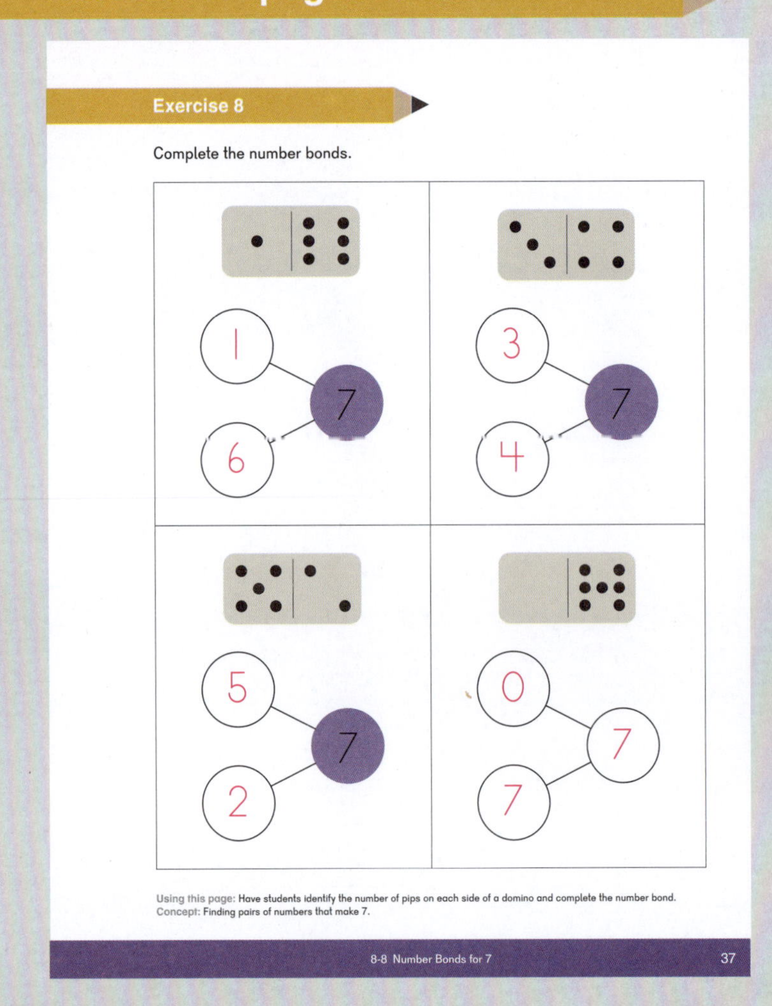

Exercise 8

Complete the number bonds.

1, 6 → 7	3, 4 → 7
5, 2 → 7	0, 7 → 7

Using this page: Have students identify the number of pips on each side of a domino and complete the number bond.
Concept: Finding pairs of numbers that make 7.

8-8 Number Bonds for 7 37

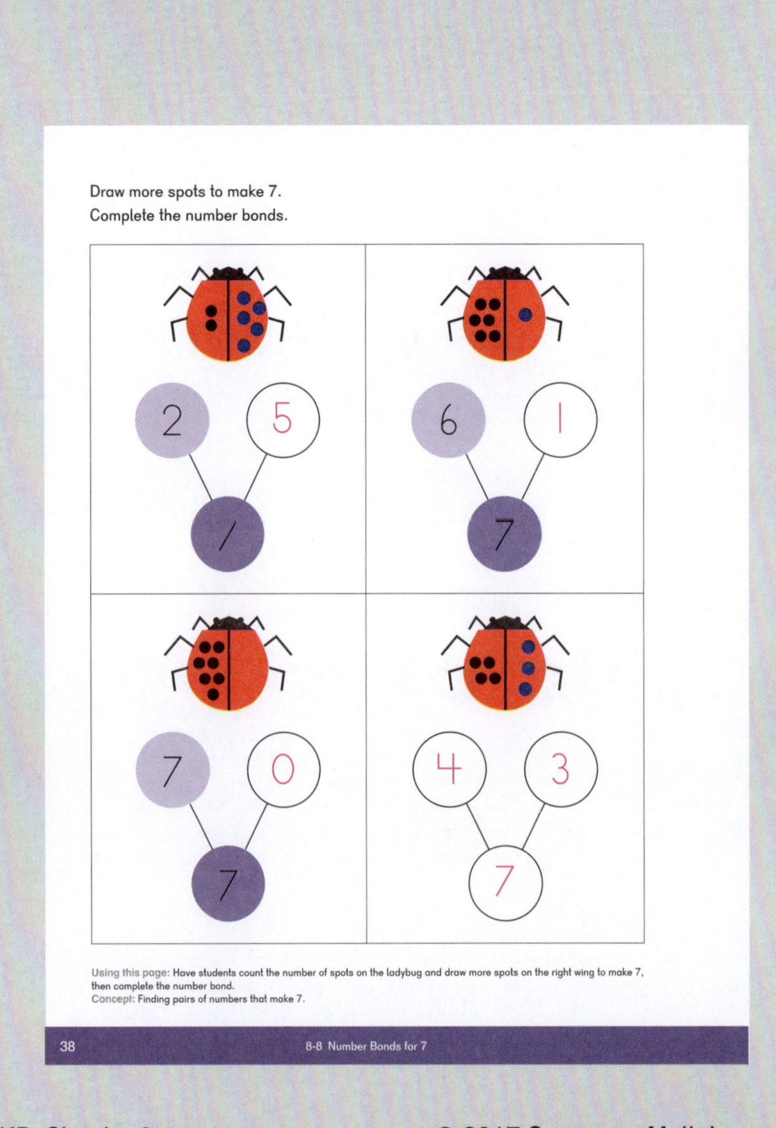

Draw more spots to make 7.
Complete the number bonds.

2, 5 → 7	6, 1 → 7
7, 0 → 7	4, 3 → 7

Using this page: Have students count the number of spots on the ladybug and draw more spots on the right wing to make 7, then complete the number bond.
Concept: Finding pairs of numbers that make 7.

38 8-8 Number Bonds for 7

Exercise 9 • pages 39–40

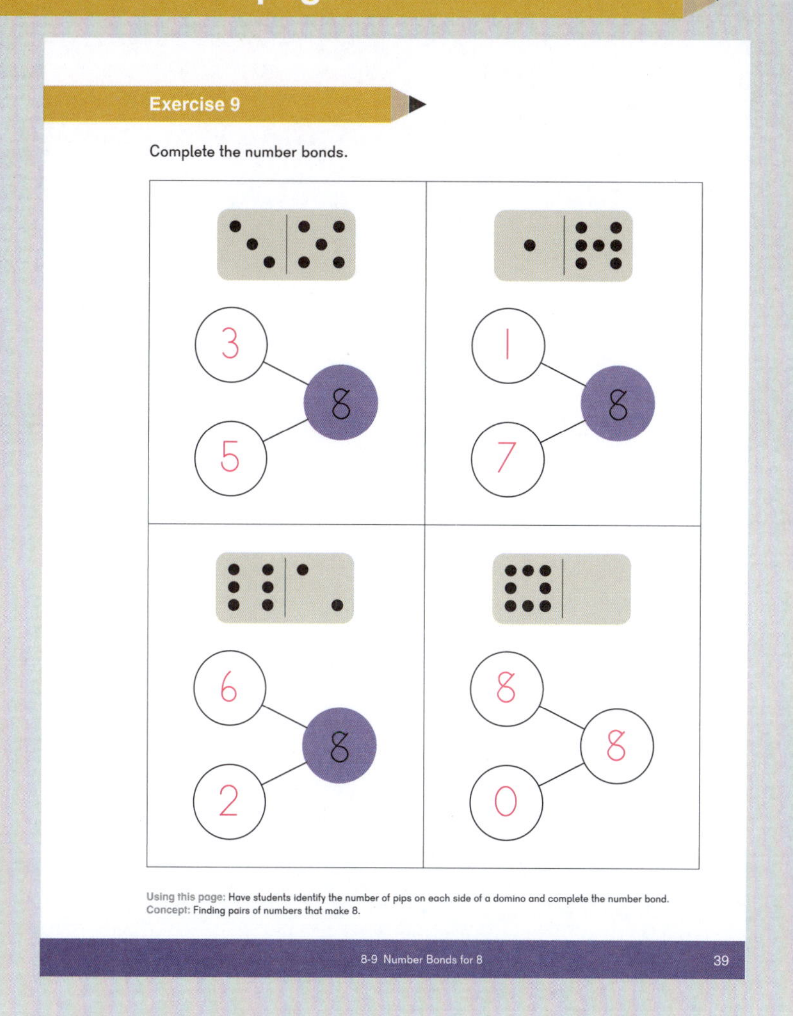

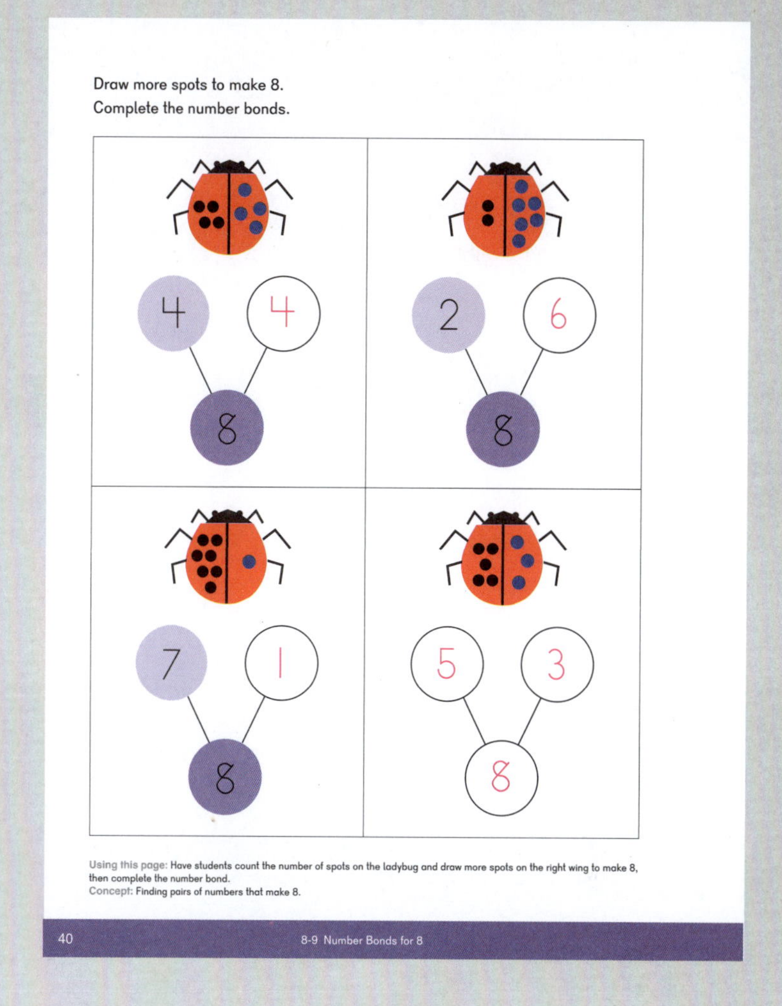

Exercise 10

Complete the number bonds.

9: 5, 4

9: 6, 3

9: 7, 2

9: 8, 1

Using this page: Have students count the number of objects in each part and complete the number bond for 9.
Concept: Finding pairs of numbers that make 9.

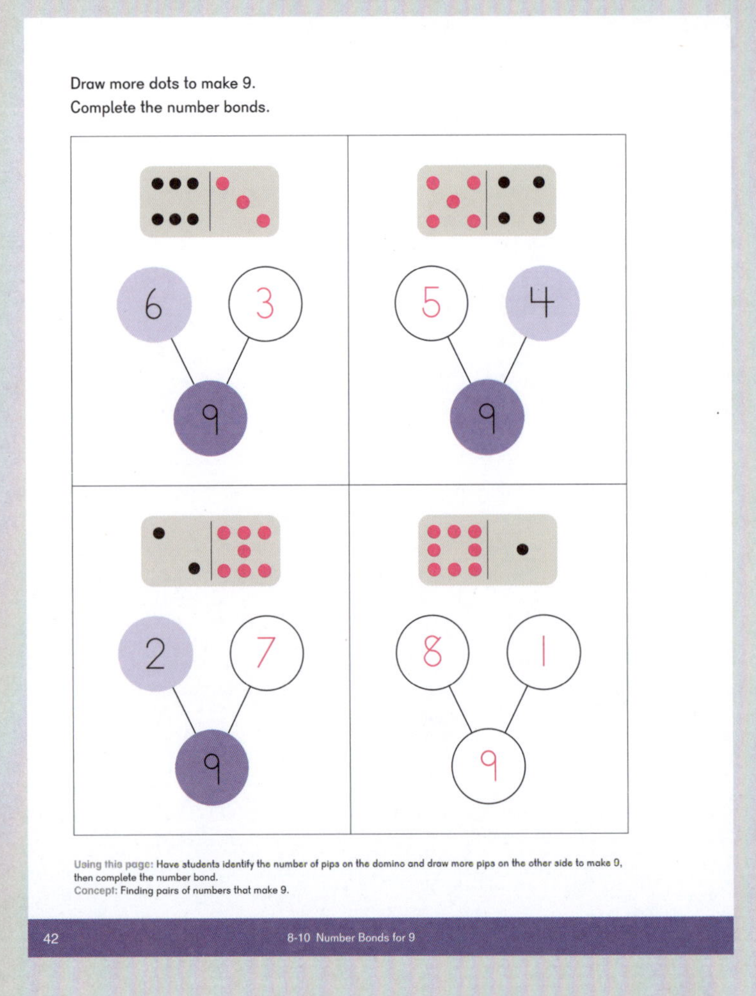
Draw more dots to make 9.
Complete the number bonds.

6, 3: 9

5, 4: 9

2, 7: 9

8, 1: 9

Using this page: Have students identify the number of pips on the domino and draw more pips on the other side to make 9, then complete the number bond.
Concept: Finding pairs of numbers that make 9.

42 8-10 Number Bonds for 9

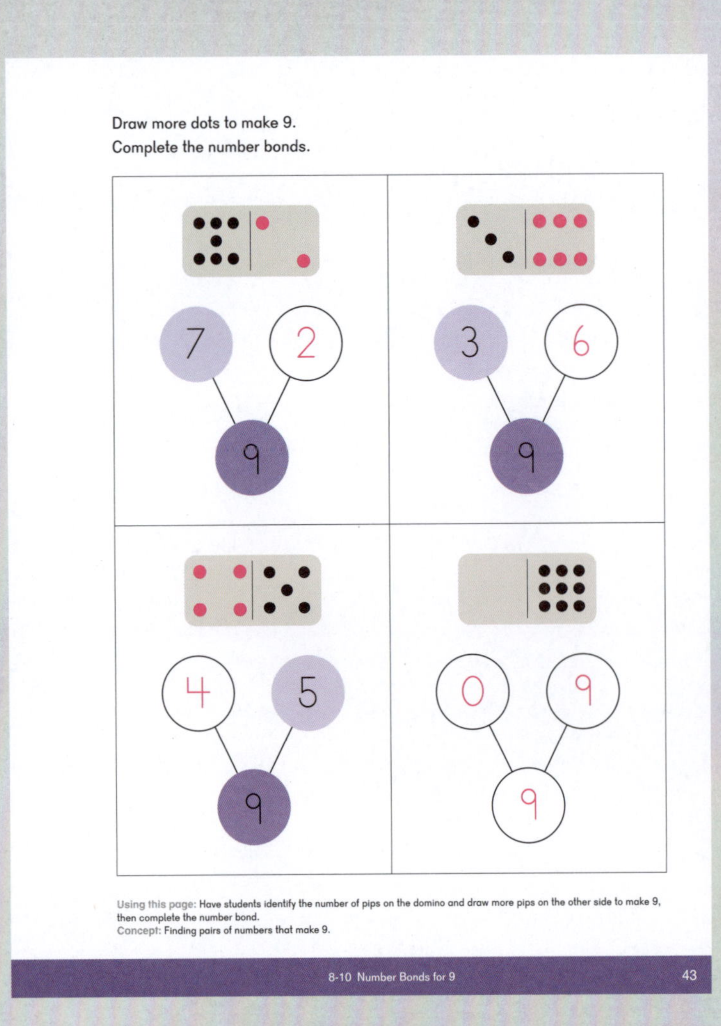
Draw more dots to make 9.
Complete the number bonds.

7, 2: 9

3, 6: 9

4, 5: 9

0, 9: 9

Using this page: Have students identify the number of pips on the domino and draw more pips on the other side to make 9, then complete the number bond.
Concept: Finding pairs of numbers that make 9.

8-10 Number Bonds for 9 43

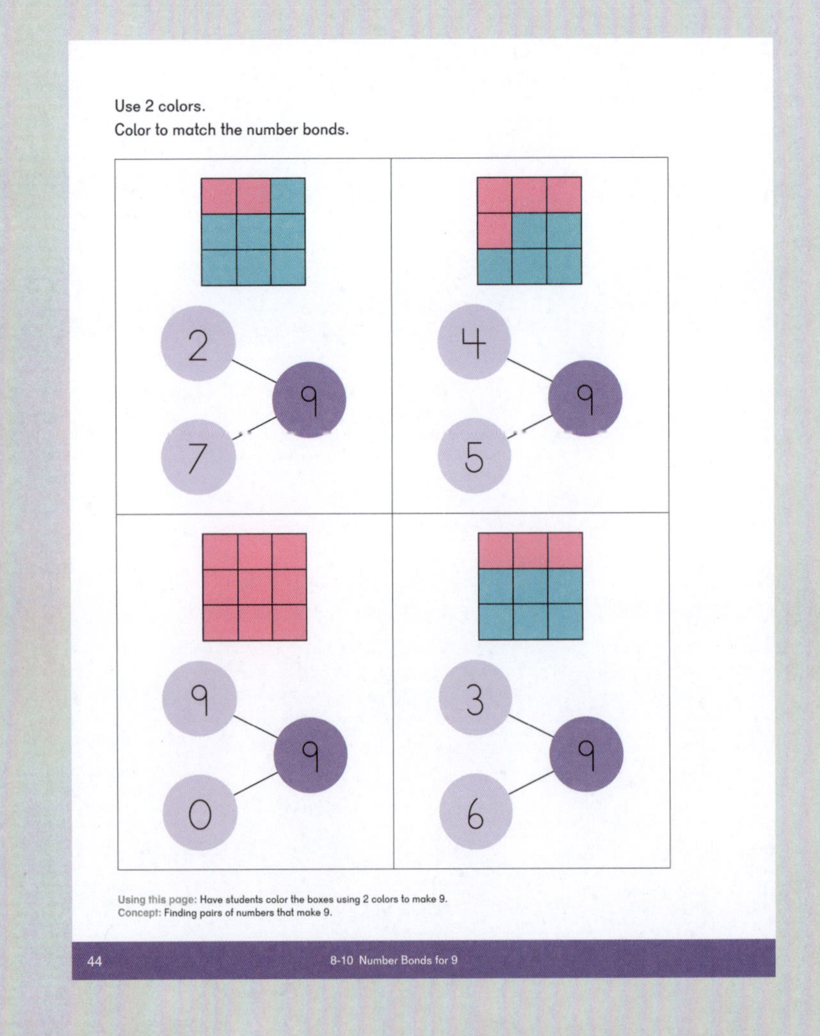
Use 2 colors.
Color to match the number bonds.

2, 7: 9

4, 5: 9

9, 0: 9

3, 6: 9

Using this page: Have students color the boxes using 2 colors to make 9.
Concept: Finding pairs of numbers that make 9.

44 8-10 Number Bonds for 9

Exercise 11

Complete the number bonds.

4, 6 → 10

2, 8 → 10

5, 5 → 10

7, 3 → 10

Using this page: Have students identify the number of objects in each part and complete the number bond for 10.
Concept: Finding pairs of numbers that make 10.

Draw more dots to make 10.
Complete the number bonds.

5, 5 → 10

3, 7 → 10

8, 2 → 10

6, 4 → 10

Using this page: Have students identify the number of pips on the domino and draw more pips on the right to make 10, then complete the number bond.
Concept: Finding pairs of numbers that make 10.

Draw more dots to make 10.
Complete the number bonds.

6, 4 → 10

1, 9 → 10

0, 10 → 10

3, 7 → 10

Using this page: Have students identify the number of pips on the domino and draw more pips on the left to make 10, then complete the number bond.
Concept: Finding pairs of numbers that make 10.

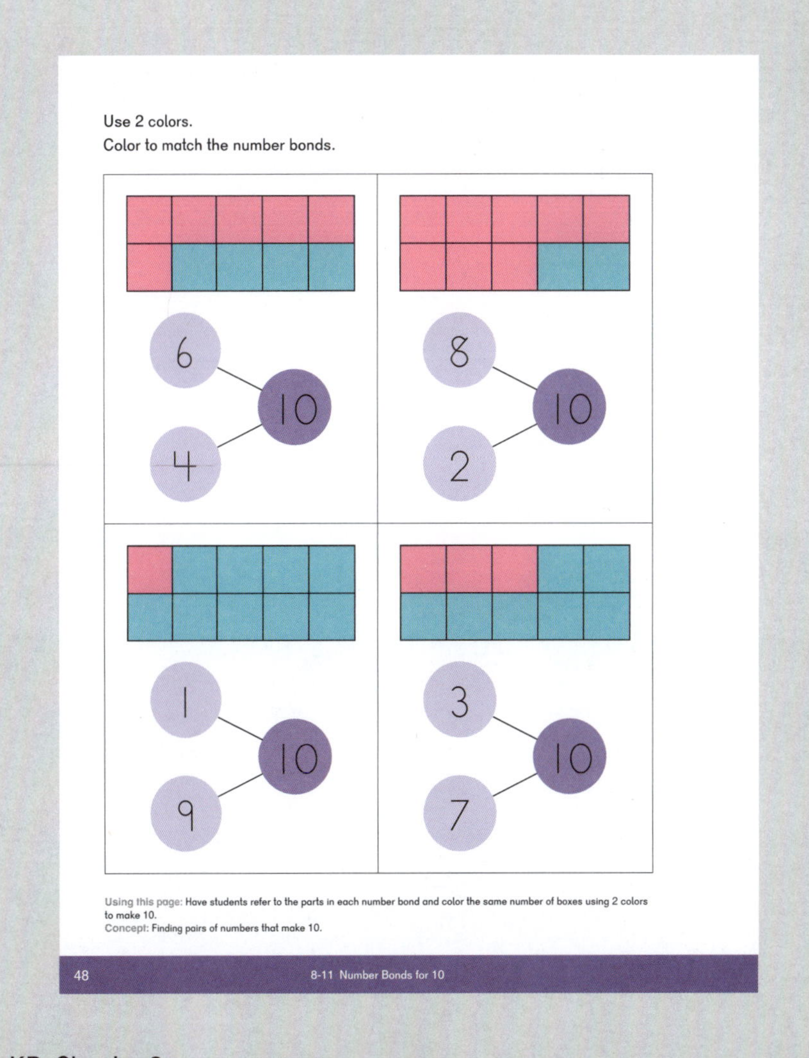

Use 2 colors.
Color to match the number bonds.

6, 4 → 10

8, 2 → 10

1, 9 → 10

3, 7 → 10

Using this page: Have students refer to the parts in each number bond and color the same number of boxes using 2 colors to make 10.
Concept: Finding pairs of numbers that make 10.

Exercise 12 • pages 49–50

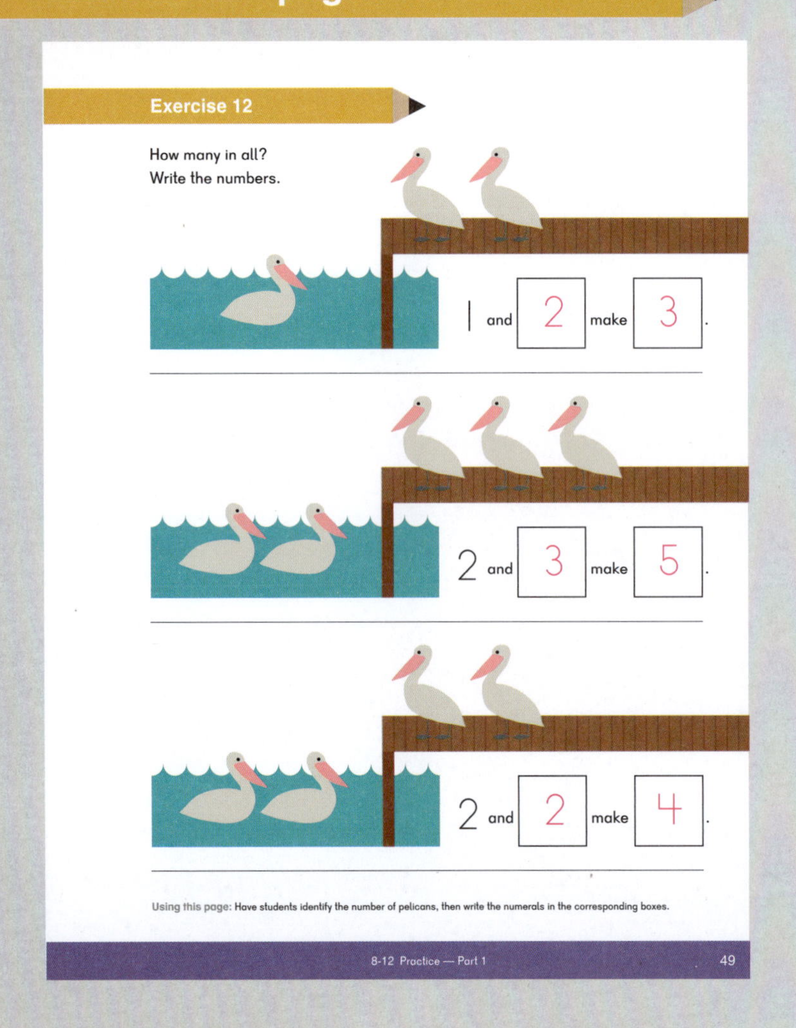

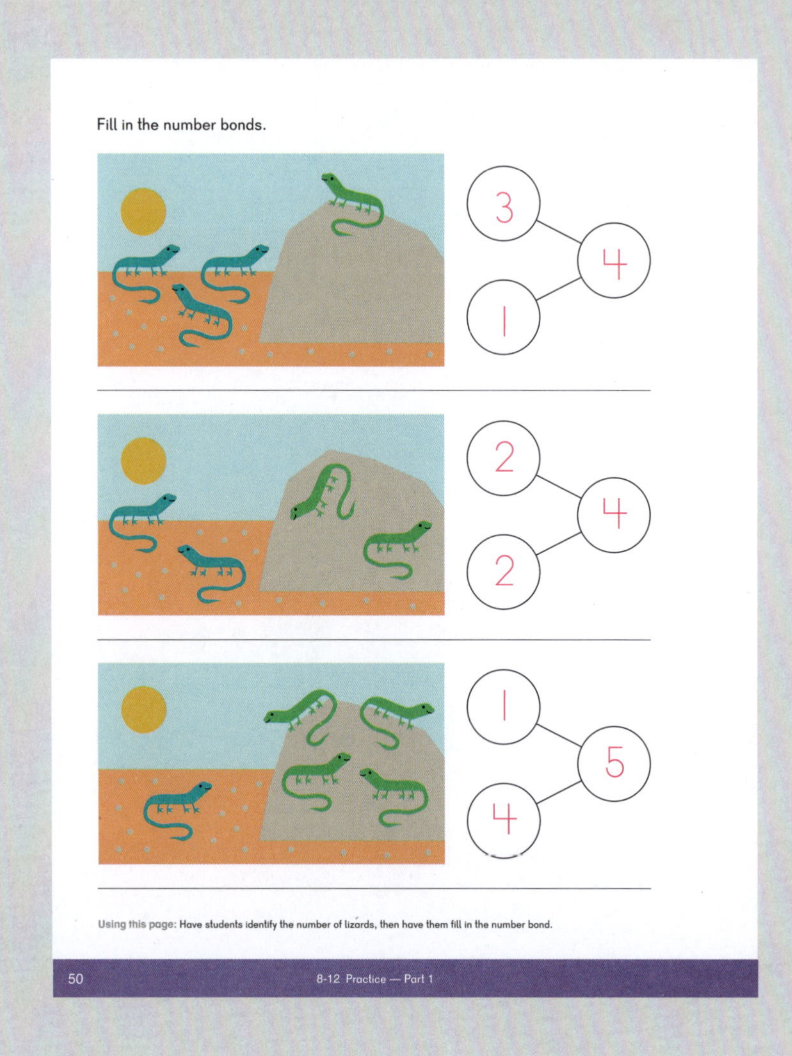

Exercise 13 • pages 51–52

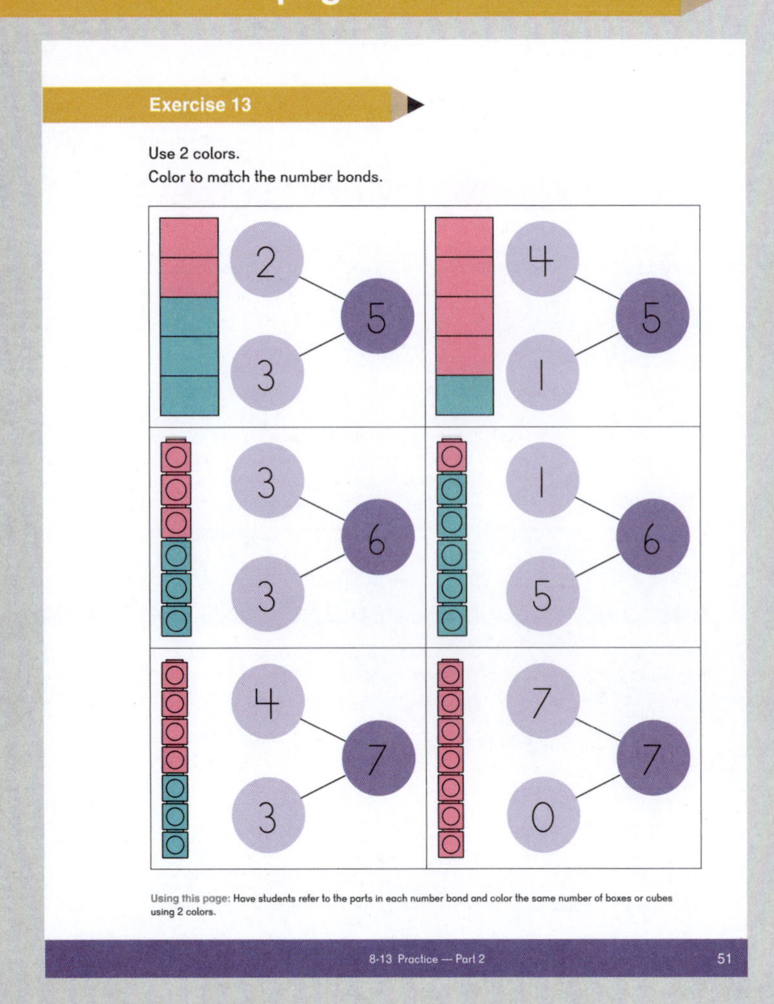

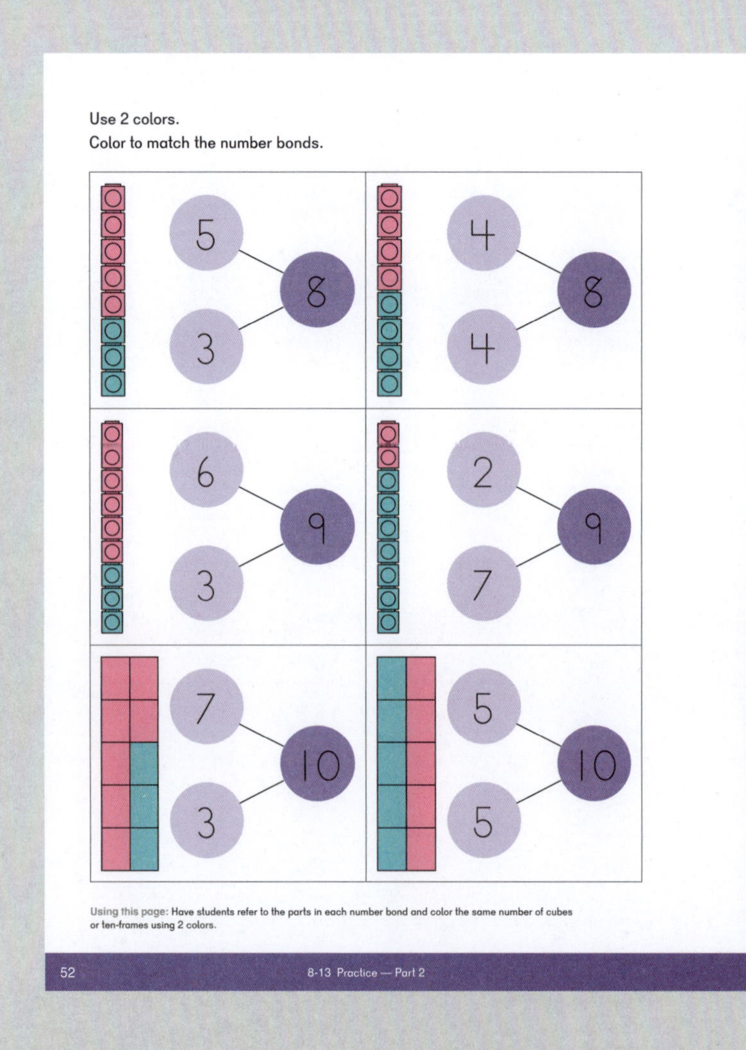

Exercise 14 • pages 53–54

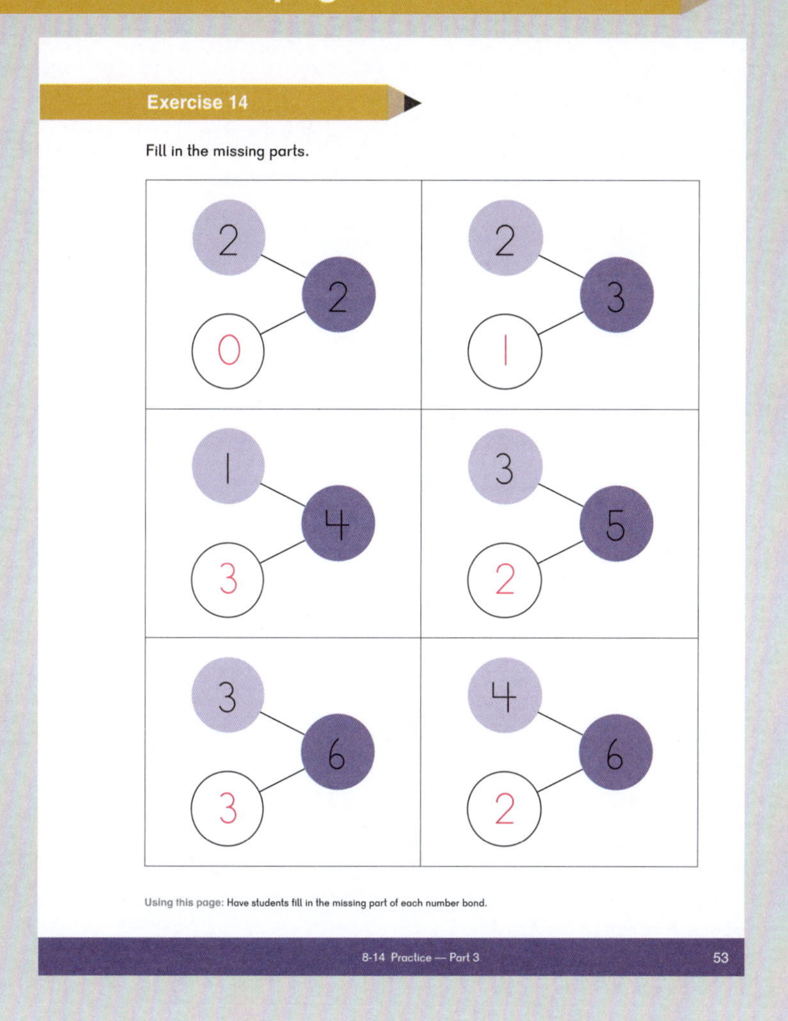

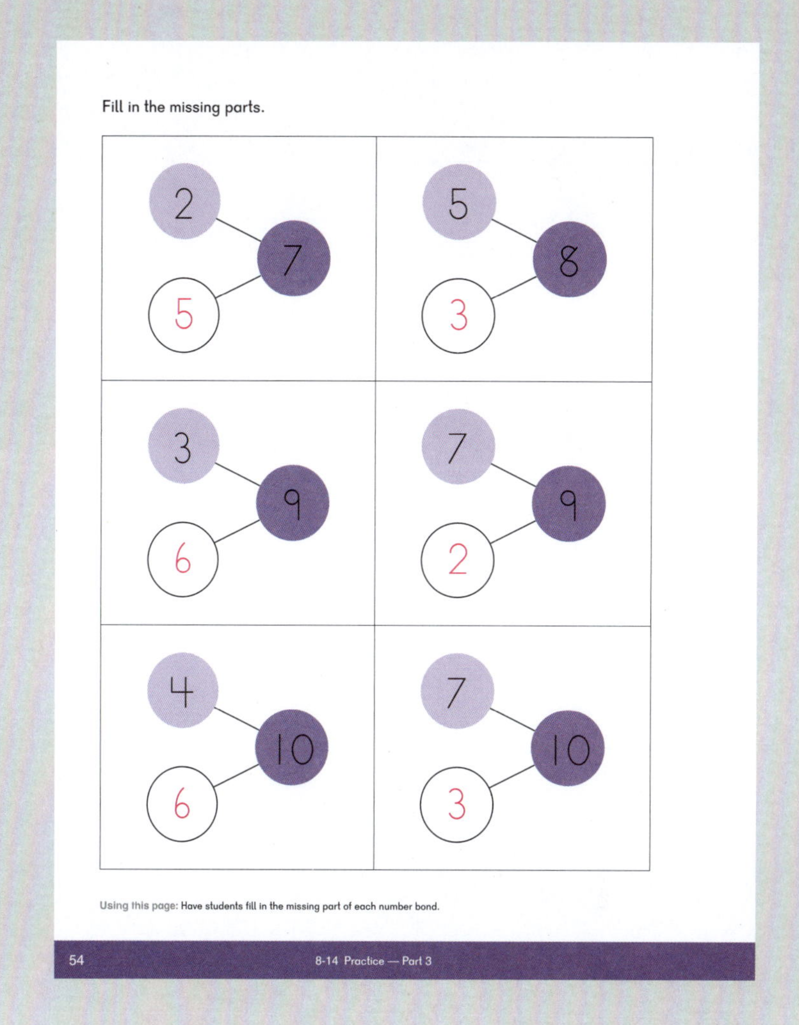

Notes

Suggested number of class periods: 12–13

	Lesson	Page	Resources	Objectives
	Chapter Opener	p. 87	TB: p. 71	
1	Introduction to Addition — Part 1	p. 88	TB: p. 72 WB: p. 55	Recognize the addition and equals symbols. Understand addition as joining two quantities together.
2	Introduction to Addition — Part 2	p. 90	TB: p. 74 WB: p. 57	Recognize the relationship between number bonds and addition sentences.
3	Introduction to Addition — Part 3	p. 93	TB: p. 77 WB: p. 59	Recognize patterns in addition of one more.
4	Addition	p. 95	TB: p. 79 WB: p. 61	Put quantities together to add.
5	Count On — Part 1	p. 97	TB: p. 81 WB: p. 63	Count on 1, 2, or 3 to add.
6	Count On — Part 2	p. 99	TB: p. 83 WB: p. 65	Count on 1, 2, or 3 to add on a number path.
7	Add Up to 3 and 4	p. 101	TB: p. 85 WB: p. 67	Write addition sentences up to a sum of 4.
8	Add Up to 5 and 6	p. 103	TB: p. 87 WB: p. 69	Write addition sentences up to a sum of 6.
9	Add Up to 7 and 8	p. 105	TB: p. 89 WB: p. 71	Write addition sentences up to a sum of 8.
10	Add Up to 9 and 10	p. 107	TB: p. 91 WB: p. 73	Write addition sentences up to a sum of 10.
11	Addition Practice	p. 109	TB: p. 93 WB: p. 75	Use number bonds and number sentences for addition.
12	Practice	p. 111	TB: p. 95 WB: p. 77	Practice skills from the chapter.
	Workbook Solutions	p. 114		

Notes

In **Chapter 8: Number Bonds**, students learned that a number, or the whole, can be split into two parts. In this chapter they will extend this idea to addition equations. Students will be introduced to the symbols "+" and "=."

Students will first relate number bonds to addition equations (number sentences with the symbols + and =) and then solve addition problems by:

- Counting all objects
- Counting on 1, 2, and 3 more
- Finding all addition combinations to 10

The correct use of math vocabulary, once introduced, should be emphasized. Students are adding, not "plus-ing," and subtracting (in **Chapter 10: Subtraction**) not "minus-ing." The equals sign also should be stated correctly, as "equals," in addition to the previously used convention, "makes."

In this curriculum, students will use the phrase "count on" rather than "count up," and "count back" rather than "count down."

Using "count up" or "count down" to indicate an increase or decrease is often confusing to students because on a typical number chart, the numbers increase going from left to right or to the next line down.

When students count on or back using a number path, the number of steps (or "hops") is what is counted, which does not include the starting square.

Dimensions Mathematics® KB uses the terms "addition sentence," "number sentence," and "subtraction sentence," instead of "equation." This is intended to help students recognize that the symbols are conveying a complete thought, the way a sentence is a complete thought.

Since students in early grades are often asked to "find the answer" for equations (written in this form: $3 + 4 = ?$), they might think equations always mean "find the answer." However, the equals sign simply means that the expressions on either side of it are the same. Thus, $3 + 4 = 7$ is the same thing as $7 = 3 + 4$.

At this level, students will not be required to write the entire equations (including the symbols), but instead will see equations with boxes where they need to write in the missing numbers. They will see some equations where the total is on the right side of the equals sign, and some equations where the total is on the left side of the equals sign.

It is assumed that all students will have access to recording tools. When a lesson refers to a whiteboard, any writing materials can be used.

Materials

- Art materials such as dot painters, stickers, or stamps
- Chalk
- Classroom game board
- Counters
- Crayons or dry erase markers
- Cup
- Deck of playing cards
- Dice
- Large cube-shaped box
- Linking cubes
- Modified die with sides labeled 0, 0, 1, 1, 2, 2
- Number bond mats
- Painter's tape
- Paper plates
- Pattern blocks
- Recording device
- Storybooks or magazines
- Two-color counters
- Whiteboards

Blackline Masters

- Addition Facts Cards
- Addition Template
- Addition to 10 Cover-Up Cards
- Animal-shaped counters from classroom or Animal Counters
- Blank Number Bond Template
- Dot Cards
- Equation Symbol Cards
- Five-frame Cards
- Number Bond Cards
- Number Cards
- Number Cards — Large
- Number Path
- Shake and Count Addition
- Ten-frame Cards

Storybooks

- Any picture book provides opportunities to create a number story from images on the page

Letters Home

- Chapter 9 Letter

Notes

Lesson Materials

- Two-color counters
- Blank Number Bond Template (BLM)

Ask students to tell number stories about the ducks on page 71. As stories are told, have students show number bonds for each, using two-color counters and a Blank Number Bond Template (BLM). Examples:

- There are 3 brown ducks flying. There are 2 white ducks flying. There are 5 ducks flying in all.

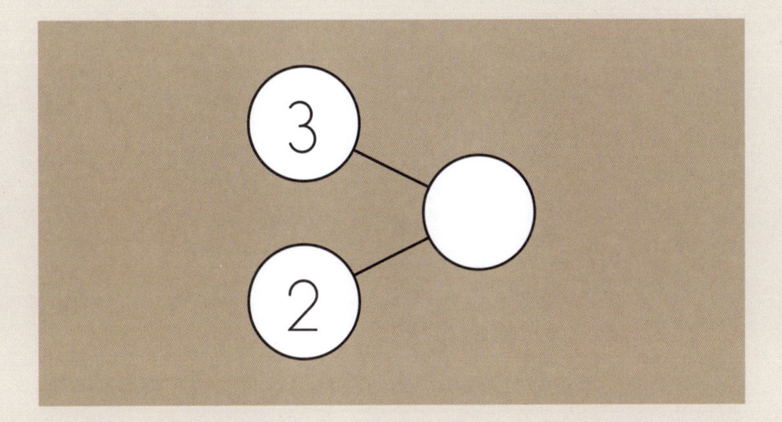

- 3 ducklings are on land. 1 duckling is in the water. There are 4 ducklings in all.

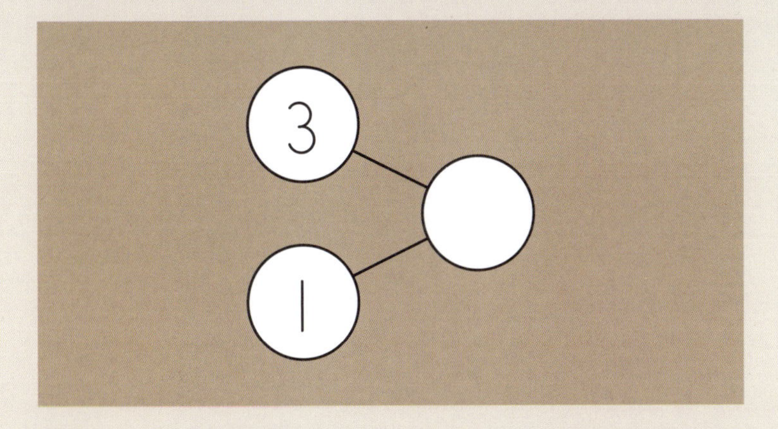

71

Extend

★ Number Story Pictures

Materials: Number Cards (BLM) 0 to 9

Students choose 2 cards and illustrate a joining or addition story using the numbers chosen.

For example, the numbers 5 and 3 are chosen. A student may illustrate an addition story by drawing 5 black puppies and 3 spotted puppies. Although not introduced, the student may also illustrate a subtraction story by drawing 5 ducks, 3 in the pond and 2 out of the pond.

Have students show the number bond for each situation.

Lesson 1 Introduction to Addition — Part 1

Objectives

- Recognize the addition and equals symbols.
- Understand addition as joining two quantities together.

Lesson Materials

- Linking cubes, 5 per student

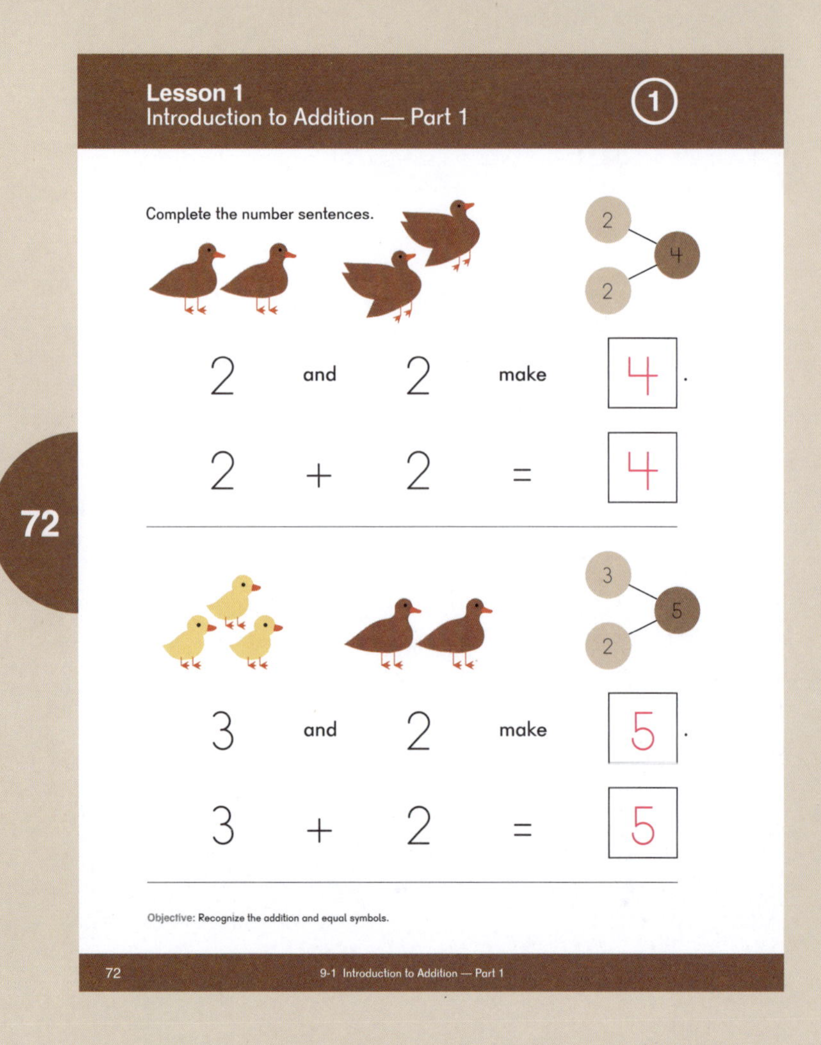
72

Lesson 1
Introduction to Addition — Part 1 ①

Complete the number sentences.

2 and 2 make 4.

2 + 2 = 4

3 and 2 make 5.

3 + 2 = 5

Objective: Recognize the addition and equal symbols.

72 9-1 Introduction to Addition — Part 1

Explore

Tell a story about 3 ducklings and 2 adult ducks. Have students use linking cubes to explore how they could find out how many ducks in all. Have students share the number bond they would use to show how many ducks in all.

Learn

Tell students that instead of writing "and" and "make," we can write "+" and "=." Introduce the symbols:

- The addition symbol, +, stands in for the word "and." It is read "plus." Say, "3 plus 2."
- The equals sign, =, stands in for the word "make." It is read "equals." Say, "3 plus 2 equals 5."

Write the addition sentences (or equations) for stories and number bonds found in the **Chapter Opener**. Tell students when we are joining amounts together, it is called "adding." We can say that $3 + 2 = 5$ is an addition sentence. Tell students that when we know the parts and find the total, we are adding the parts together.

Whole Group Activity

▲ Human Number Sentences

Select students to come to the front of the class based on some obvious differences. Ask students to notice what is the same and what is different about the students. Ask students to say addition sentences in two ways.

Examples:

- 2 students wearing pants and 3 students wearing shorts make 5 students altogether. $2 + 3 = 5$
- 3 students wearing shorts and 2 students wearing pants make 5 students altogether. $3 + 2 = 5$

Small Group Activities

Textbook Pages 72–73. Discuss and complete the different representations: pictures, number bonds and number sentences.

▲ Shake and Count Addition

Materials: 10 two-color counters, cup, Shake and Count Addition (BLM)

Students shake up to 10 counters in the cup, dump them out and complete the addition sentence on Shake and Count Addition (BLM).

▲ Addition Sentences

Materials: + and = cards from Equation Symbol Cards (BLM), counters

Player 1 creates an addition sentence using counters and the + and = cards from Equation Symbol Cards (BLM). Player 2 tells the answer. Players switch roles and play continues.

Exercise 1 • page 55

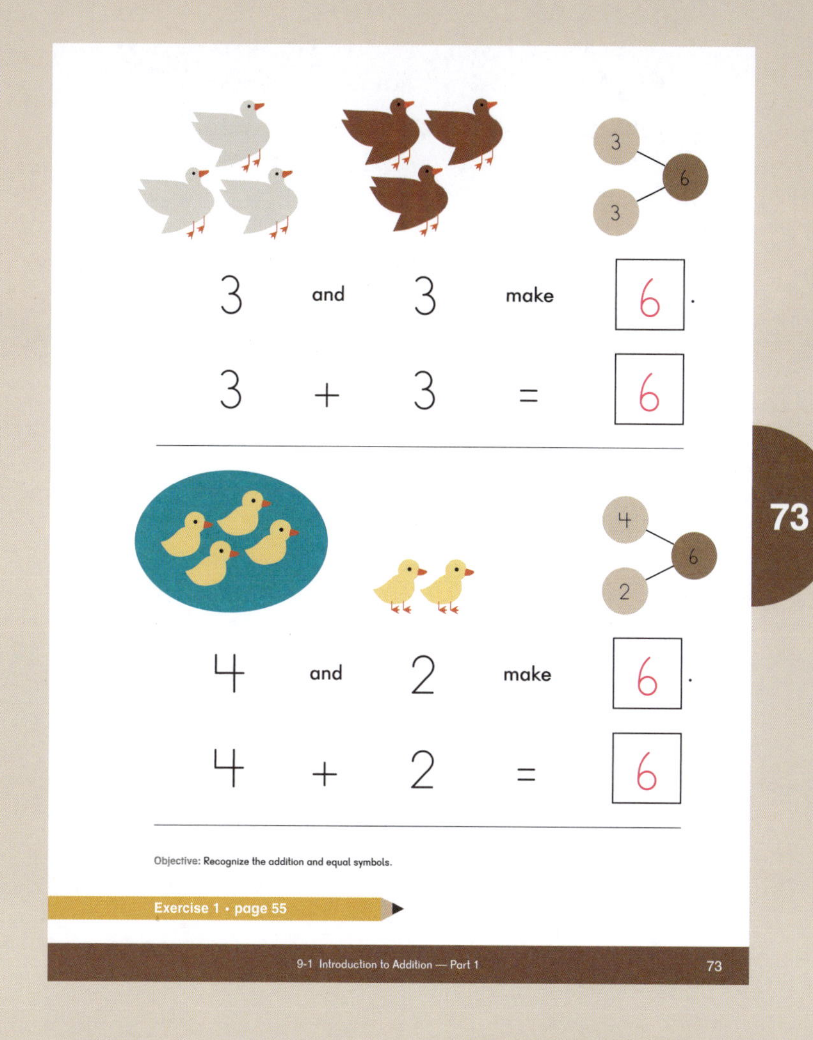

73

Extend

★ Greatest Whole

Materials: Number bond mats, 4 of each Number Card (BLM) 0 to 5

Players shuffle the Number Cards (BLM). Each player draws 2 cards from the deck. Students place the cards in the "parts" of the number bond and find the whole. The player with the greatest whole wins the cards.

To extend, use number cards up to 9.

Objective

- Recognize the relationship between number bonds and addition sentences.

Lesson Materials

- Blank Number Bond Template (BLM)
- Two-color counters, at least 8 per student
- Pattern blocks

74

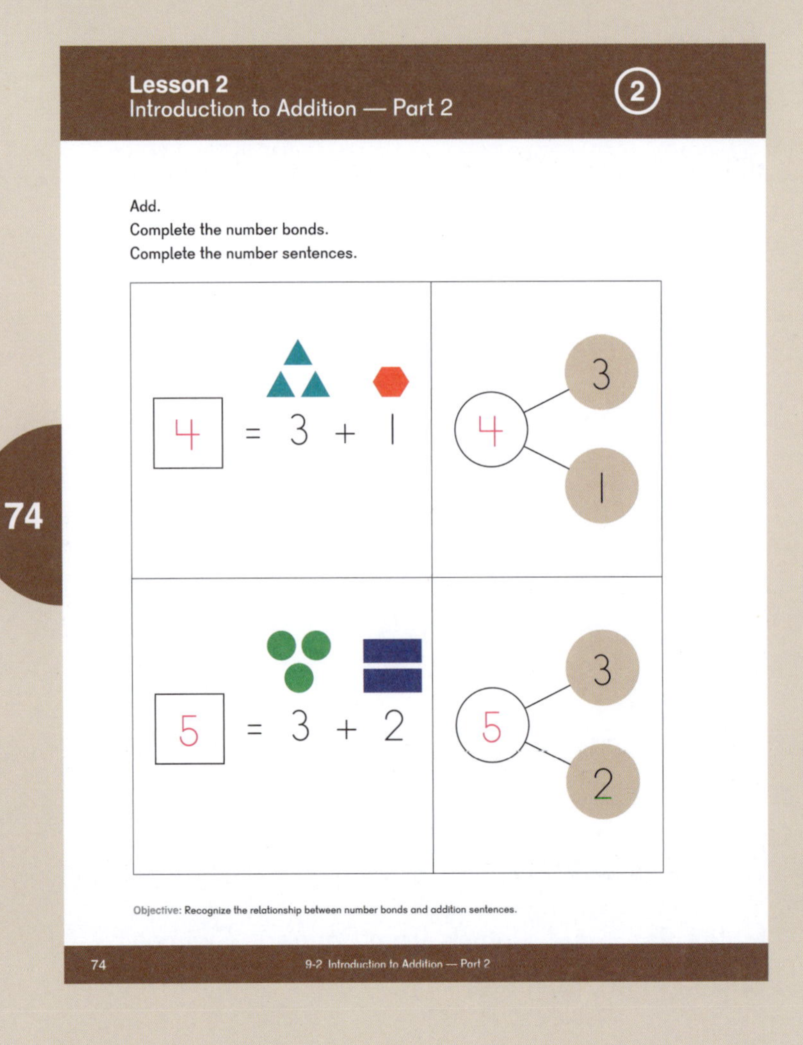
Lesson 2
Introduction to Addition — Part 2 ②

Add.
Complete the number bonds.
Complete the number sentences.

4 = 3 + 1

5 = 3 + 2

Objective: Recognize the relationship between number bonds and addition sentences.

74 9-2 Introduction to Addition — Part 2

Explore

Provide each student with a Blank Number Bond Template (BLM) and at least 8 counters, and have them show a number bond with two parts. Ask students how many different ways they can say what is on the number bond.

For example:

- 4 and 1 make 5
- 1 and 4 make 5
- 5 is 4 and 1
- 5 is 1 and 4

Learn

Tell students that the equals sign (or symbol) means the value of what is on both sides of it is the same. The word "equals" means "is the same as."

Model the addition sentence on page 74 with pattern blocks or other counters.

☐ = 3 triangles + 1 hexagon can be read as,

"4 shapes is the same as 3 triangles plus 1 hexagon," or, "4 equals 3 plus 1."

Give several other examples and encourage students to see that "equals" means they are worth the same amount.

Whole Group Activity

▲ Number Sentence Flash

Materials: Addition to 10 Cover-Up Cards (BLM)

Show students an Addition to 10 Cover-Up Card (BLM), using a finger to cover up the total. Have students say the total.

To extend, use a finger to cover up one of the parts in the addition sentence and have students say the missing part.

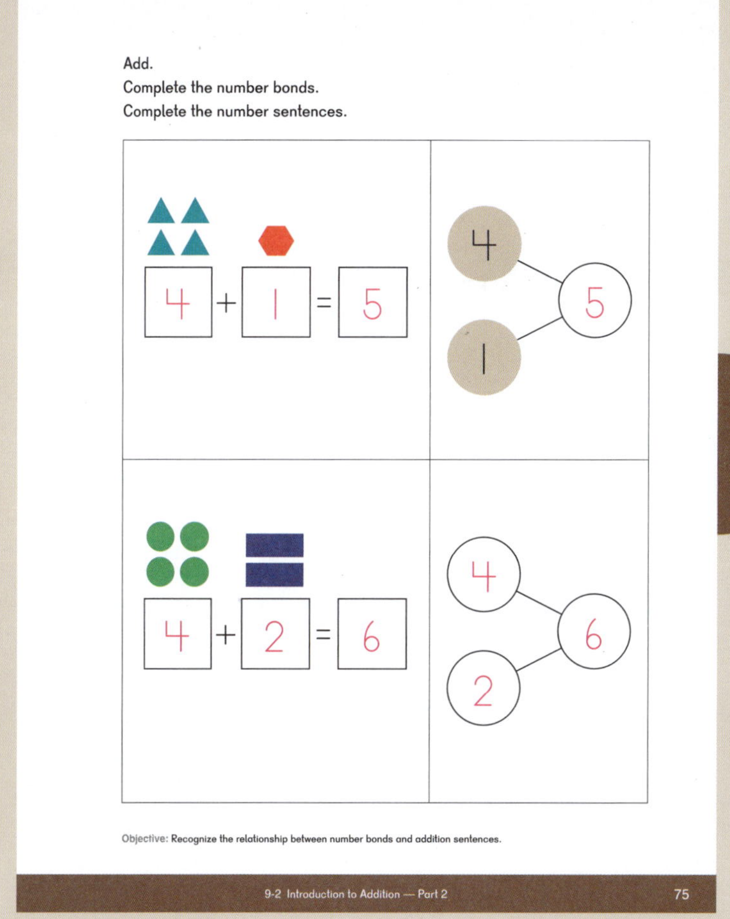

75

Small Group Activities

Textbook Pages 74–76. Discuss and complete.

▲ Ten-frame Card Addition

Materials: 4 sets of Ten-frame Cards (BLM) 0 to 5, + and = cards from Equation Symbol Cards (BLM), Optional: Number Cards (BLM)

Give each pair of students 4 sets of Ten-frame Cards (BLM) 0 to 5 and the + and = from Equation Symbol Cards (BLM).

Each player draws two cards and places them in the "parts" of the equation along with the + and = cards. The player with the greater total wins the cards. Shuffle the cards and repeat.

To extend this game:

- Draw 3 cards and try to make the greatest or least total.
- Play with Number Cards (BLM) instead of Ten-frame Cards (BLM).
- Use Number Cards (BLM) or Ten-frame Cards (BLM) up to 9.

Exercise 2 • page 57

Extend

★ How Many Ways?

Materials: Addition Template (BLM)

Students can cut the numbers out from the bottom of the Addition Template (BLM), then use them to create as many correct number sentences as possible using each digit only once in each sentence.

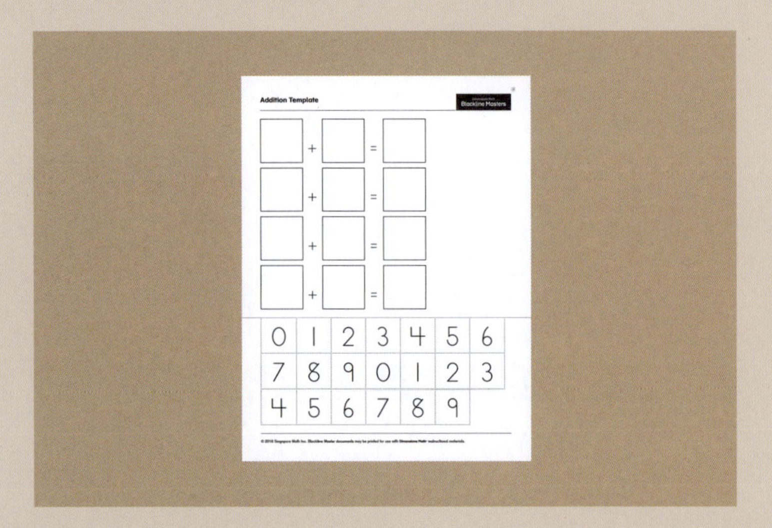

Addition Template

Blackline Masters

□ + □ = □

□ + □ = □

□ + □ = □

□ + □ = □

0	1	2	3	4	5	6
7	8	9	0	1	2	3
4	5	6	7	8	9	

76

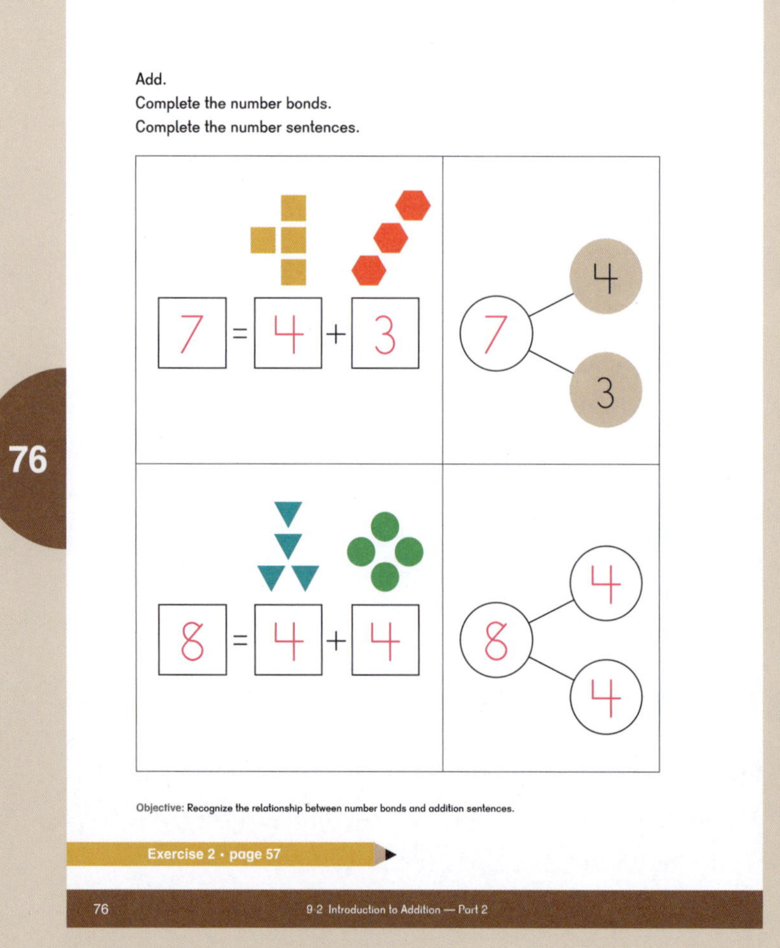

Add.
Complete the number bonds.
Complete the number sentences.

7 = 4 + 3

7 — 4, 3

8 = 4 + 4

8 — 4, 4

Objective: Recognize the relationship between number bonds and addition sentences.

Exercise 2 • page 57

76 9-2 Introduction to Addition — Part 2

Objective

- Recognize patterns in addition of one more.

Lesson Materials

- Linking cubes, 21 per student or pair of students

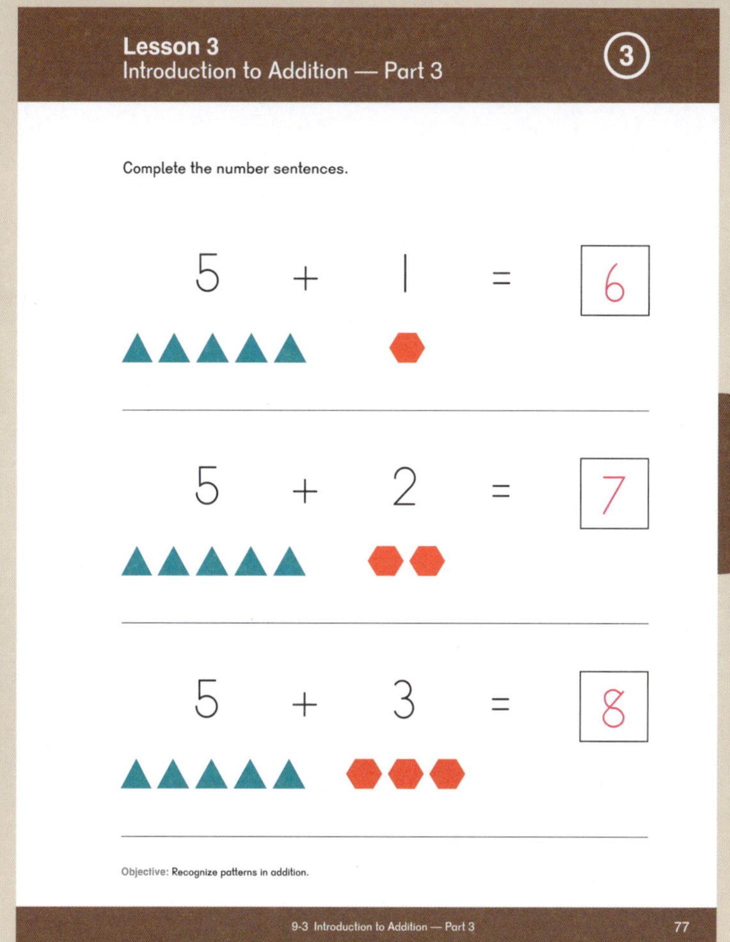

77

Explore

Provide students with linking cubes and have them show, then build, 1 + 1 on their whiteboards arranged as shown below. Once they know that 1 + 1 is 2 cubes, have them build 1 + 2 and lay it below the other two rows. The students should build a staircase in the "total" column showing one more than the previous number.

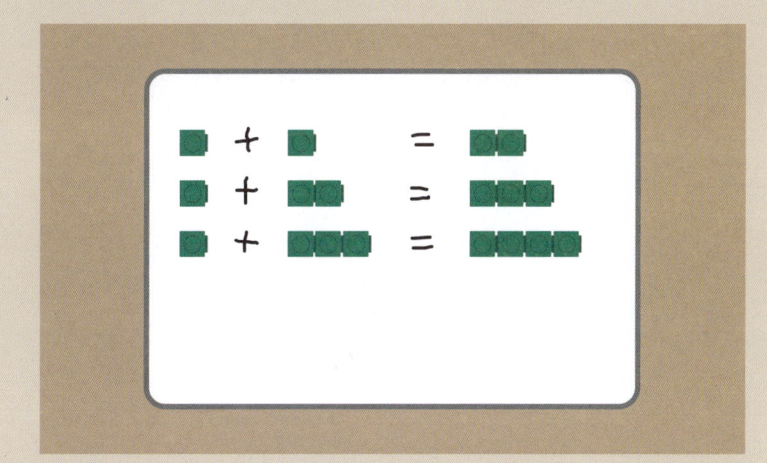

Learn

Ask students to use the pattern of adding one more to determine the answers for a few examples.

- If 3 + 1 = 4, then what will 3 + 2 be?
- If 6 + 2 = 8, then what will 6 + 3 be?

Some students may need to manipulate cubes to find the answer. Encourage students to use the pattern to find the answer rather than counting all of the cubes.

Whole Group Activity

▲ Group Up

Write number sentence adding 1, 2, or 3 on the board. Students work together to form groups of the total. For example, the teacher writes "4 + 2 =" and students form groups of 6.

Since not all students will be included in groups every time, play several rounds.

Small Group Activities

Textbook Pages 77–78. Discuss the pattern.

▲ Same or 1 More

Materials: 4 sets of Number Cards (BLM) 0 to 9 per group

Have students form groups of 3 or 4, then give each group 4 sets of Number Cards (BLM) 0 to 9. Each student is dealt 5 cards. Place the remaining cards facedown in a deck. One card from the deck is flipped faceup and placed in the middle to make the play pile.

Students take turns playing a card that is either the same or 1 more than the card showing. If a player doesn't have a card to play, he must draw one from the deck until he does. The first player to play all of his cards wins the game.

To extend, play **Same or 2 More**.

▲ Next Number Snap

Materials: At least 2 sets of Number Cards (BLM) 1 to 9, or a deck of playing cards with face cards removed

Deal all cards facedown to up to 4 players. Players take turns turning over their top cards and saying the numbers aloud. They put those cards into a discard pile.

If the new card is one more than the top card on the discard pile, players say, "Snap." The first person to do so collects the discard pile.

The game ends when a player is out of cards. The player with the most cards wins.

Exercise 3 • page 59

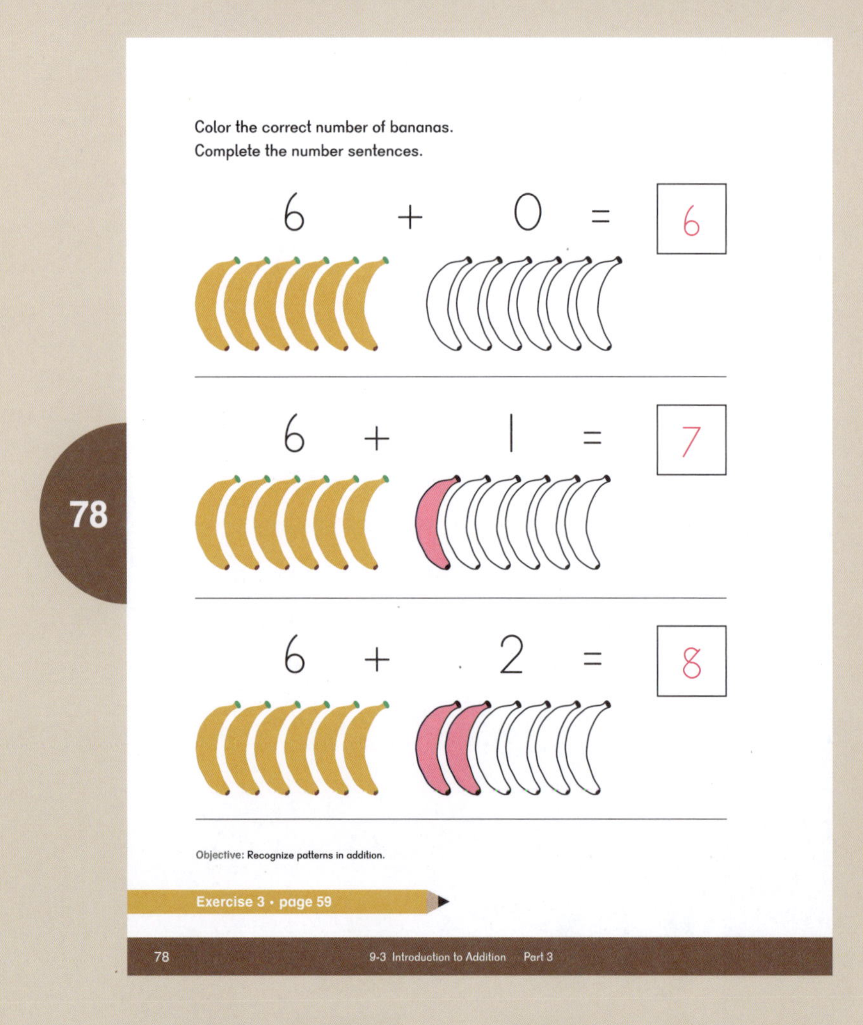

Extend

★ Counting On Dice

Materials: 1 regular die, 1 modified die with sides labeled: 0, 0, 1, 1, 2, and 2, any classroom game board

Almost any game where students need to count on to move their pieces will work with the modified dice. Players take turns rolling the two dice and adding the numbers together to determine their move.

Objective

- Put quantities together to add.

Lesson Materials

- Two-color counters, 10 per student

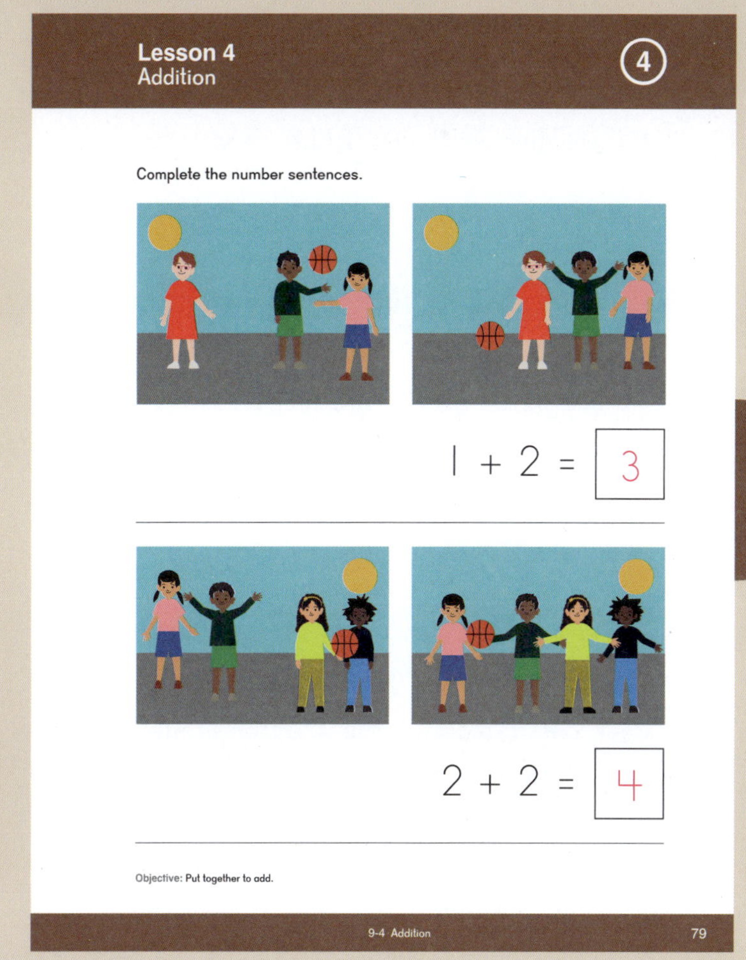

79

Explore

Tell an addition story and ask students to act the story out with counters. For example, "There are 5 red apples and 2 green apples in a basket. How many apples are there altogether?"

Ask students to tell addition stories (using numbers up to 5) while all students act out the stories with counters.

Learn

Have students discuss the two scenarios on page 79. Discuss what was the same about the two sets of pictures. Possible answers:

- In the top pictures, Alex and Mei were playing basketball and Emma joined them. And then in the bottom pictures, Sofia and Dion were playing basketball and Mei and Alex joined them.
- More friends played basketball in both sets of pictures.
- They were both addition stories.

Whole Group Activity

▲ Five-frame Flash

Materials: Five-frame Cards (BLM) 1 to 5, and additional Five-frame Cards (BLM) for 1, 2, and 3

Show students a Five-frame Card (BLM) for 5, and Five-frame Cards (BLM) showing 1, 2, or 3. Have students call out the total.

Alternatively, students could show the answer with their fingers or write the answer on a whiteboard.

Small Group Activities

Textbook Pages 79–80

▲ Addition Pictures

Materials: Number Cards (BLM) 0 to 5, and art materials such as dot painters, stickers, or stamps

Provide dot painters, stickers, stamps, or other art materials, and Number Cards (BLM) 0 to 5. Have students draw two cards and illustrate addition pictures with two different colors or stamps. After they have created the picture, have students write the corresponding addition sentence.

▲ Total Up!

Materials: Dot Cards (BLM) 0 to 5, Ten-Frame Cards (BLM) 0 to 5, Number Cards (BLM) 0 to 5

Make a deck of cards comprised of 4 each of cards 0 to 5, using either Dot Cards (BLM), Ten-frame Cards (BLM), or Number Cards (BLM), and deal all the cards equally between 2 players.

Each player flips a card at the same time and adds the 2 cards together. The first player to say the total of the two numbers on the cards collects the cards. If a player loses all of his cards, the game is over.

Reshuffle the cards and play again.

To extend this activity, have students play in groups of 3, or combine and use cards from the three different sets.

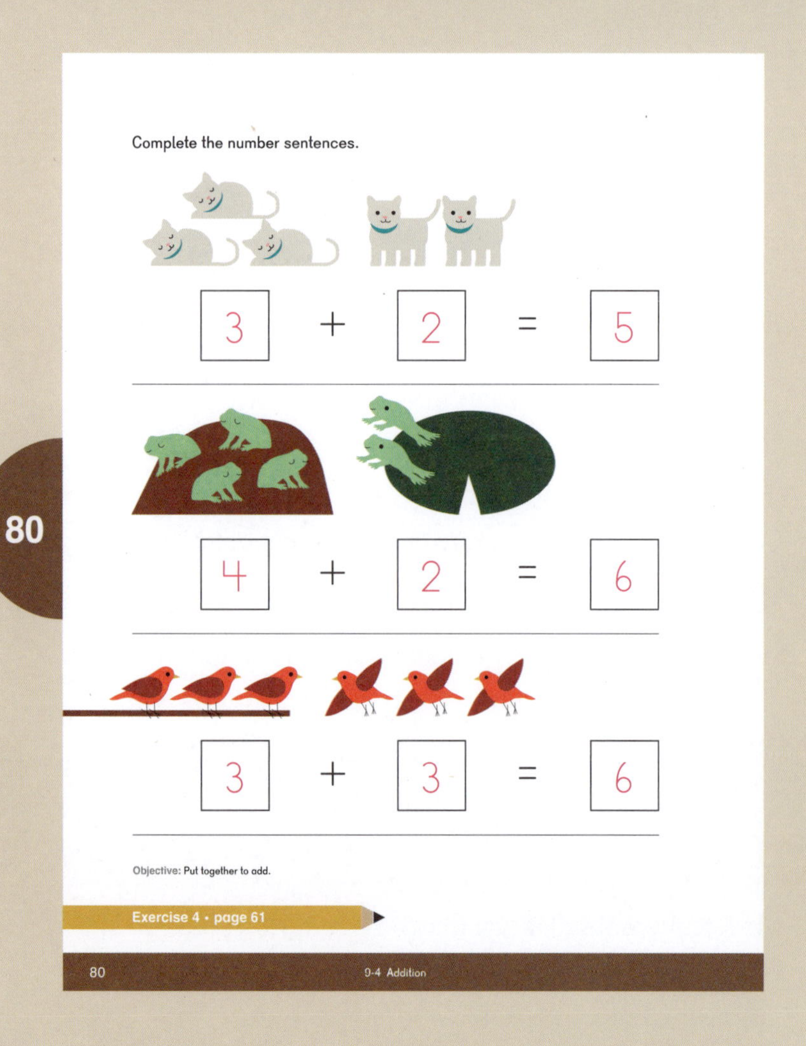

▲ Rock-Paper-Scissors-Math!

Similar to Rock-Paper-Scissors, students work in pairs and tap their fists on their hand while saying, "Rock-Paper-Scissors-Math!"

On "Math!" each student holds up 0 to 5 fingers on one hand. Students then add all the fingers together. Students can take turns saying the total. Alternatively, both students can find the total and the first player to say it out loud wins the round.

Exercise 4 • page 61

Extend

★ Roll and Add

Materials: 2 dice with sides 0, 1, 2, 3, 4, and 5

Students take turns rolling the dice. On each turn, students write an addition sentence using the numbers shown on each die as a part on a whiteboard. The partner checks the addition sentence and if it is correct, the player who rolled gets a point.

Objective

- Count on 1, 2, or 3 to add.

Lesson Materials

- Number Cards — Large (BLM) 4, 5, and 6
- Counters
- Cup

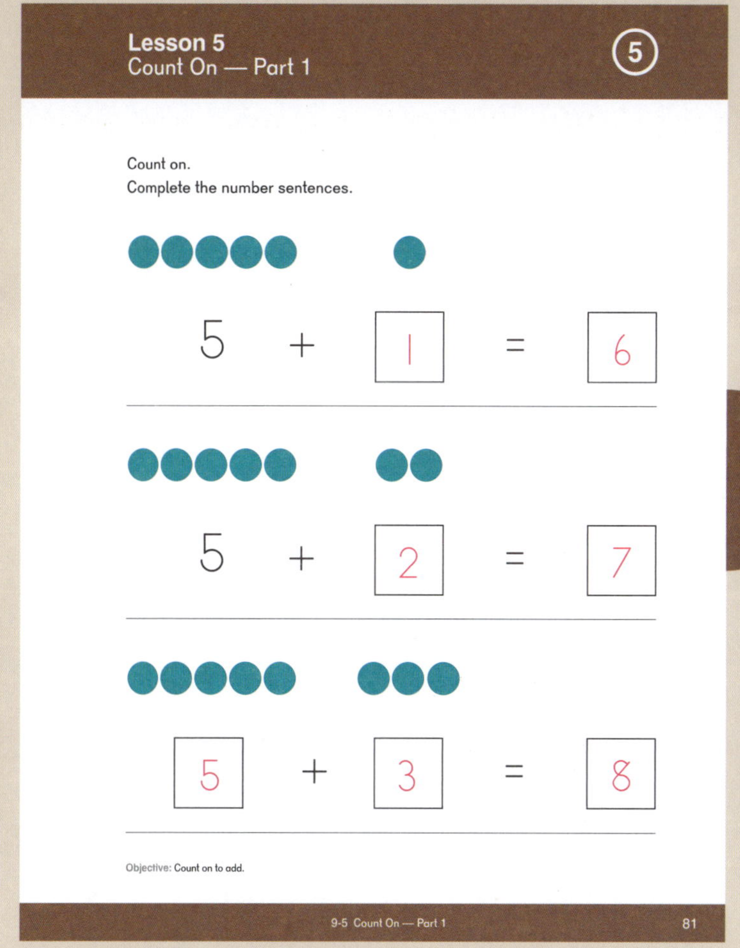

Explore

Invite 4 students to come to the front of the classroom. Count the students. Have one of them hold a Number Card — Large (BLM) for 4. Invite one more student to join the group. Students say, "4," and then count on to add. Repeat with other groups of students or classroom objects, adding only up to 3 more each time.

Learn

Tell students that when adding 1, 2, or 3, we can count on from the starting number.

Write 4 + 2 on the board. Put 4 counters into a cup and label it "4." Ask students to say, "4," and count on 2 more ("5, 6") as you drop counters one by one into the cup.

Practice with a few other examples, adding up to 3 counters to the cup at a time.

Whole Group Activity

▲ Shout and Clap

Have students take turns choosing a starting number from 5 to 9 and adding 1, 2, or 3. Chorally, all students shout the number and clap to count on.

For example, a student chooses the number 7 and wants to add on 2. Students shout, "7," then clap and say, "8, 9."

For variation, the student can also choose a different action, such as hopping, snapping, marching, etc.

Small Group Activities

Textbook Pages 81–82

Activities from previous lessons can be extended to count on by 1, 2, or 3.

▲ Face-Off!

Materials: 4 of each Number Card (BLM) or Ten-frame Card (BLM) 0 to 9, 3 each of Dot Cards (BLM) 1, 2, and 3

Place the Number Cards (BLM) in one deck and the Dot Cards (BLM) in another deck. Each player draws 1 card from each deck. Students say the number on the Number Card (BLM) and count on to add the number on the Dot Card (BLM). The player with the greatest total wins the cards.

▲ Same or 2 or 3 More

Materials: 4 sets of Number Cards (BLM) 0 to 9 per group

Play using the rules from **Same or 1 More** in Lesson 3, this time adding on 2 more or 3 more.

▲ Total Up!

Materials: Dot Cards (BLM) 0 to 5, Ten-Frame Cards (BLM) 0 to 5, Number Cards (BLM) 0 to 5

Play as described in Lesson 4.

▲ Roll and Add

Materials: 1 die with numbers 0 to 5 and 1 die with numbers 1, 1, 2, 2, 3, and 3

Play as explained in Lesson 4, with dice modified as directed here.

82

Color the ten-frame cards and complete the number sentences.
The first one has been done for you.

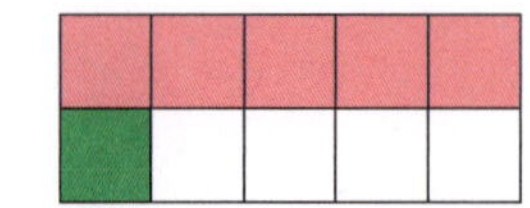

5 + 1 = 6

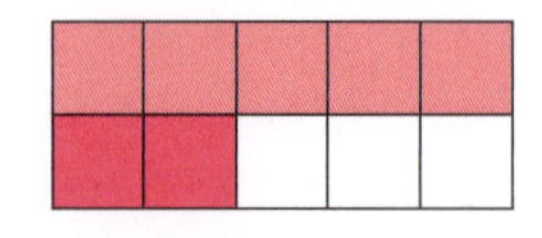

5 + 2 = 7

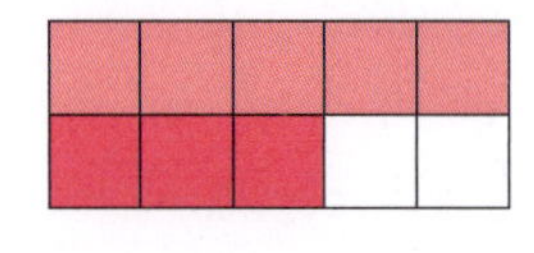

5 + 3 = 8

Objective: Count on to add.

Exercise 5 • page 63

82 9-5 Count On — Part 1

Exercise 5 • page 63

Extend

★ Speed Face-Off!

Materials: 4 sets of Number Cards (BLM) 0 to 9, die modified with sides 1, 1, 2, 2, 3, and 3 for each player

Deal all cards equally to up to 4 players. Players each flip over a card and roll their die. The first player to say his total by adding the number on the card and the die together, wins the round and collects the other players' cards.

In the case of a tie, players put their cards on the bottom of their piles and play continues.

Objective

- Count on 1, 2, or 3 to add on a number path.

Lesson Materials

- Number Path (BLM)
- Animal-shaped counters from classroom or Animal Counters (BLM)

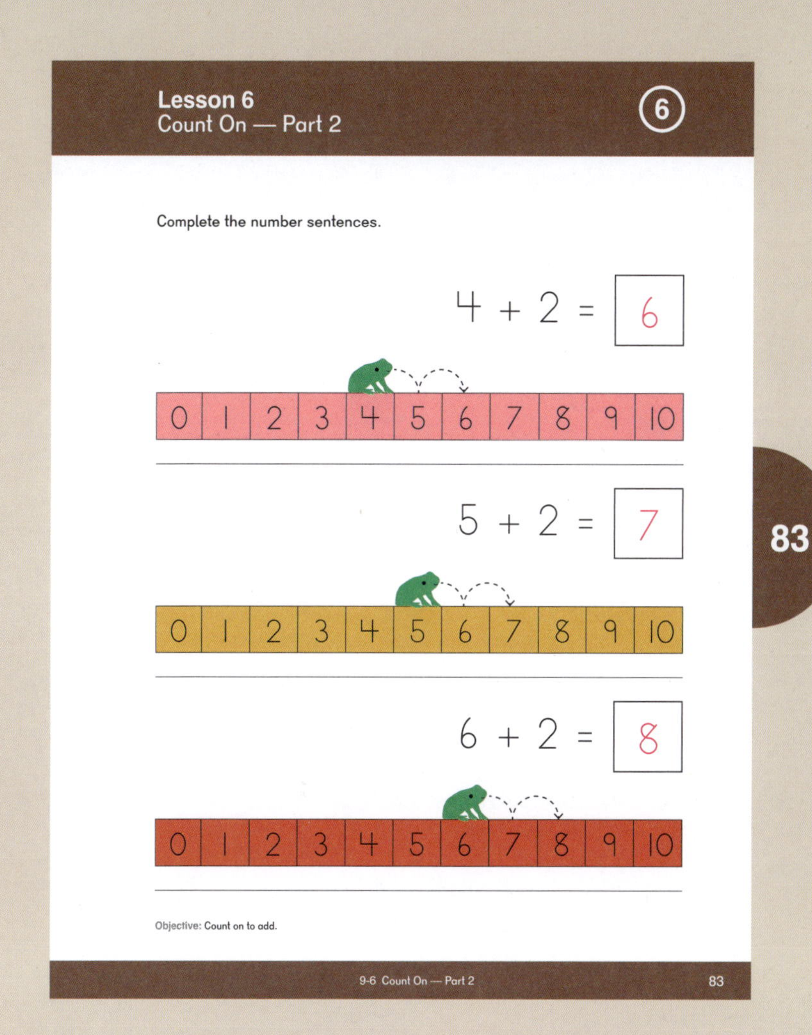

Explore

Provide each student with an Number Path (BLM). Each student places an Animal Counter (BLM) or his finger on the number 4. Ask students to add 1 on the number path by hopping one number to the right on the path. Ask them what number their counters are on, then give a number sentence with "4 + 1 =" and have students tell the total.

Learn

Repeat the activity in **Explore**, having students add on 1, 2, or 3. Emphasize that we start from the number that we are on, and the number of hops are what we are adding on.

Write the number sentences prompting students to provide the parts. For example:

- What number did we start on?
- How many did we add?
- What does five + two equal? (Use various numbers)
- How many does ___ + ___equal? Where does that number belong in the addition sentence?

Whole Group Activity

▲ Shout and Clap

Play as directed in Lesson 5.

Small Group Activities

Textbook Pages 83–84

Activities from previous lessons can be extended to count on by 1, 2, or 3.

▲ Same or 2 or 3 More

Materials: 4 sets of Number Cards (BLM) 0 to 9 per group

Play using the rules from **Same or 1 More** in Lesson 3, this time adding on 2 more or 3 more.

▲ Total Up!

Materials: Dot Cards (BLM) 0 to 5, Ten-Frame Cards (BLM) 0 to 5, Number Cards (BLM) 0 to 5

Play as described in Lesson 4.

▲ Roll and Add

Materials: Number Path (BLM) 1 die with numbers 1, 1, 2, 2, 3, and 3

On a number path, use one die with numbers 1, 1, 2, 2, 3, and 3 and beginning on 0. The first to 10 wins.

Take it Outside

▲ Next Number Hop: Add

Materials: Chalk, painter's tape, paper plates, Number Cards (BLM) 0 to 9, large cube-shaped box

Create large number paths outside with chalk, or inside with painter's tape. You could also use paper plates with numbers on them:

84

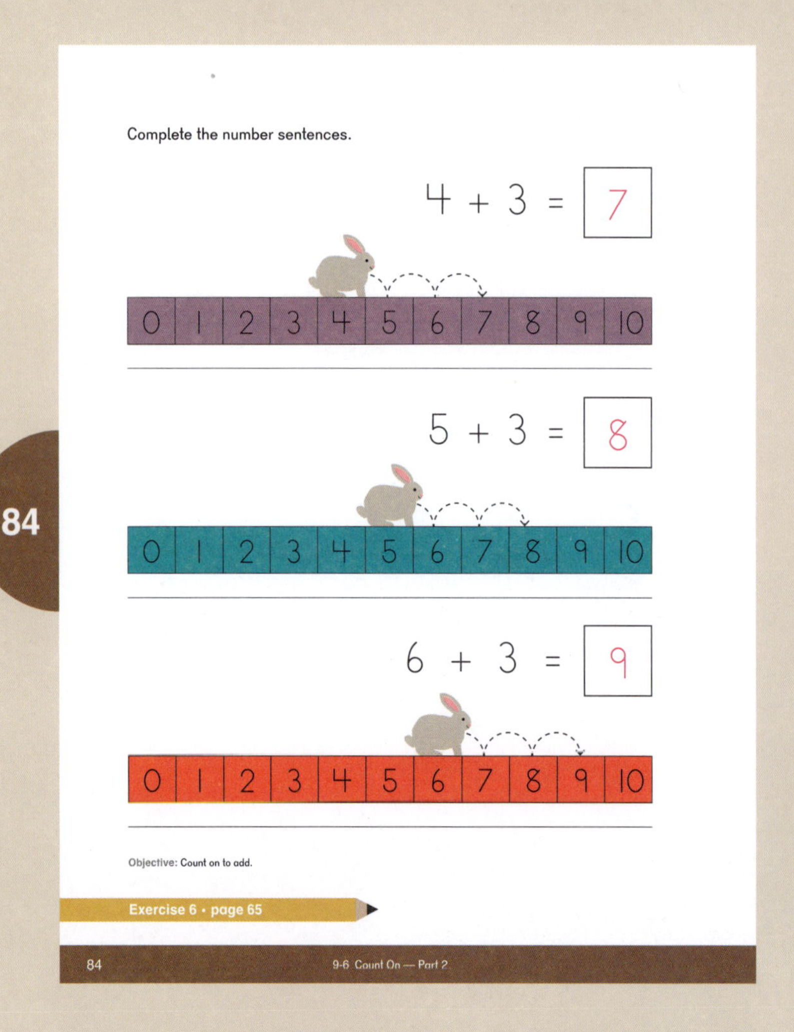

Call a number from 0 to 9 and have students stand on that number. Ask them to add 0, 1, or 2 to that number and walk to the new number.

To extend, make a large die out of a box with sides of 0, 0, 1, 1, 2, and 2. Have players take turns drawing from a stack of Number Cards (BLM) 0 to 9 and directing their partners to that starting spot on the number path.

The same player that drew the card then rolls the die and tells her partner how many steps to take. The partner walks that many steps on the path and says the new number.

Exercise 6 • page 65

Objective

- Write addition sentences up to a sum of 4.

Lesson Materials

- Counters, 4 per student
- Number bond mat

Explore

Ask students to put 3 counters in the whole of their number bonds and then to make as many different arrangements as they can, breaking apart 3 into 2 parts.

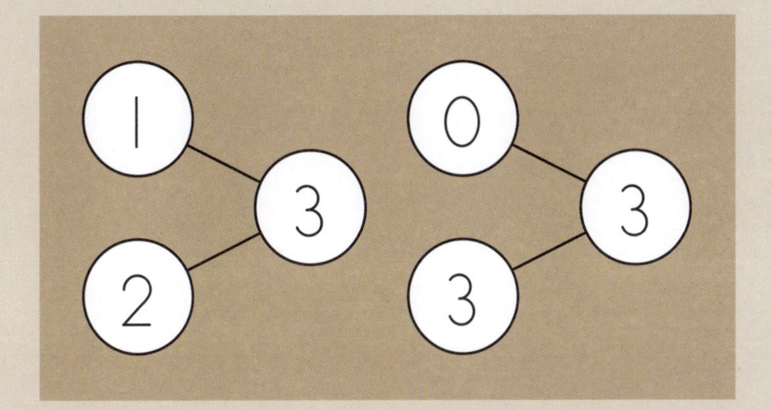

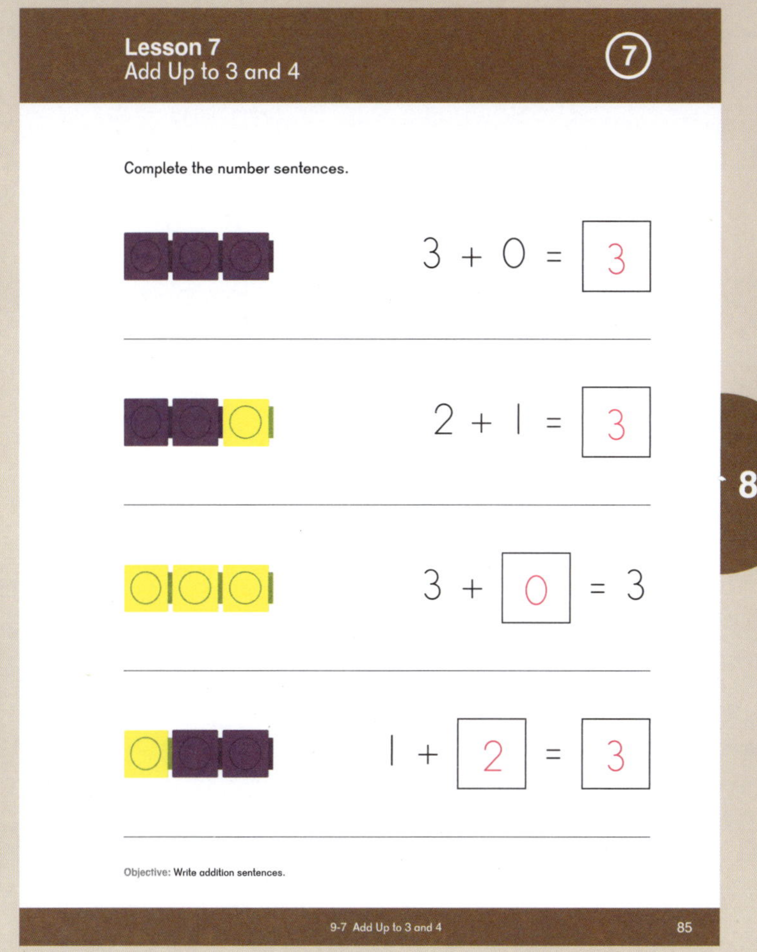
Lesson 7
Add Up to 3 and 4 ⑦

Complete the number sentences.

3 + 0 = 3

2 + 1 = 3

3 + 0 = 3

1 + 2 = 3

Objective: Write addition sentences.

9-7 Add Up to 3 and 4 85

85

Learn

As students find ways to make 3, write the addition sentence on the board. Have students discuss why 2 and 1 makes 3 is the same as 1 and 2 makes 3.

Repeat with 4 counters and ways to make 4.

Whole Group Activities

● Complete the Bond

Tell the class that you are going to find bonds for 3. The teacher or leader holds up fingers (representing a part), and students complete the bond by showing the other part using their fingers. Repeat for bonds for 4.

▲ Find a Partner

Materials: Multiple sets of Number Cards (BLM) 0 to 4

Pass out Number Cards (BLM) 0 to 3, or 0 to 4, so that each student has a card. Have students find partners to make 3 or 4, depending on what number they are adding to.

Small Group Activities

Textbook Pages 85–86

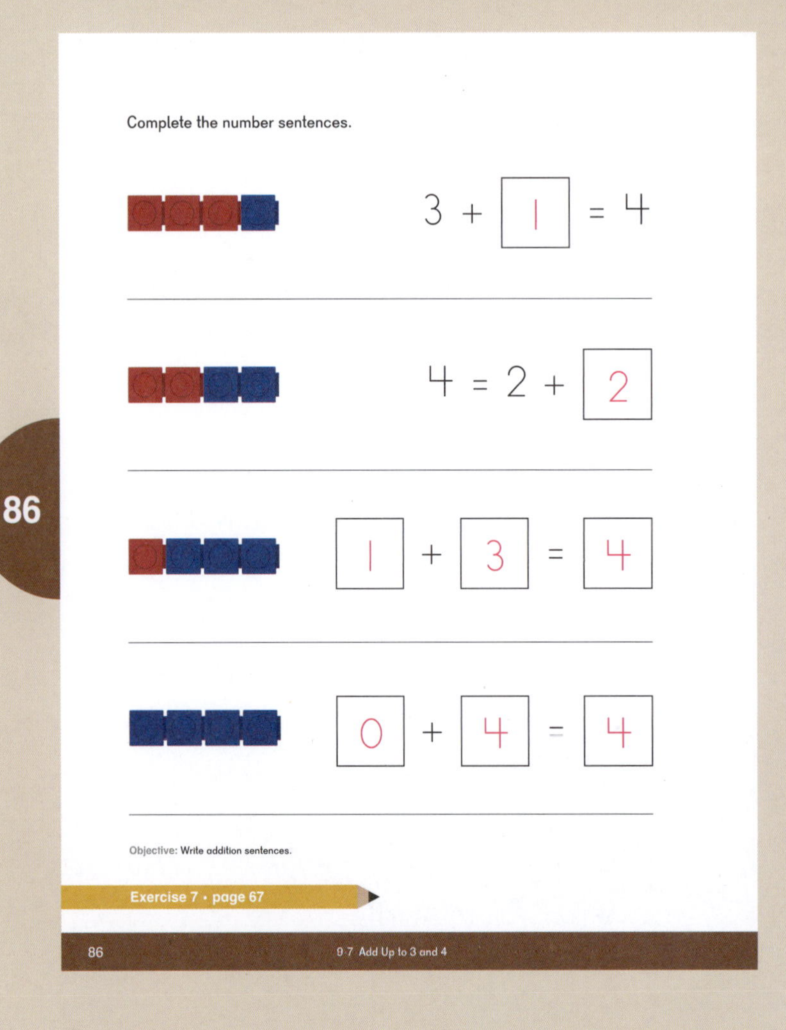

● Match

Materials: 2 sets of Ten-frame Cards (BLM) 0 to 4

Students arrange the cards faceup in a grid. Students take turns finding two cards that go together to make 3 or 4.

▲ Memory

Materials: 2 sets of Ten-frame Cards (BLM) 0 to 4

Students arrange the cards facedown in a grid. Students take turns finding two cards that go together. If the total is 3 or 4, they have a match.

▲ Go Fish

Materials: 4 sets of either Ten-frame Cards (BLM) or Number Cards (BLM) 0 to 4 per small group

Shuffle 4 sets of either Ten-frame Cards (BLM) or Number Cards (BLM), using cards from 0 to 4. Deal 4 cards to each player. Players play **Go Fish** with the cards, where a pair is a set of cards that together make 4.

Use Ten-frame Cards (BLM) for students who are still counting. Number Cards (BLM) can be used for students who understand the abstract number without counting.

Exercise 7 • page 67

Extend

★ Addition Stories

Materials: Storybooks or magazines, optional: recording device

Use a storybook or a magazine and find pictures where addition stories could be made. Have students tell the addition story to a friend or record it. Students then draw a picture of the addition story and write the number sentence.

Objective

- Write addition sentences up to a sum of 6.

Lesson Materials

- Two-color counters, 6 per student
- Blank Number Bond Template (BLM)

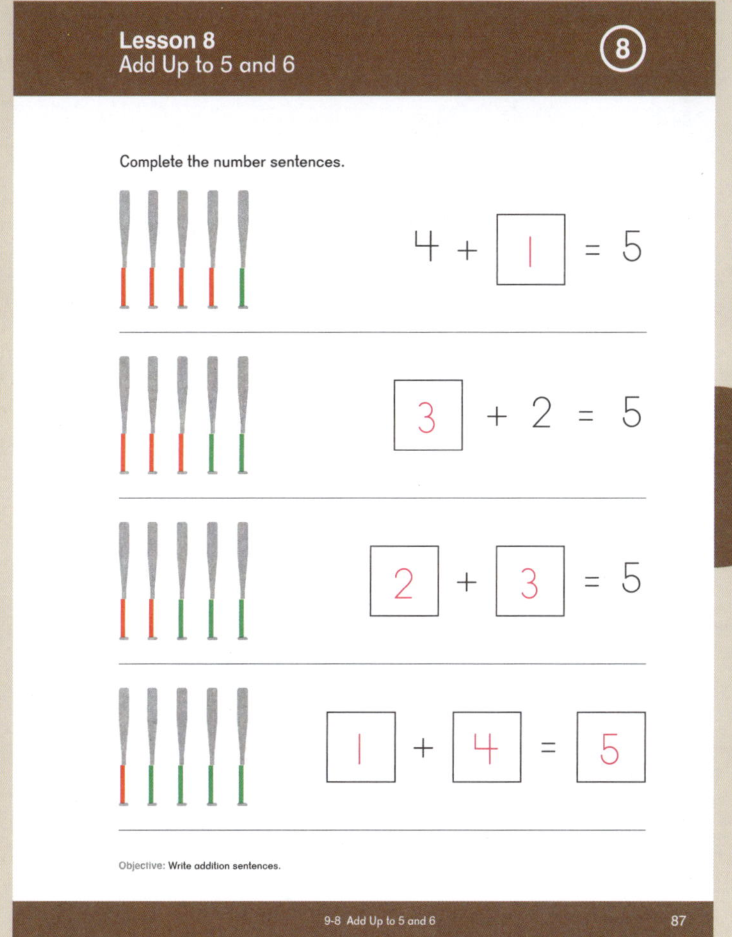

87

Explore

Ask students to put 5 counters in the whole of their Blank Number Bond Template (BLM) and then make as many different arrangements as they can, breaking apart 5 into 2 parts.

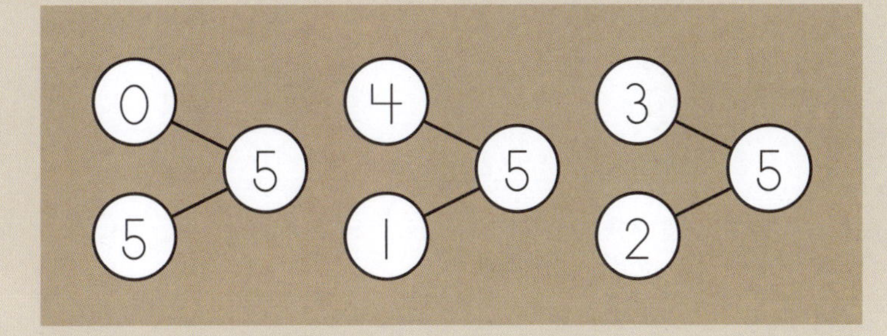

Learn

As students find ways to make 5, have them write the addition sentences on a whiteboard.

Write the number sentences prompting students to provide the parts. For example:

- How many are one color?
- How many did we add?
- How many does ___ + ___ equal? Where does that number belong in the addition sentence?

Repeat with 6 counters and ways to make 6.

Whole Group Activity

▲ Find a Partner

Materials: Multiple sets of Number Cards (BLM) 0 to 6

Play as instructed in Lesson 7, having students find a partner to make an addition sentence for 5 or 6.

Small Group Activities

Textbook Pages 87–88

● **Match**

Materials: 2 sets of Ten-frame Cards (BLM) 0 to 6

Students arrange the cards faceup in a grid. Students take turns finding two cards that go together to make 5 or 6.

▲ **Memory**

Materials: 2 sets of Ten-frame Cards (BLM) 0 to 6

Play as described in Lesson 7. If the total is 5 or 6, they have a match.

▲ **Snap!**

Materials: 4 sets of either Ten-frame Cards (BLM) or Number Cards (BLM) 0 to 6

Shuffle the cards and deal the whole deck equally between 2 players. Players flip their top card at the same time and if the total of the 2 cards is 5 or 6, the first player to say, "Snap!" wins the cards. The game is over when one player has no more cards to play.

▲ **How Many Ways?**

Materials: Addition Template (BLM), Number Cards (BLM) 0 to 6

Students place Number Cards (BLM) on the Addition Template (BLM) to create as many correct number sentences as possible using each digit only once in each sentence.

Exercise 8 • page 69

88

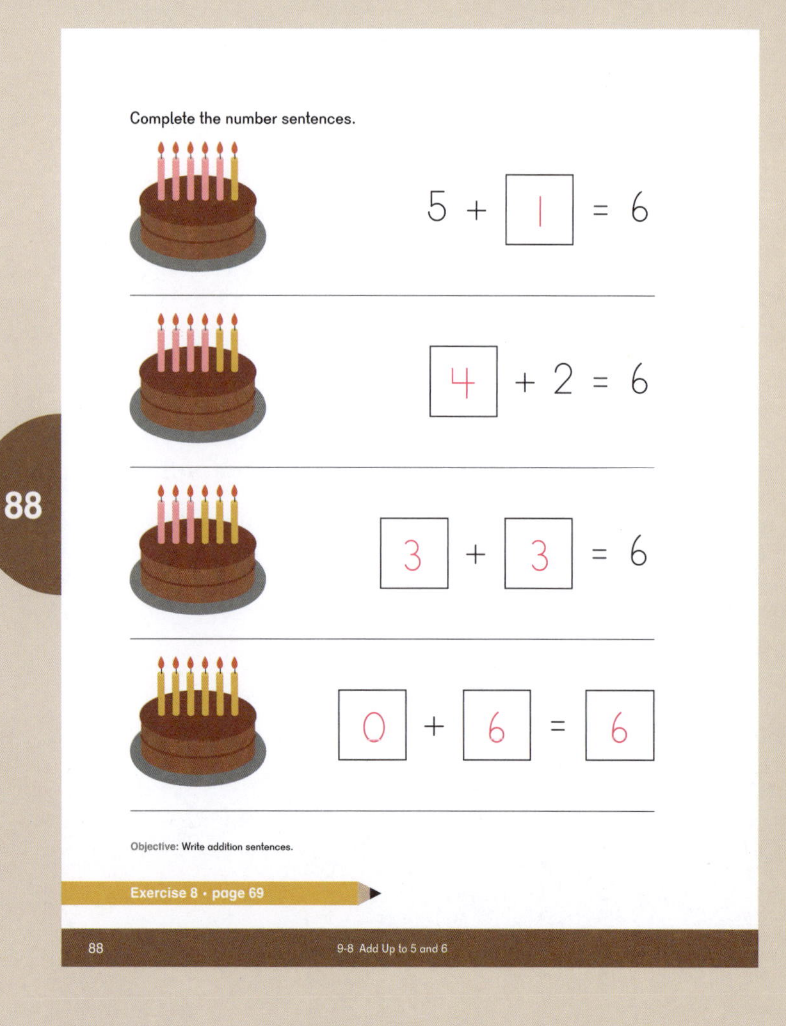

Complete the number sentences.

5 + 1 = 6

4 + 2 = 6

3 + 3 = 6

0 + 6 = 6

Objective: Write addition sentences.

Exercise 8 • page 69

88 9-8 Add Up to 5 and 6

Extend

★ **How Many Other Ways?**

Have students record different ways to make 6. Possible student responses:

- 1 more than 5
- 1 less than 7
- 3 + 3
- 2 + 2 + 2

Objective

- Write addition sentences up to a sum of 8.

Lesson Materials

- 20 linking cubes, 10 each of 2 different colors for each pair of students

Explore

Provide each pair of students with 20 linking cubes, 10 each of 2 different colors. Have the partners make 7 with the different color cubes.

Learn

Discuss Alex's example on textbook page 89. Have students share how they made 7 and write the addition sentences on the board.

Ask students to do the same activity for 8. Have them write the addition sentences on their whiteboards while they make them with the cubes.

Whole Group Activities

▲ Complete the Bond

Play as instructed in Lesson 7, finding bonds for numbers up to 8.

▲ Find a Partner

Materials: Multiple sets of Number Cards (BLM) 0 to 8

Play as instructed in Lesson 7, having students find a partner to make an addition sentence for 7 or 8.

Small Group Activities

Textbook Page 90

● **Match**

Materials: 2 sets of Ten-frame Cards (BLM) 0 to 8

Students arrange the cards faceup in a grid. Students take turns finding two cards that go together to make 7 or 8.

▲ **Memory**

Materials: 2 sets of Ten-frame Cards (BLM) 0 to 8

Play as described in Lesson 7. If the total is 7 or 8, they have a match.

▲ **Go Fish**

Materials: 4 sets of either Ten-frame Cards (BLM) or Number Cards (BLM) per small group

Play as described in Lesson 7 with Number Cards (BLM) for 0 to 8, where a pair is a set of cards that make 7 or 8. Use Ten-frame Cards (BLM) for students who are still counting.

▲ **Face-Off!**

Materials: 4 of each Number Card (BLM) or Ten-frame Card (BLM) 0 to 8

Shuffle the cards and deal all cards equally between 2 players. Players both flip 2 cards off the top of their decks, add the two together, and compare the total to their opponent's total. The player with the greatest total wins the cards.

Exercise 9 • page 71

90

Complete the number sentences.

6 + 1 = 7

7 = 5 + 2

5 + 3 = 8

8 = 4 + 4

Objective: Write addition sentences.

Exercise 9 • page 71

90 9-9 Add Up to 7 and 8

Extend

★ **What's Special About a Die?**

Have students look at the pips on a die.

Ask:

- Do you see any patterns?
- How many ways can you add the pips?
- What's the lowest sum? Greatest sum?
- How many numbers can you make if you add the pips?

Students may notice that opposite faces always add up to 7.

Objective

- Write addition sentences up to a sum of 10.

Lesson Materials

- 20 linking cubes, 10 each of 2 different colors for each pair of students
- 2 colors of crayons or dry erase markers

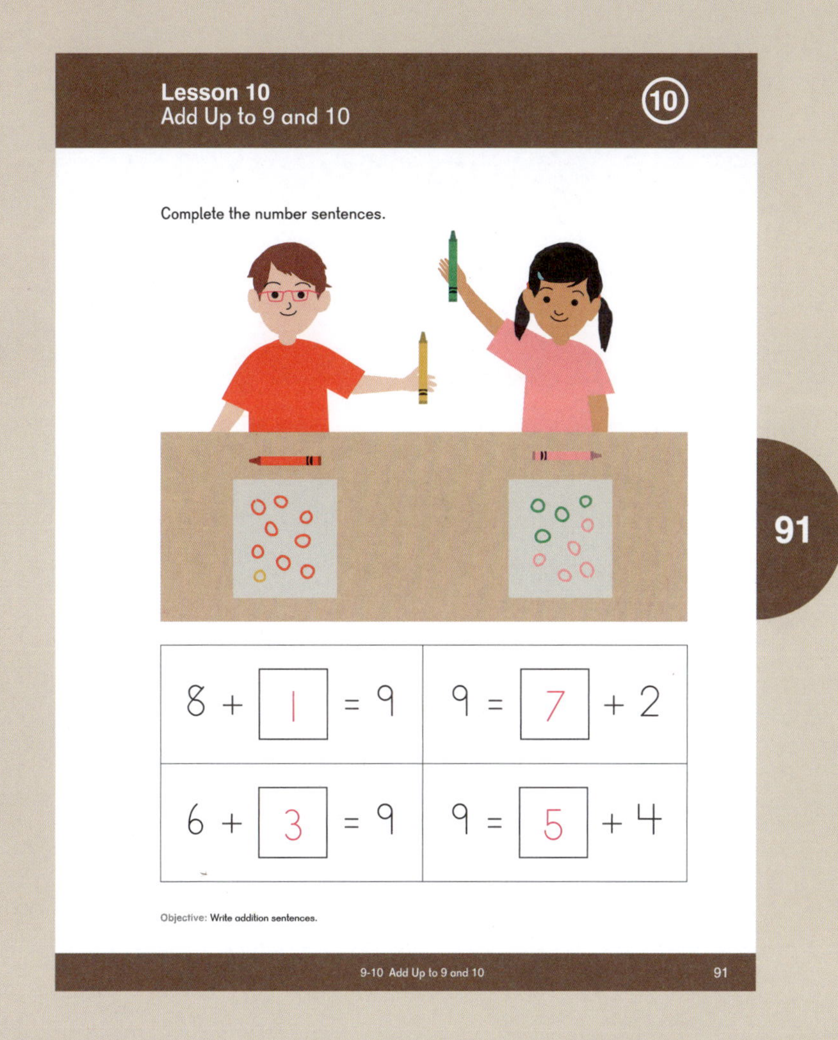

91

Explore

Similar to the previous lesson, provide each pair of students with 20 cubes, 10 of each of 2 colors. Have partners make 9 with the different color cubes.

Learn

Discuss what Emma and Mei are doing on textbook page 91.

Have students make a simple picture using two colors, with a total of 9 or 10, similar to Emma and Mei's pictures. Have students share their pictures. As students share, have other students tell the addition sentences they see in each picture.

Whole Group Activity

▲ Find a Partner

Materials: Multiple sets of Number Cards (BLM) 0 to 10

Play as instructed in Lesson 7, having students find a partner to make an addition sentence for 9 or 10.

Small Group Activities

Textbook Page 92

▲ Go Fish

Materials: 4 sets of either Ten-frame Cards (BLM) or Number Cards (BLM) per small group

Play as described in Lesson 7 with Number Cards (BLM) for 0 to 10, where a pair is a set of cards that make 9 or 10. Use Ten-frame Cards (BLM) for students who are still counting.

▲ Snap!

Materials: 4 sets of either Ten-frame Cards (BLM) or Number Cards (BLM) 0 to 10

Shuffle the cards and deal the whole deck equally between 2 players. Players flip their top cards at the same time and if the total of the 2 cards is 10, the first player to say, “Snap!” wins the cards. The game is over when one player has no more cards to play.

▲ Face-Off!

Materials: 4 of each Number Card (BLM) or Ten-frame Card (BLM) 0 to 10

Shuffle the cards and deal all cards equally between 2 players. Players both flip 2 cards off the top of their deck, add the two together, and compare the total to their opponent’s total. The player with the greatest total wins the cards.

92

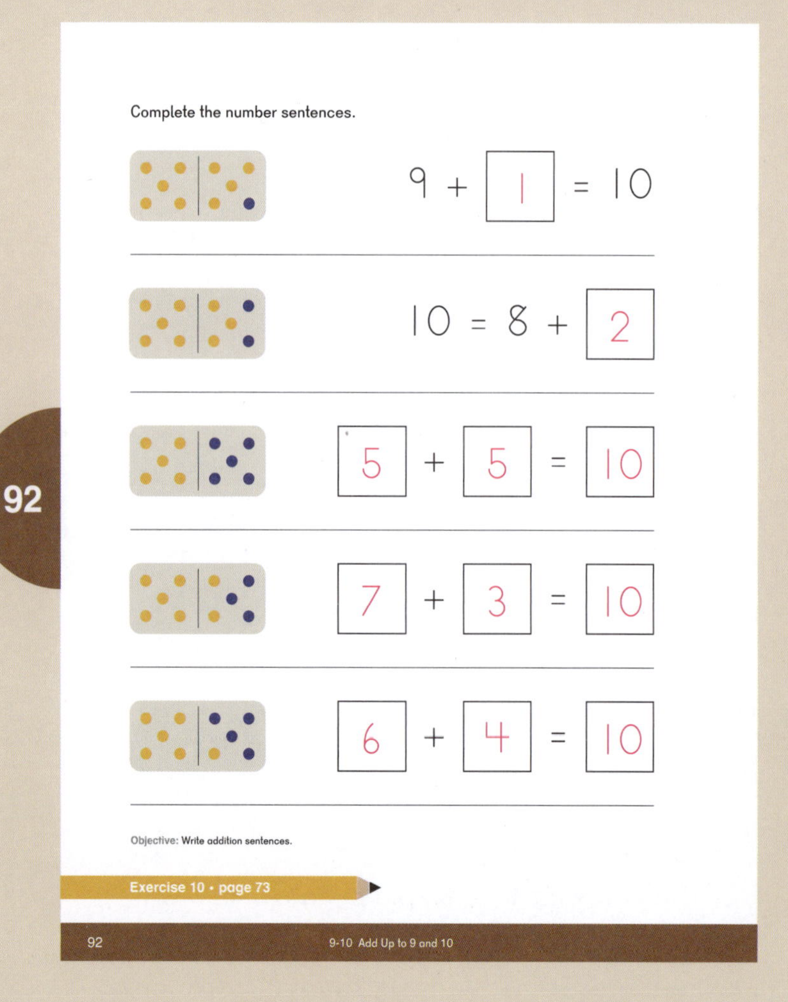
Complete the number sentences.

9 + 1 = 10

10 = 8 + 2

5 + 5 = 10

7 + 3 = 10

6 + 4 = 10

Objective: Write addition sentences.

Exercise 10 • page 73

92 9-10 Add Up to 9 and 10

Exercise 10 • page 73

Objective

- Use number bonds and number sentences for addition.

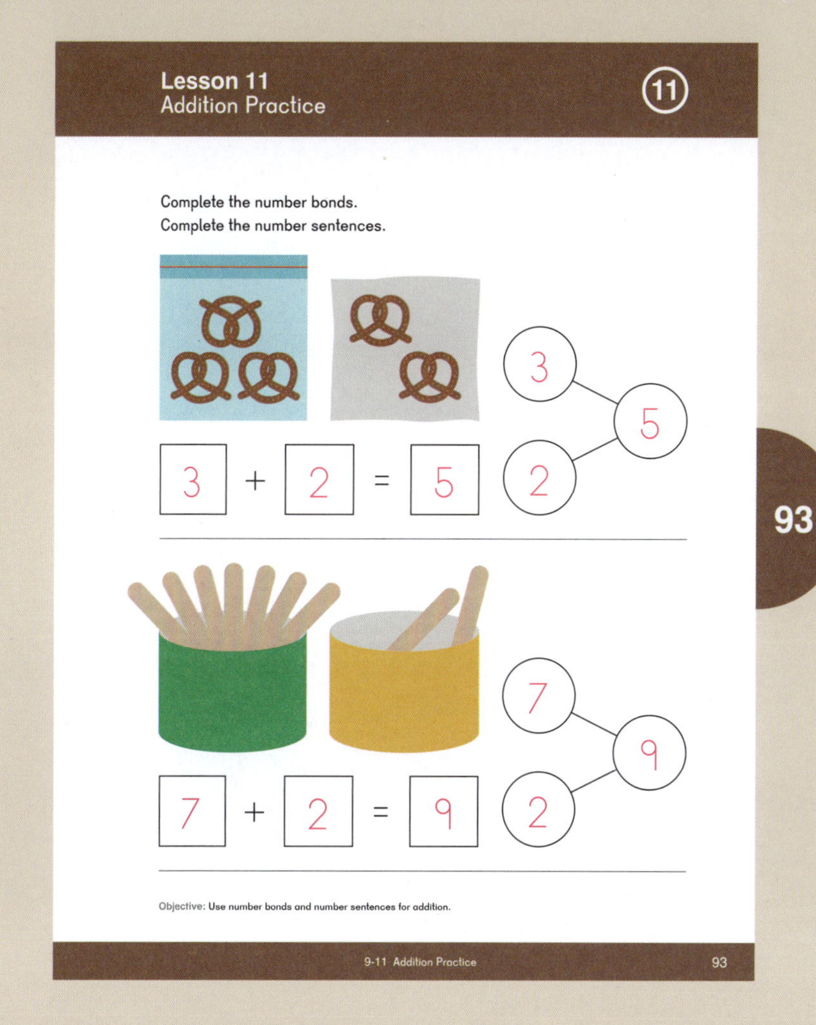

93

Practice lessons are designed for further practice and assessment as needed.

Students can complete the textbook and workbook pages as practice and/or as assessment.

Use activities and extensions from the chapter for additional review and practice.

Small Group Activity

▲ Draw and Write — Addition

Materials: Number Bond Cards (BLM)

Students draw a Number Bond Card (BLM) and write the corresponding addition equation on their whiteboards. Students can also illustrate a picture that represents the number bond on paper.

Exercise 11 • page 75

Extend

★ Addition Sort

Materials: Addition Facts Cards (BLM), Number Cards (BLM) 1 to 9

Students draw two Number Cards (BLM) at random. These cards become the target sums for the game. The Addition Fact Cards (BLM) will be sorted under these number cards.

Students then draw an Addition Fact Card (BLM) and find the total. If the total is the same as one of the Number Cards (BLM), they put it under the Number Card (BLM). If the total is not equal to either of the Number Cards (BLM), the fact card is discarded.

To add challenge: Play head-to-head. The winner is the first player to collect 5 fact cards under each Number Card (BLM).

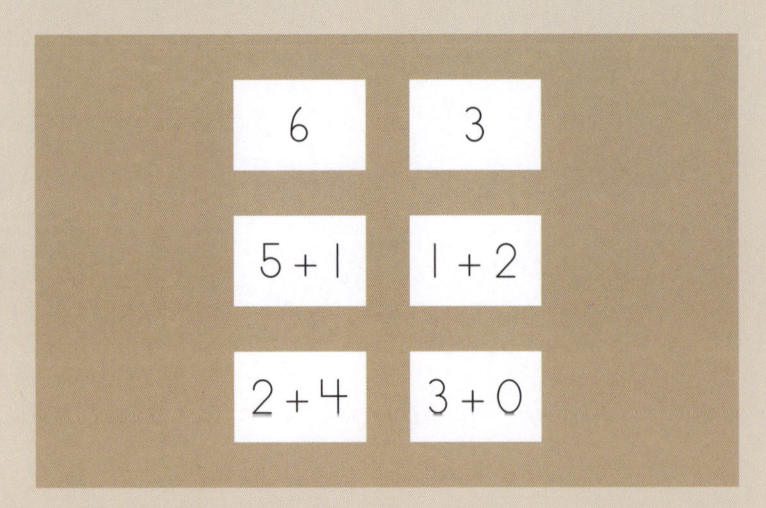

94

Complete the number sentences.

5 + 2 = 7

7 = 3 + 4

7 = 1 + 6

7 = 0 + 7

Objective: Use number bonds and number sentences for addition.

Exercise 11 • page 75

94 9-11 Addition Practice

Objective

- Practice skills from the chapter.

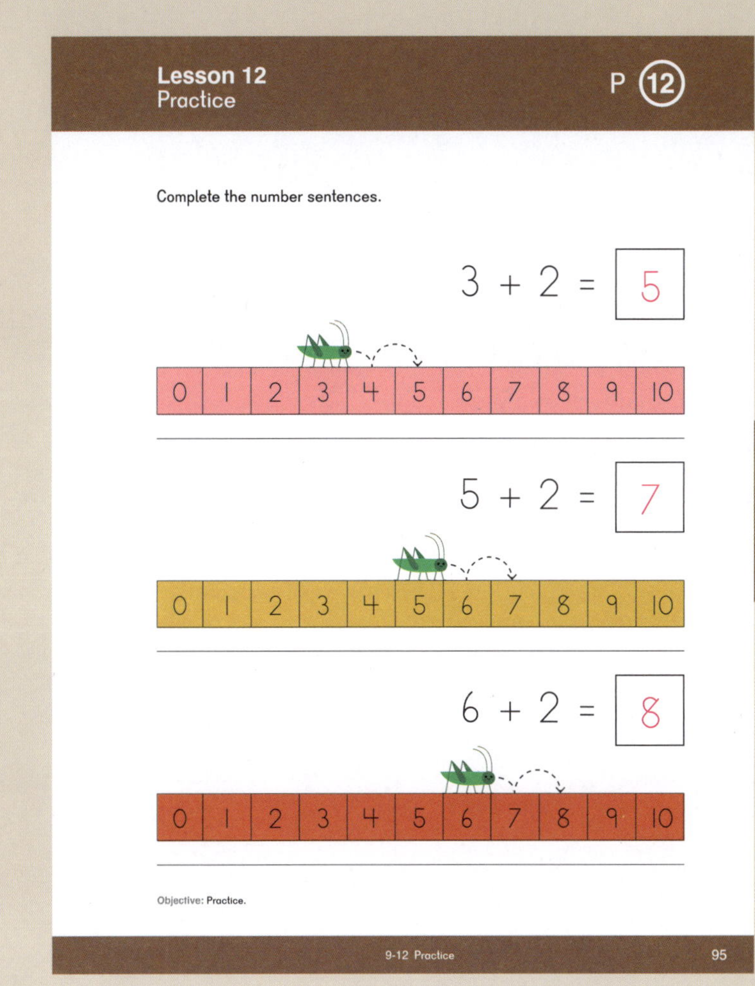

95

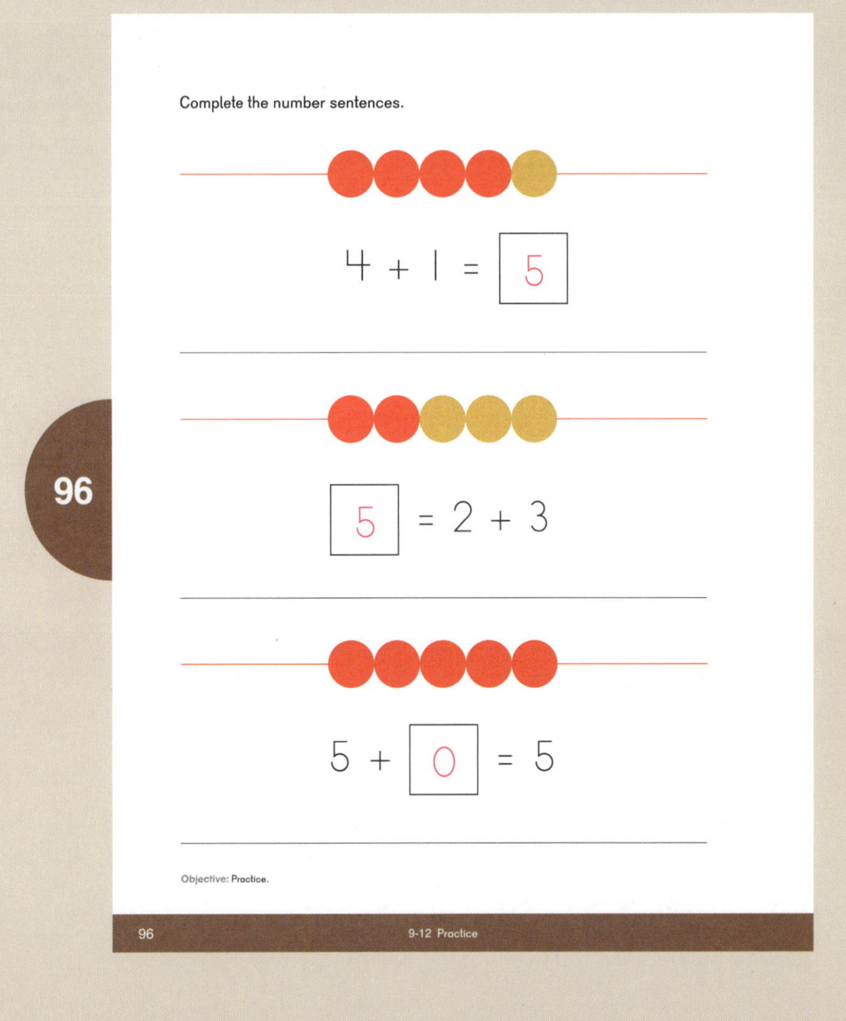

96

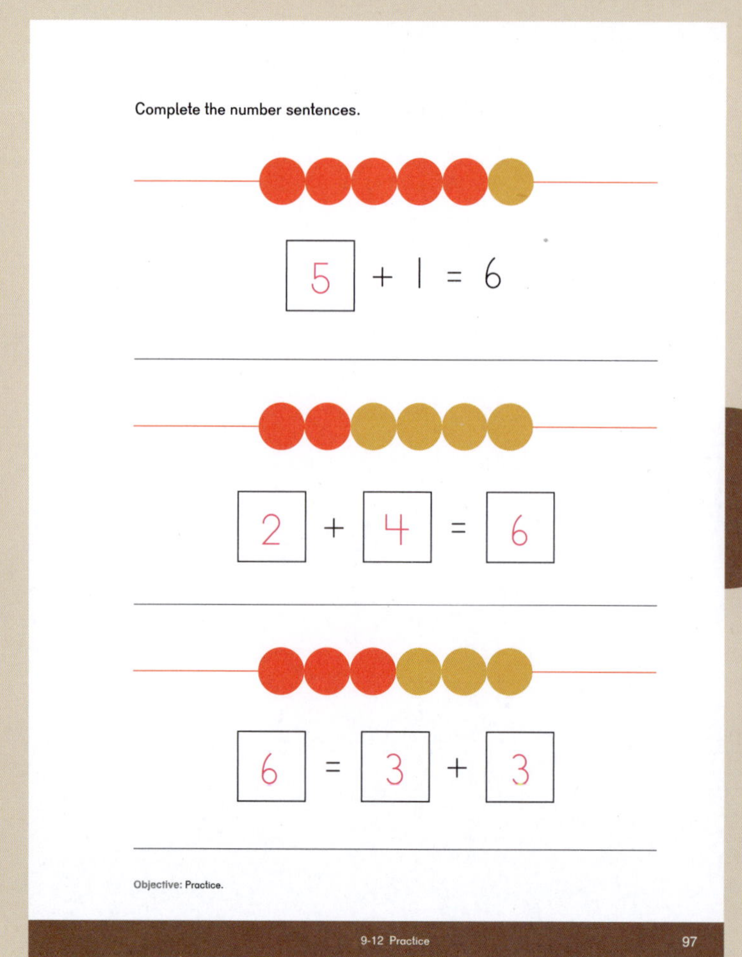
Complete the number sentences.

5 + 1 = 6

2 + 4 = 6

6 = 3 + 3

Objective: Practice.

9-12 Practice 97

97

98

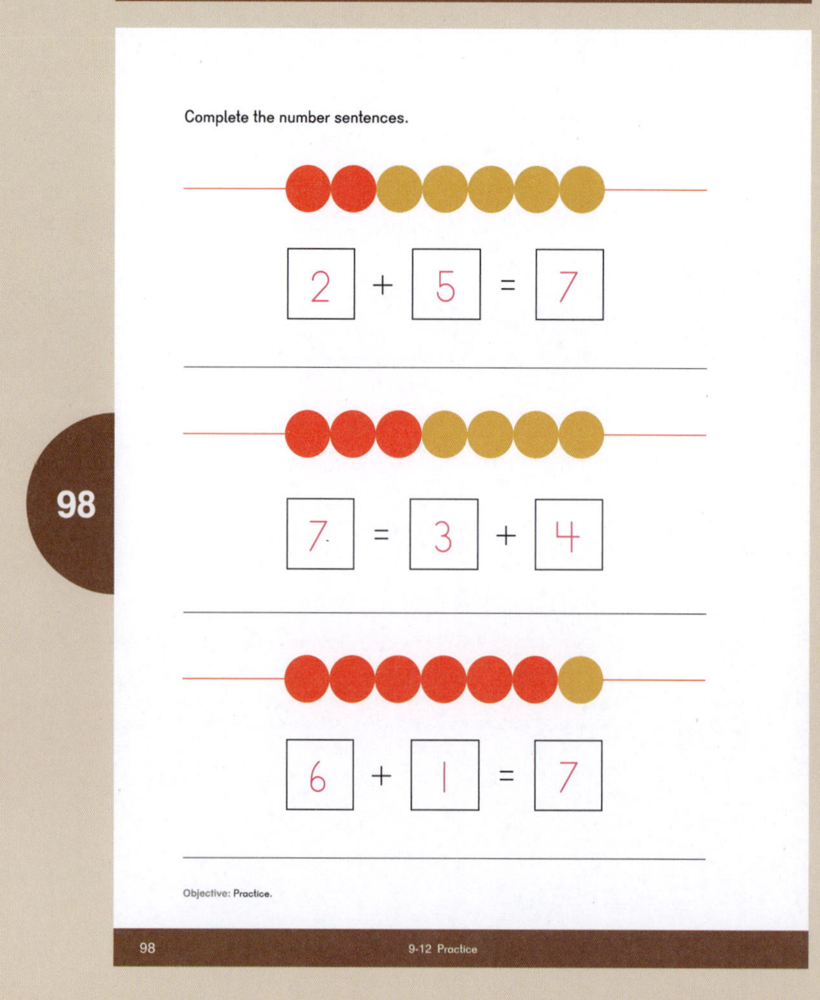
Complete the number sentences.

2 + 5 = 7

7 = 3 + 4

6 + 1 = 7

Objective: Practice.

98 9-12 Practice

Exercise 12 • page 77

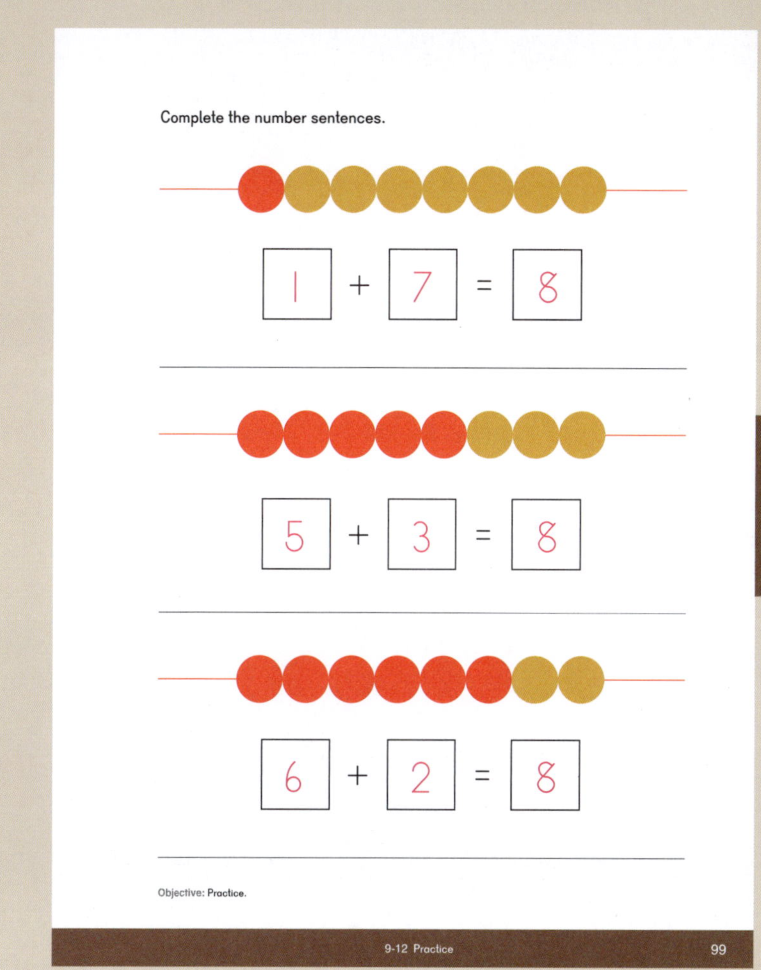

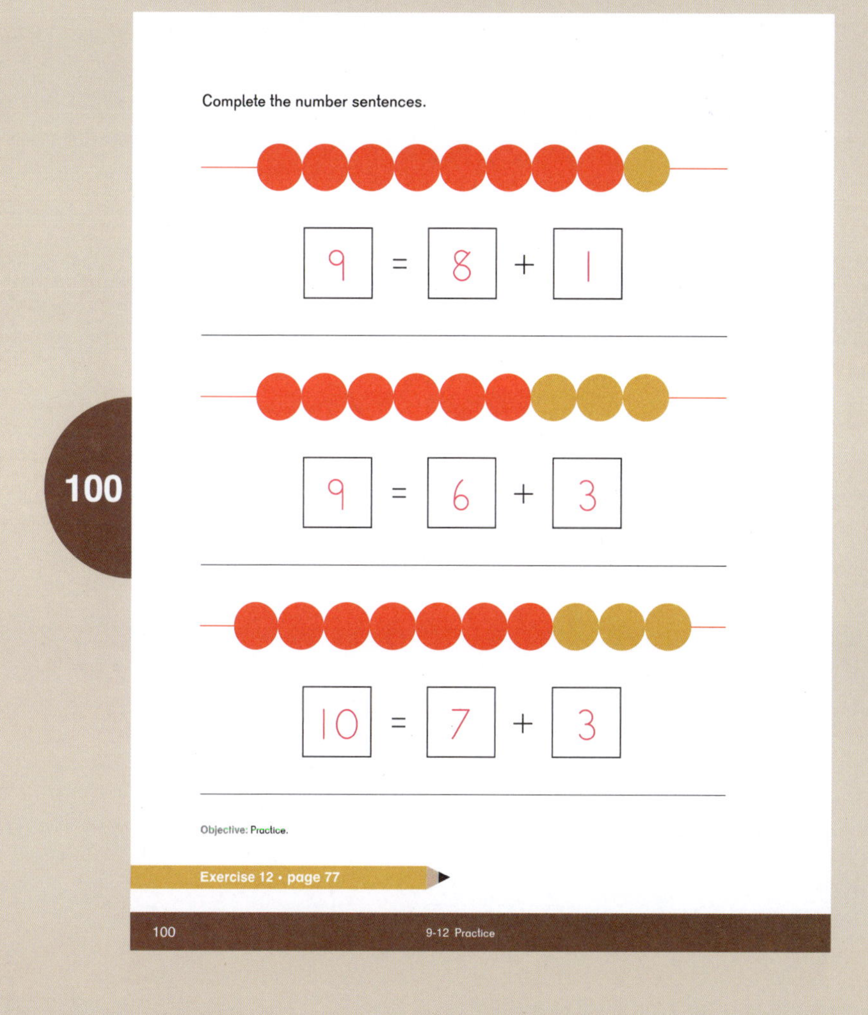

Exercise 1 • pages 55–56

Chapter 9 Addition

Exercise 1

Complete the number bonds.
Complete the number sentences.

2 and 1 make 3.

2 + 1 = 3

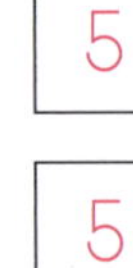

1 and 4 make 5.

1 + 4 = 5

Using this page: Have students find how many rabbits there are in all and fill in the numerals for the number bond and number sentences.
Concept: Recognizing the addition and equal symbols.

Complete the number bonds.
Complete the number sentences.

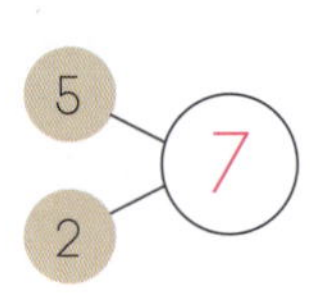

5 and 2 make 7.

5 + 2 = 7

4 and 4 make 8.

4 + 4 = 8

Using this page: Have students find how many rabbits there are in all and fill in the numerals for the number bond and number sentences.
Concept: Recognizing the addition and equal symbols.

Exercise 2 • pages 57–58

Exercise 2

Add.
Complete the number bonds.
Complete the number sentences.

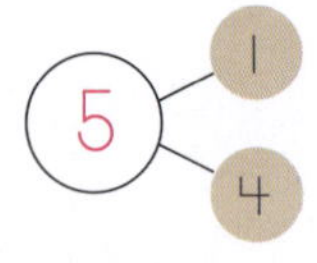

5 = 1 + 4

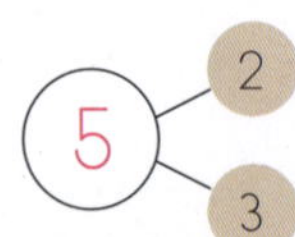

5 = 2 + 3

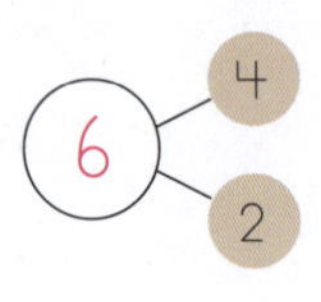

6 = 4 + 2

Before using this page: Give each student 10 linking cubes.
Using this page: Have students take the corresponding number of linking cubes for each vegetable or fruit, then put them together to find the whole. Have them write the numeral to complete the number bond and number sentence.
Concept: Recognizing the relationship between number bonds and addition sentences.

Add.
Complete the number bonds.
Complete the number sentences.

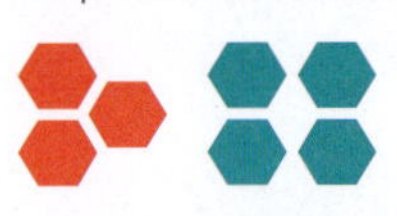
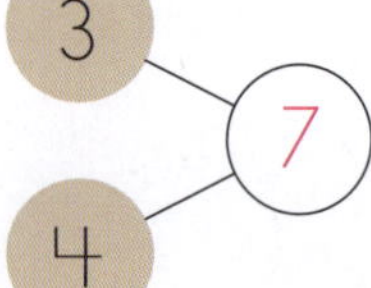

3 + 4 = 7

5 + 3 = 8

3 + 6 = 9

Before using this page: Give each student 10 linking cubes.
Using this page: Have students take the corresponding number of linking cubes for each shape, then put them together and find the total number of shapes. Have them write the numeral to complete the number bond and number sentence.
Concept: Recognizing the relationship between number bonds and addition sentences.

Exercise 3 • pages 59–60

Exercise 3

Color 1 more dot blue.
Complete the number sentences.

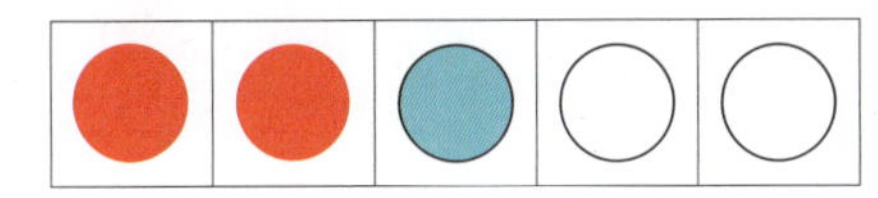

$2 + 1 = \boxed{3}$

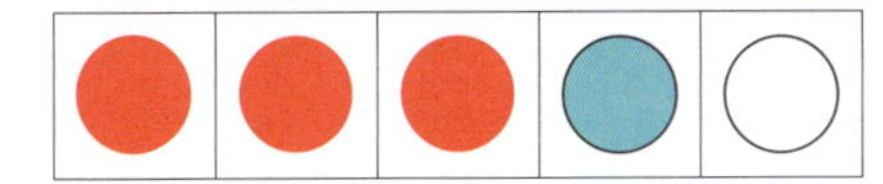

$3 + 1 = \boxed{4}$

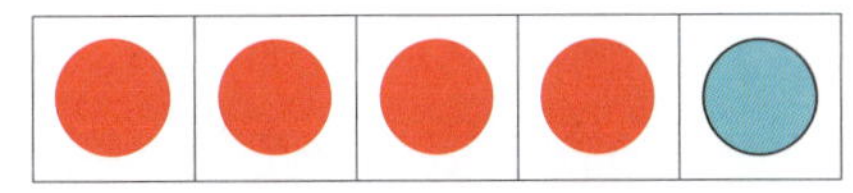

$4 + 1 = \boxed{5}$

Using this page: Have students color 1 more dot blue on the five-frame, then count on from the number of red dots to determine how many in all and write the total in the box.
Concept: Recognizing patterns in addition.

Color 2 more eggs pink.
Complete the number sentences.

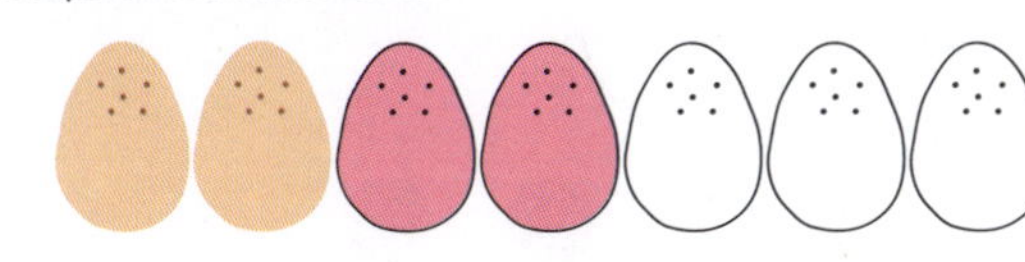

$2 + 2 = \boxed{4}$

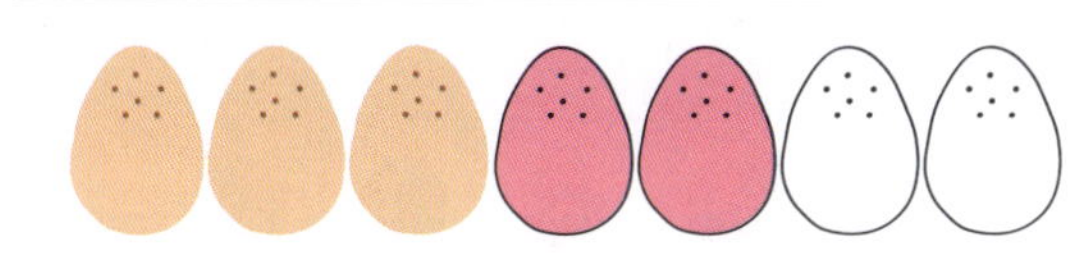

$3 + 2 = \boxed{5}$

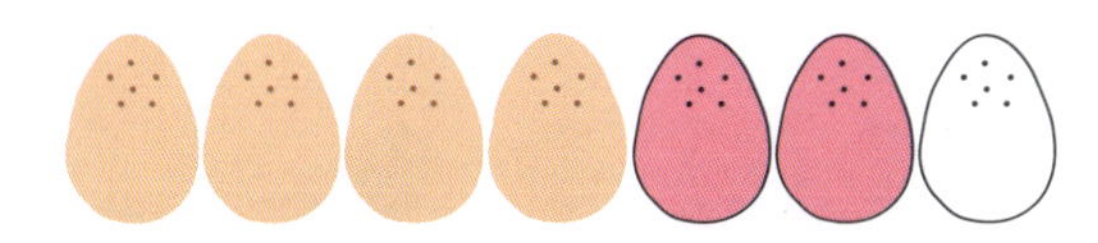

$4 + 2 = \boxed{6}$

Using this page: Have students color 2 more eggs pink, then count on from the beige eggs to determine how many eggs there are in all and write the total in the box.
Concept: Recognizing patterns in addition.

Exercise 4 • pages 61–62

Exercise 4

Complete the number sentences.

$2 + \boxed{2} = \boxed{4}$

$3 + \boxed{3} = \boxed{6}$

$4 + \boxed{3} = \boxed{7}$

Before using this page: Distribute 10 linking cubes to students.
Using this page: Have students place a cube on each insect. Next, have them count and write the missing part. Then count all the cubes and write the total.
Concept: Putting together to add.

Complete the number sentences.

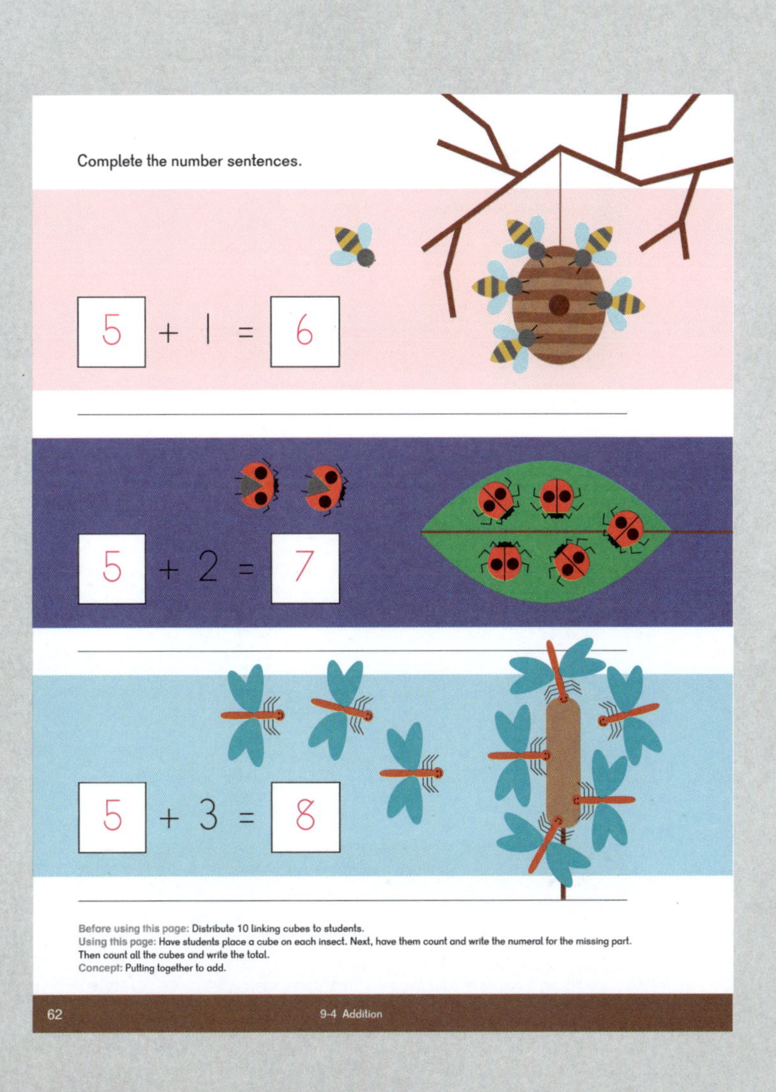

$\boxed{5} + 1 = \boxed{6}$

$\boxed{5} + 2 = \boxed{7}$

$\boxed{5} + 3 = \boxed{8}$

Before using this page: Distribute 10 linking cubes to students.
Using this page: Have students place a cube on each insect. Next, have them count and write the numeral for the missing part. Then count all the cubes and write the total.
Concept: Putting together to add.

Exercise 5 • pages 63–64

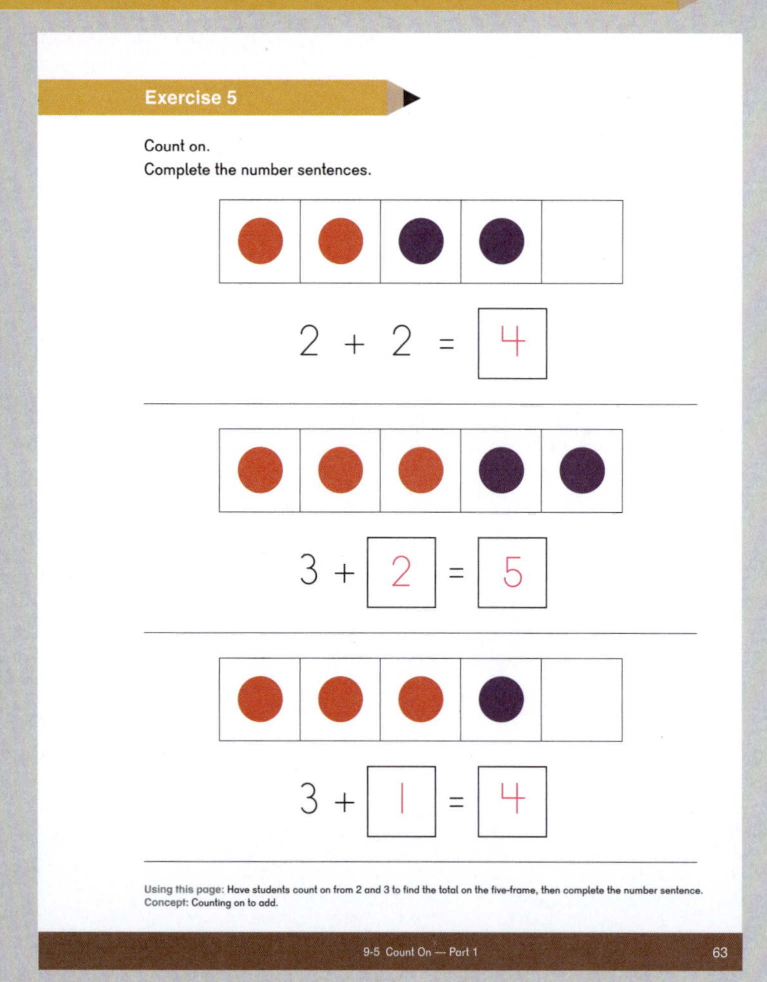

Exercise 5

Count on.
Complete the number sentences.

$2 + 2 = 4$

$3 + 2 = 5$

$3 + 1 = 4$

Using this page: Have students count on from 2 and 3 to find the total on the five-frame, then complete the number sentence.
Concept: Counting on to add.

9-5 Count On — Part 1 63

Count on.
Complete the number sentences.

$4 + 1 = 5$

$4 + 2 = 6$

$4 + 3 = 7$

Using this page: Have students count on from 4 to find the total, then complete the number sentence.
Concept: Counting on to add.

64 9-5 Count On — Part 1

Exercise 6 • pages 65–66

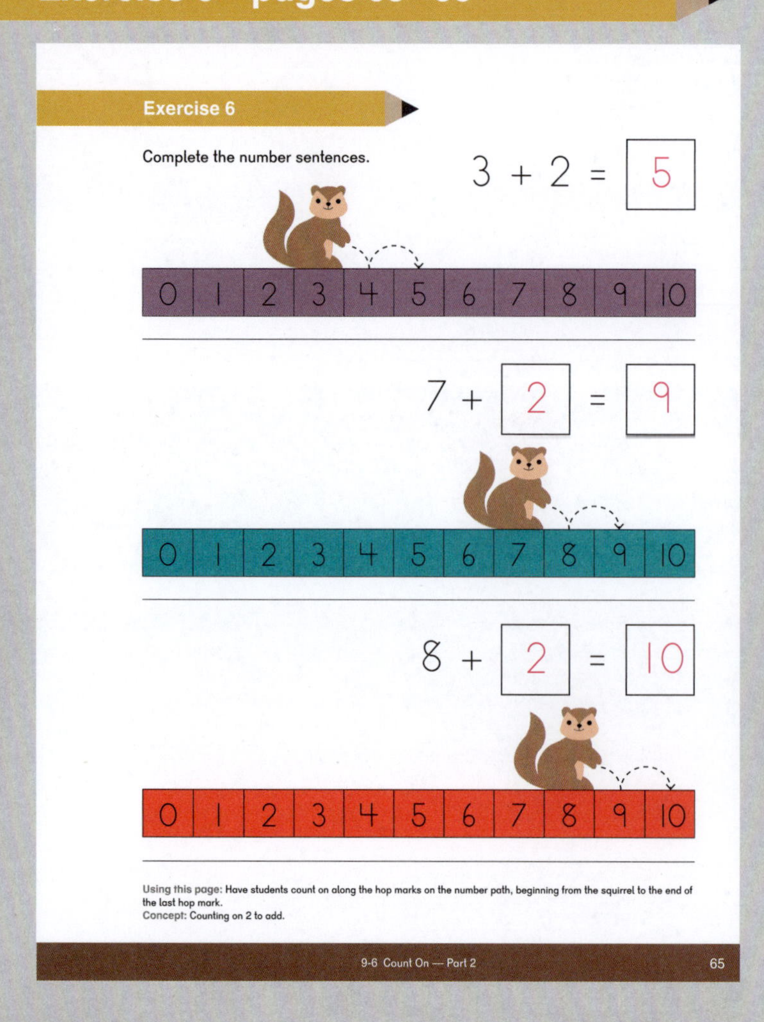

Exercise 6

Complete the number sentences.

$3 + 2 = 5$

$7 + 2 = 9$

$8 + 2 = 10$

Using this page: Have students count on along the hop marks on the number path, beginning from the squirrel to the end of the last hop mark.
Concept: Counting on 2 to add.

9-6 Count On — Part 2 65

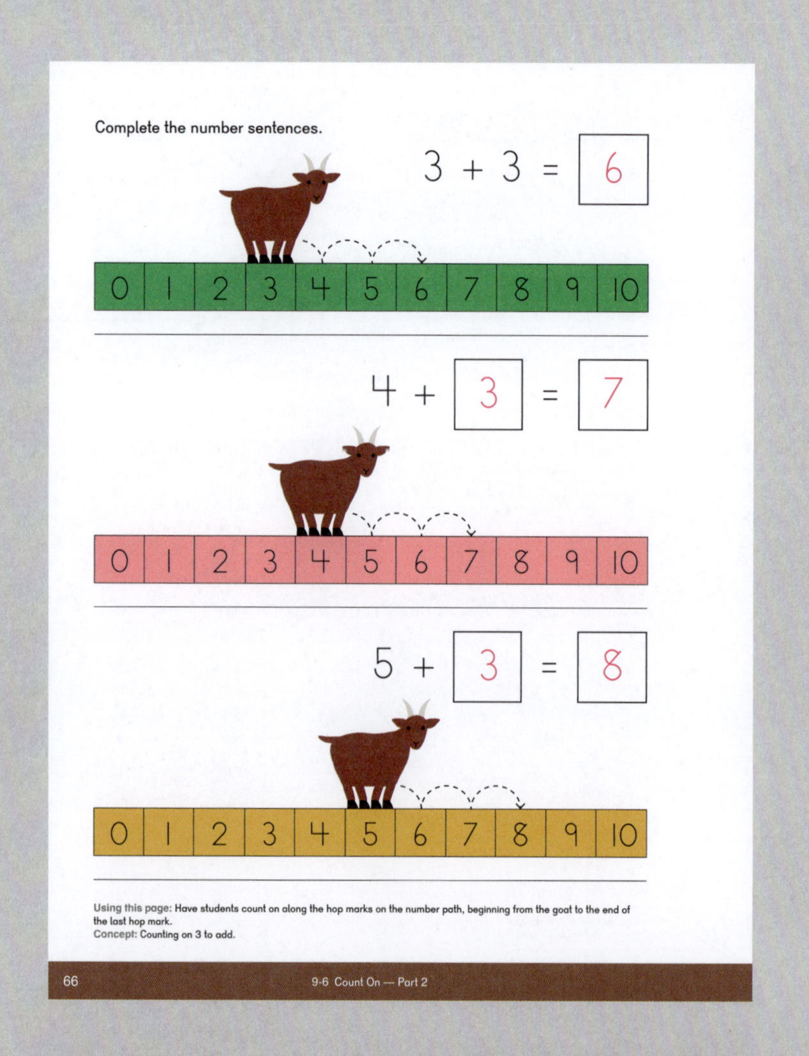

Complete the number sentences.

$3 + 3 = 6$

$4 + 3 = 7$

$5 + 3 = 8$

Using this page: Have students count on along the hop marks on the number path, beginning from the goat to the end of the last hop mark.
Concept: Counting on 3 to add.

66 9-6 Count On — Part 2

Exercise 7 • pages 67–68

Exercise 7

Complete the number sentences.

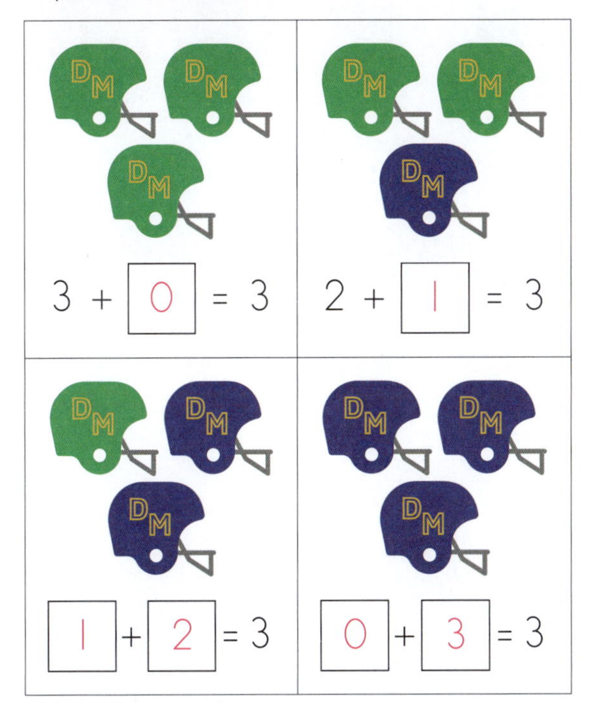

Using this page: Have students count the green helmets, then the blue helmets and complete the number sentence accordingly for addition up to 3.
Concept: Writing addition sentences.

9-7 Add Up to 3 and 4 67

Complete the number sentences.

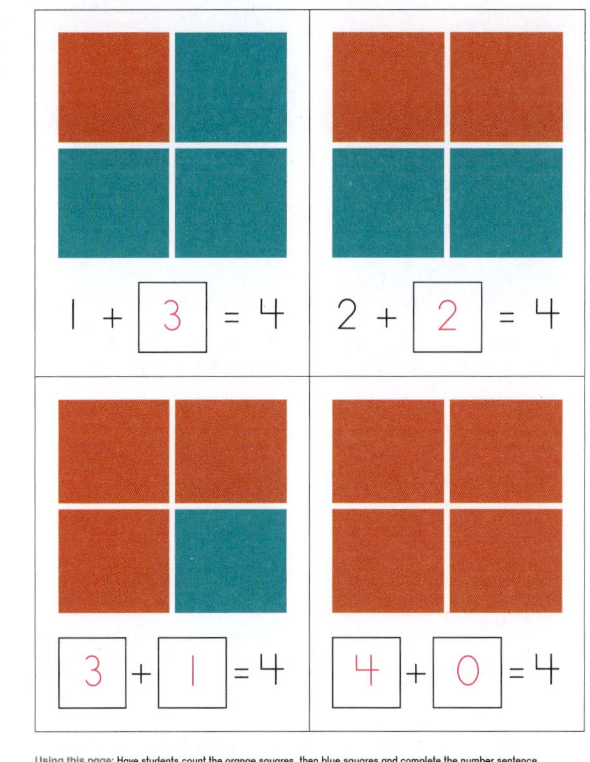

Using this page: Have students count the orange squares, then blue squares and complete the number sentence accordingly for addition up to 4.
Concept: Writing addition sentences.

68 9-7 Add Up to 3 and 4

Exercise 8 • pages 69–70

Exercise 8

Complete the number sentences.

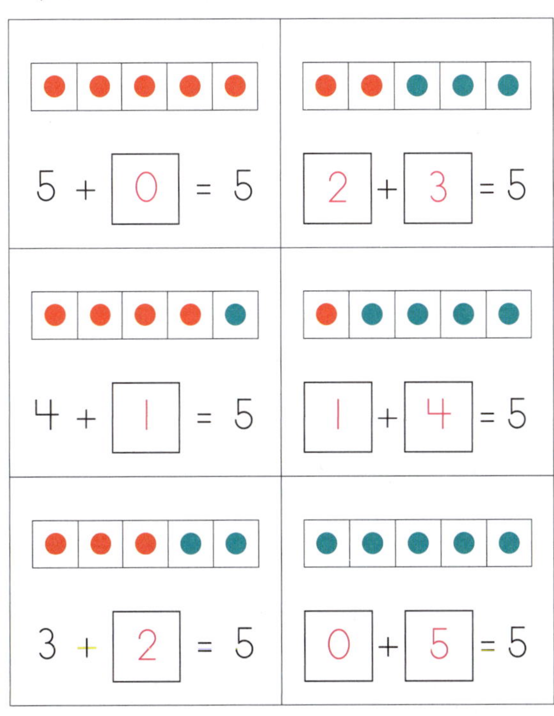

Using this page: Have students complete the number sentence for addition up to 5.
Concept: Writing addition sentences.

9-8 Add Up to 5 and 6 69

Complete the number sentences.

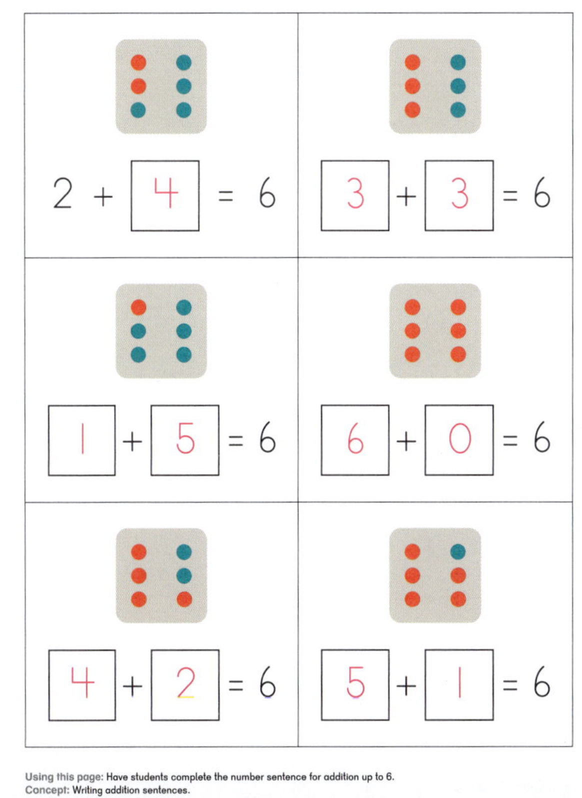

Using this page: Have students complete the number sentence for addition up to 6.
Concept: Writing addition sentences.

70 9-8 Add Up to 5 and 6

Exercise 9 • pages 71–72

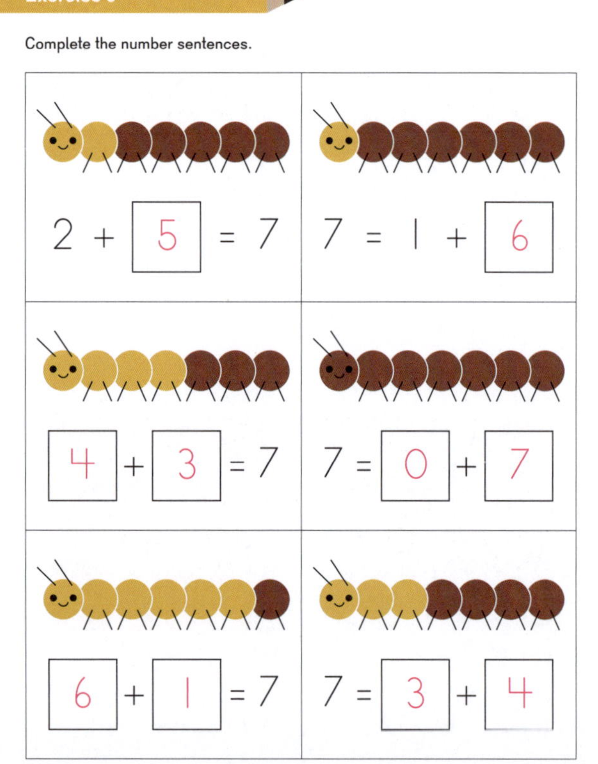
Exercise 9

Complete the number sentences.

2 + 5 = 7 | 7 = 1 + 6
4 + 3 = 7 | 7 = 0 + 7
6 + 1 = 7 | 7 = 3 + 4

Using this page: Have students complete the number sentence for addition up to 7.
Concept: Writing addition sentences.

9-9 Add Up to 7 and 8 71

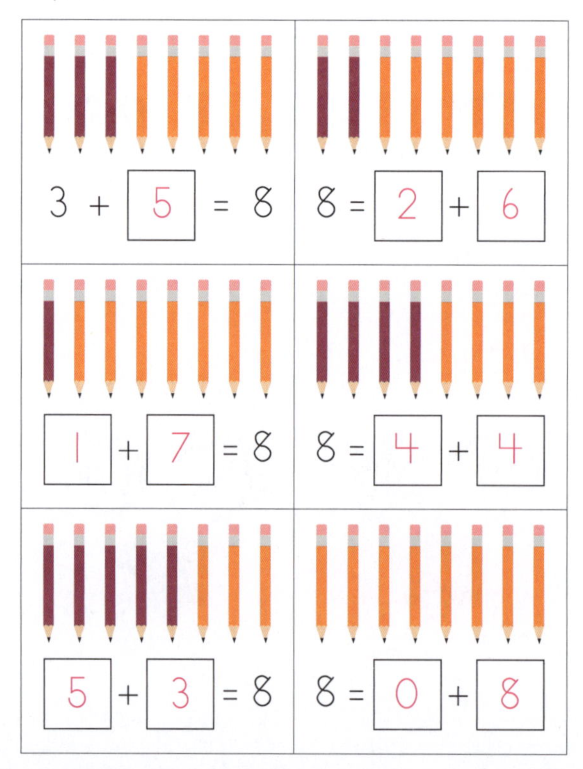
Complete the number sentences.

3 + 5 = 8 | 8 = 2 + 6
1 + 7 = 8 | 8 = 4 + 4
5 + 3 = 8 | 8 = 0 + 8

Using this page: Have students complete the number sentence for addition up to 8.
Concept: Writing addition sentences.

72 9-9 Add Up to 7 and 8

Exercise 10 • pages 73–74

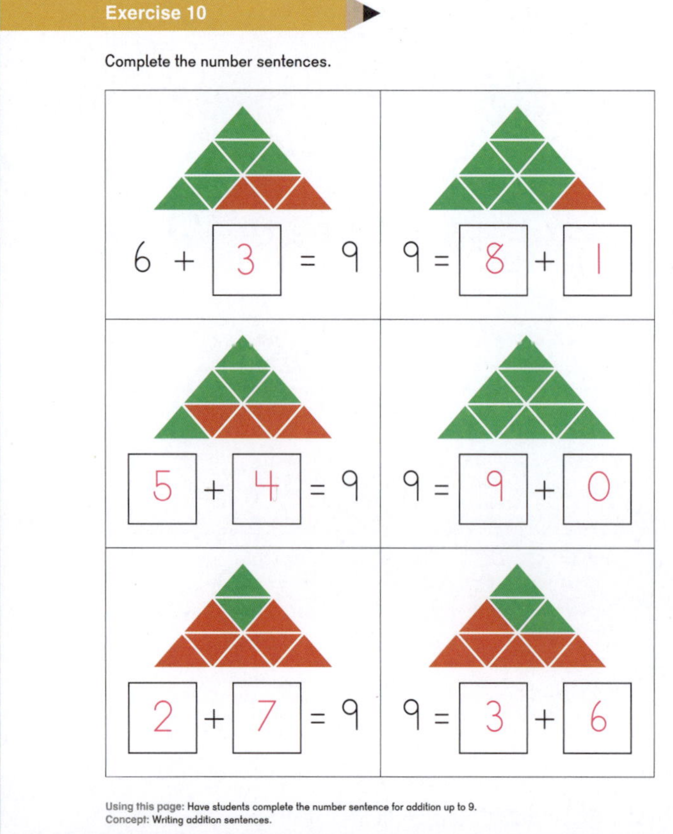
Exercise 10

Complete the number sentences.

6 + 3 = 9 | 9 = 8 + 1
5 + 4 = 9 | 9 = 9 + 0
2 + 7 = 9 | 9 = 3 + 6

Using this page: Have students complete the number sentence for addition up to 9.
Concept: Writing addition sentences.

9-10 Add Up to 9 and 10 73

Complete the number sentences.

8 + 2 = 10 | 10 = 7 + 3
6 + 4 = 10 | 10 = 10 + 0
9 + 1 = 10 | 10 = 5 + 5

Using this page: Have students complete the number sentence for addition up to 10.
Concept: Writing addition sentences.

74 9-10 Add Up to 9 and 10

Exercise 11 • pages 75–76

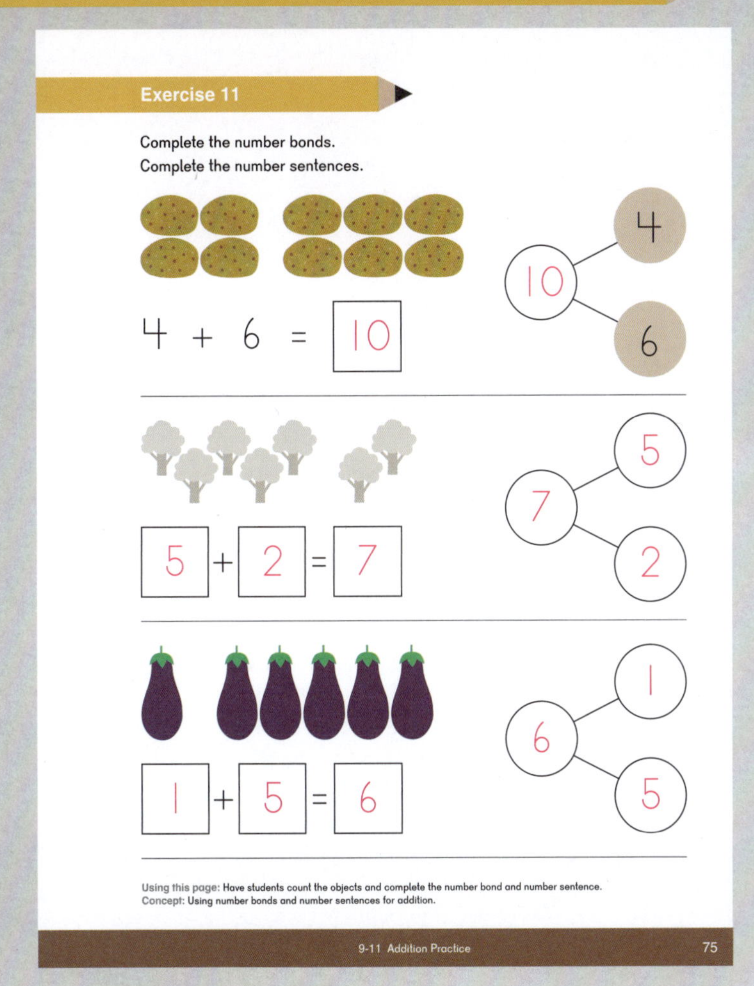

Exercise 11

Complete the number bonds.
Complete the number sentences.

4 + 6 = 10 (number bond: 10 → 4, 6)

5 + 2 = 7 (number bond: 7 → 5, 2)

1 + 5 = 6 (number bond: 6 → 1, 5)

Using this page: Have students count the objects and complete the number bond and number sentence.
Concept: Using number bonds and number sentences for addition.

9-11 Addition Practice 75

Complete the number bonds.
Complete the number sentences.

3 + 5 = 8 (number bond: 8 → 3, 5)

4 + 3 = 7 (number bond: 7 → 4, 3)

2 + 4 = 6 (number bond: 6 → 2, 4)

Using this page: Have students count the pips on the dominoes and complete the number bond and number sentence.
Concept: Using number bonds and number sentences for addition.

76 9-11 Addition Practice

Exercise 12 • pages 77–78

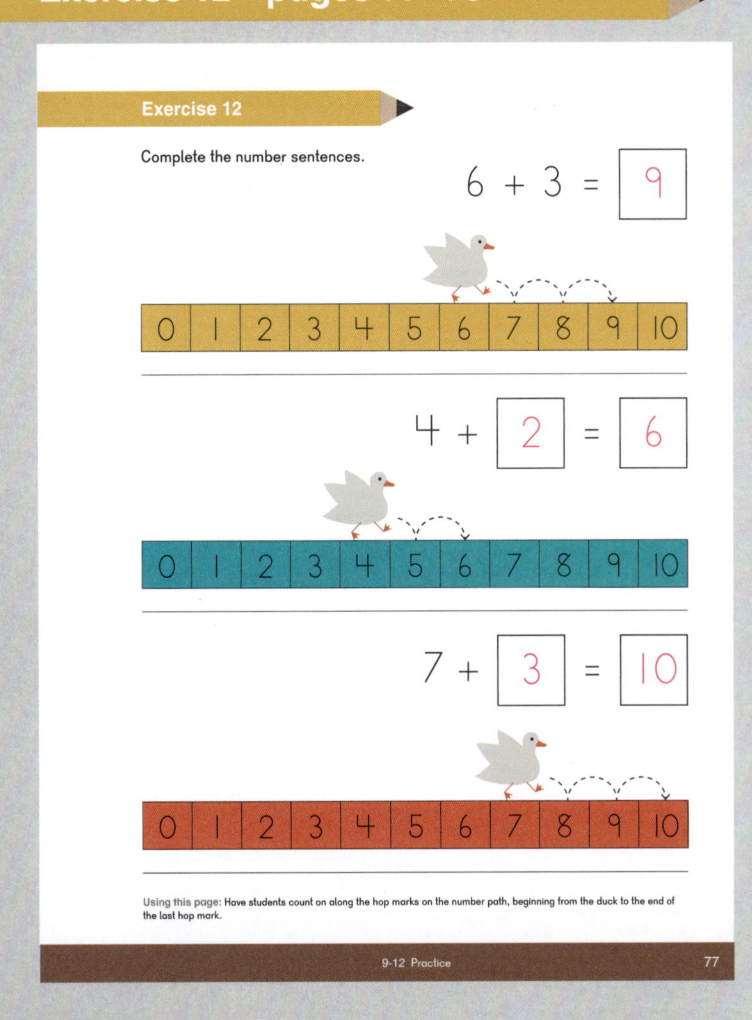

Exercise 12

Complete the number sentences.

6 + 3 = 9

0	1	2	3	4	5	6	7	8	9	10

4 + 2 = 6

0	1	2	3	4	5	6	7	8	9	10

7 + 3 = 10

0	1	2	3	4	5	6	7	8	9	10

Using this page: Have students count on along the hop marks on the number path, beginning from the duck to the end of the last hop mark.

9-12 Practice 77

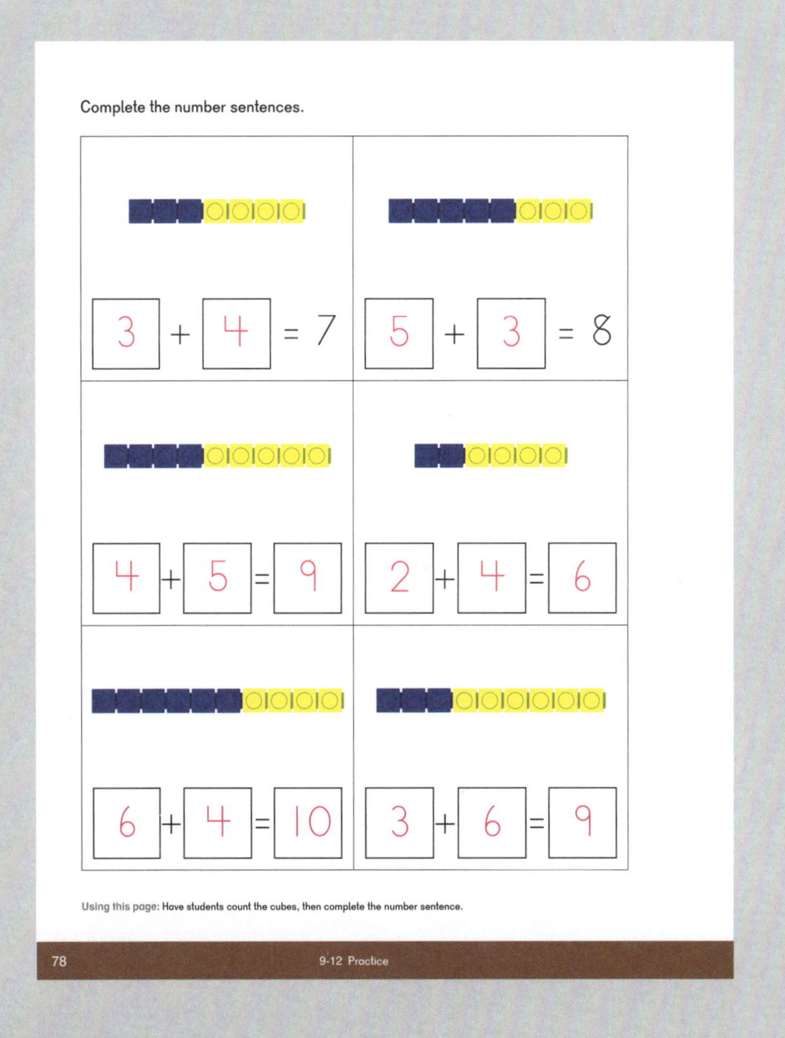

Complete the number sentences.

3 + 4 = 7	5 + 3 = 8
4 + 5 = 9	2 + 4 = 6
6 + 4 = 10	3 + 6 = 9

Using this page: Have students count the cubes, then complete the number sentence.

78 9-12 Practice

Notes

Suggested number of class periods: 10–11

	Lesson	Page	Resources	Objectives
	Chapter Opener	p. 125	TB: p. 101	
1	Take Away to Subtract — Part 1	p. 126	TB: p. 102 WB: p. 79	Recognize subtraction as taking away from a whole.
2	Take Away to Subtract — Part 2	p. 128	TB: p. 104 WB: p. 81	Recognize the subtraction symbol. Take away to subtract.
3	Take Away to Subtract — Part 3	p. 130	TB: p. 106 WB: p. 83	Write subtraction sentences.
4	Take Apart to Subtract — Part 1	p. 132	TB: p. 108 WB: p. 85	Recognize subtraction as separating a known quantity into two parts, given one part, and having to find the other part.
5	Take Apart to Subtract — Part 2	p. 134	TB: p. 111 WB: p. 87	Recognize the relationship between number bonds and subtraction sentences.
6	Count Back	p. 136	TB: p. 113 WB: p. 91	Count back 1, 2, or 3 to subtract on a number path.
7	Subtract Within 5	p. 138	TB: p. 115 WB: p. 93	Subtract from 3, 4, and 5.
8	Subtract Within 10 — Part 1	p. 140	TB: p. 117 WB: p. 95	Subtract from 6 through 10. Write subtraction sentences.
9	Subtract Within 10 — Part 2	p. 142	TB: p. 119 WB: p. 97	Subtract from 6 through 10. Write subtraction sentences.
10	Practice	p. 144	TB: p. 121 WB: p. 99	Practice subtraction within 10.
	Workbook Solutions	p. 146		

Subtraction is often a more difficult concept for students to grasp than addition. Students will first look at subtraction through the relationships expressed in number bonds.

Students will encounter a variety of situations that can be interpreted as subtraction. In each, the whole quantity and one part is known and the student needs to find the other part. The other part could be:

- one of two quantities present, or
- the quantity remaining after a part is taken away, or
- the part taken away

In **Dimensions Math® 1B**, students will learn to interpret comparison situations, i.e., how many more or fewer one quantity is than another using subtraction.

Just as with addition, students need a large variety of examples to relate a subtraction equation to the part-part-whole representation of a number bond. In beginning subtraction, students start with the "whole." Avoid using the terms "bigger" and "smaller" with students as they can lead to later confusion. For example, which is bigger:

3 or 7?

In this curriculum, students will use the phrase "count back" rather than "count down." Typically, number charts have numbers decreasing to the left along a row, and then up to the next row of numbers. It may be confusing to students to use the word "down" to decrease when visually the decrease is to the left or up. We avoid this by using the phrase "count back."

When students count on or back on a number path, the number of steps or "hops" made is what is counted, not including the starting square.

It is assumed that all students will have access to recording tools. When a lesson refers to a whiteboard, any writing materials can be used.

Materials

- 10 objects with several attributes
- 10 small paper cups
- Bags
- Bear counters or other counters shaped like objects (bugs, dinosaurs, fruit, vehicles, etc.)
- Chalk
- Counters
- Cups
- Dice
- Dominoes
- Dry erase sleeves
- Hula hoops
- Linking cubes
- Number bond mats
- Objects to count
- Painter's tape
- Paper plates
- Partitioned plate
- Picture books
- Play dough or clay
- Sets of up to 5 objects that are the same except for their color or size
- Small ball
- Storybooks or magazines
- Two-color counters

Blackline Masters

- Blank Number Bond Template
- Blank Ten-frame
- Bowling Recording Sheet
- Domino Parking Lot
- Number Bond Cards
- Number Cards
- Number Cards — Large
- Number Path
- Picture Cards
- Subtraction Fact Cards
- Subtraction Sentences
- Subtraction Template
- Ten-frame Cards

Storybooks

- Dr. Suess books
- Nonfiction nature books
- *Ten, Nine, Eight* by Molly Bang
- *Rooster's Off to See the World* by Eric Carle
- *Pete the Cat and His Four Groovy Buttons* by Eric Litwin
- *How Many Blue Birds Flew Away?: A Counting Book With a Difference* by Paul Giganti
- *Splash* by Ann Jonas
- *Hot Rod Hamster* by Cynthia Lord
- *Elevator Magic* by Stuart J. Murphy
- *Ten Timid Ghosts* by Jennifer O'Connell
- *Skippyjon Jones books* by Judy Schachner

Letters Home

- Chapter 10 Letter

Notes

Lesson Materials

- Two-color counters

Ask students to tell number stories about the picture on page 101. As stories are told, have students represent the story with counters. They may observe that:

- There are 3 girls and 2 boys. There are 5 friends.
- There are 5 friends, and 2 friends are leaving the park.
- There are 6 flowers. Alex is holding 2 flowers and 4 flowers are in the ground.
- There are 5 ducks. 4 ducks are swimming and 1 duck is flying away.

As students share their stories, rephrase them to begin with a whole and then state the two parts. For example, if a student says, "There are 3 girls and 2 boys. There are 5 friends in all."

Ask students if we could also say, "There are 5 friends. 3 are girls and 2 are boys."

Encourage students to notice that the friends, flowers, and duck are going away.

Extend

★ Number Story Pictures

Have students illustrate a number story from a number bond where one part is "going away."

Have students talk about their picture using words that indicate that one part is taken away or leaves. For example, animals might be flying away, running away, leaving, etc.

Objective

- Recognize subtraction as taking away from a whole.

Lesson Materials

- Two-color counters, 10 per student

Lesson 1
Take Away to Subtract — Part 1 ①

Look and talk.

Objective: Take away to subtract.

102 10-1 Take Away to Subtract — Part 1

Explore

Tell a story about 3 friends playing in the park and 1 friend going home for lunch.

Have 3 students act out the situation while the remaining students model the problem with counters. Have them share how they would use counters to show how many friends remain at the park. Repeat with other numbers up to 5.

Learn

Have students discuss page 102.

Have students model the subtraction situation for the dogs and then use the **Chapter Opener** on page 101 to model the friends, ducks, and flowers. Highlight the terms that could mean "take away" or "fewer remain."

- 5 dogs were playing. 1 dog is running away. How many dogs are still playing?
- 5 friends are at the park. 2 friends go home. How many friends are still at the park?
- 5 ducks were swimming. 1 duck flew away. How many ducks are still swimming?
- There were 6 flowers. Alex picked 2 flowers for his mother. How many flowers are left?

Note: The word "left" might confuse students.

- Instead of "1 dog left the park," say, "1 dog ran away."
- Instead of "4 dogs are left," say, "4 dogs remain."

Whole Group Activity

▲ Complete the Bond

Tell the class that you are going to find bonds for 5. The teacher or leader holds up fingers (representing a part), and students complete the bond by showing the other part using their fingers.

Small Group Activities

Textbook Page 103. Note that on this page, the "whole" or number of biscuits Spot begins with is different for each problem. To help students understand, tell students that on Monday, Spot had 5 biscuits. He ate 1 biscuit. How many biscuits remain? On Tuesday, Spot had 4 biscuits and he ate 2 of them. How many remain?

▲ Subtraction Smash

Materials: Play dough or clay, dice

Partners each roll a die and make that number of play dough balls to begin. They take turns smashing 1 ball and saying how many balls remain. Partners check each other's answers.

▲ Under the Cup

Materials: 10 counters, 1 cup, linking cubes

Provide students with 5 counters to start, allowing up to 10 for extension. One student will hide some of the counters under the cup, then the other student will figure out how many are hiding based on how many are visible.

Alternatively, students can play with 5 linking cubes assembled into a tower. Player 1 breaks the tower behind his back. He shows one part of the tower to Player 2. Player 2 says how many cubes are hidden behind Player 1's back.

Exercise 1 • page 79

103

Cross off what Spot eats.
Write the number left.

Spot eats 1.	4 left.
Spot eats 2.	2 left.
Spot eats 3.	0 left.

Objective: Take away to subtract.

Exercise 1 • page 79

10-1 Take Away to Subtract — Part 1 103

Extend

★ Change It!

Materials: Counters

Player 1 grabs some counters. Player 2 counts them to determine the starting number. Player 2 hides her eyes as Player 1 makes a change to the group by taking some away and hiding them behind her back. Player 2 determines how many were taken away. Players switch roles and play continues.

Objectives

- Recognize the subtraction symbol.
- Take away to subtract.

Lesson Materials

- Counters

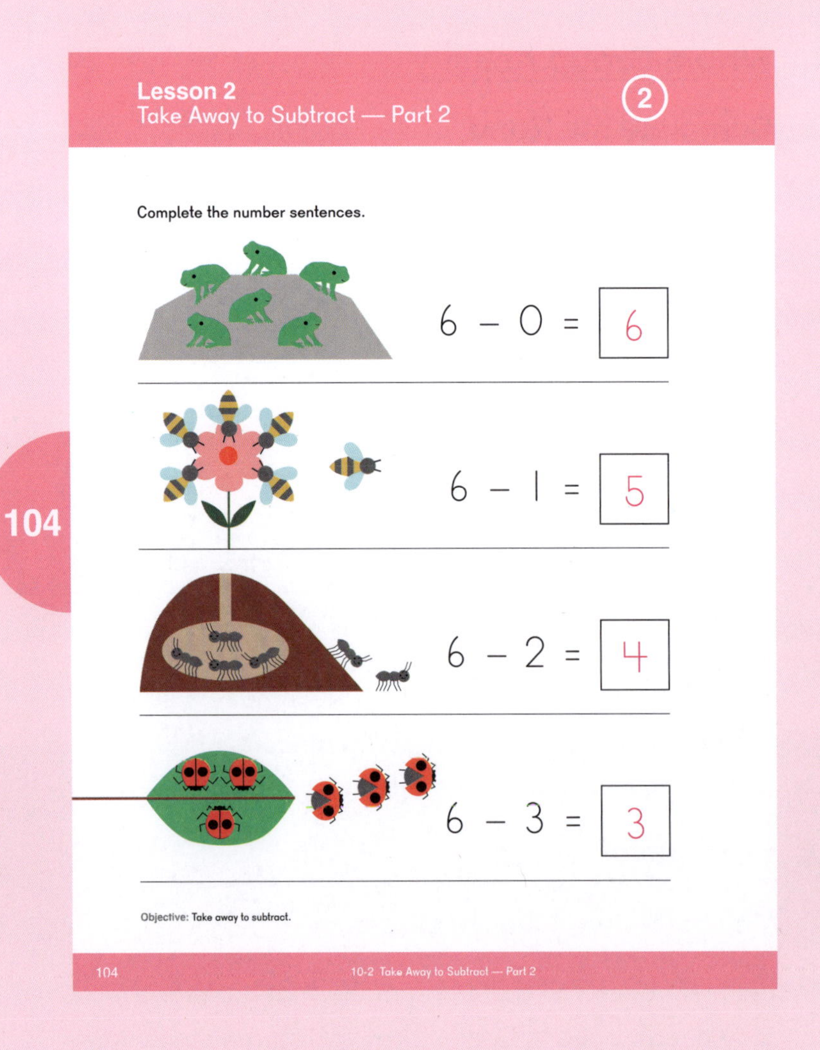

Explore

Provide students with 5 counters. Invite students to tell "take away" stories beginning with the number 5. Have students model the stories with their counters. A student may say, "There were 5 apples in my bag and I ate 3 of them. How many apples are still in the bag?"

Learn

Introduce the symbol for subtraction. Tell students that when we are "taking away," we are subtracting. We can say that 5 – 1 = 4 is a subtraction sentence.

Use the **Chapter Opener** on page 101. Write the subtraction sentence on the board as students model the situations with counters.

- There were 5 friends on the bench. 2 friends go to swing. How many friends are still on the bench?
  5 – 2 = 3
- There were 5 ducks in the pond. I duck flew away. How many ducks are still in the pond? 5 – 1 = 4
- There were 6 flowers in the park. Dion picked 2 flowers. How many flowers are still in the park?
  6 – 2 = 4

Discuss what is happening in the illustrations on page 104.

Whole Group Activity

▲ Hide and Take Away

Materials: 7 objects

Show students 7 objects, then ask students to cover their eyes. Remove some of the objects. Have students open their eyes and determine how many were taken away. Students could share stories about the situation.

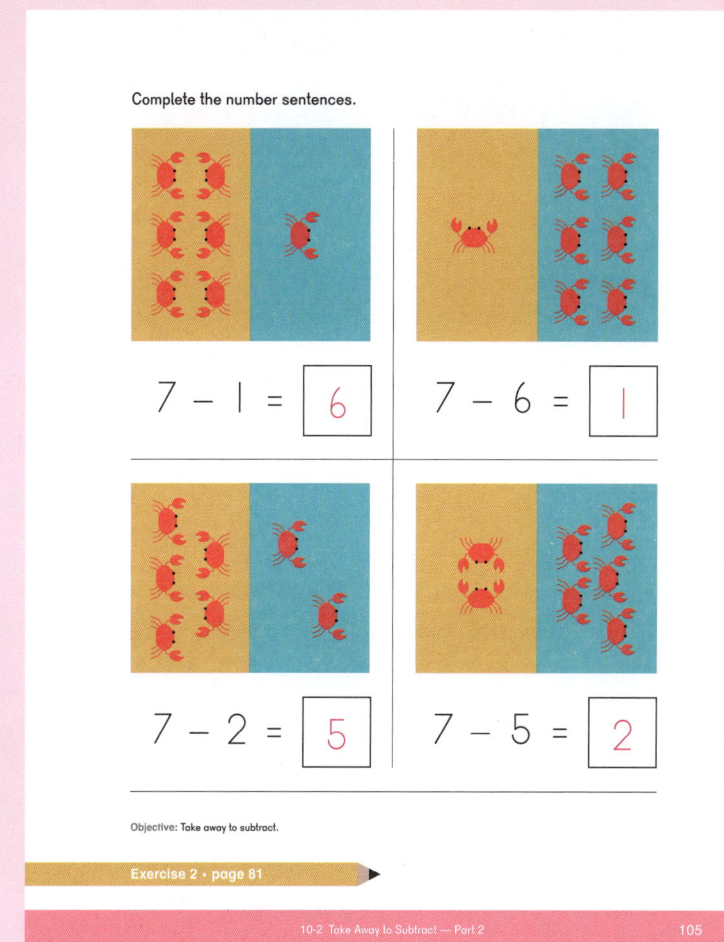

105

Small Group Activities

Textbook Pages 104–105. Students should use counters as needed. On page 105, ask students to imagine that all the crabs were on the sand at first, and then some left and went into the water. The picture shows how many left or went away to the water.

▲ Take Away From 7

Materials: 3 sets of Number Cards (BLM) 0 to 10, Picture Cards (BLM) 0 to 10, or Ten-frame Cards (BLM) 0 to 10

Students arrange the cards 0 to 7 faceup in the middle of the group. Students take turns saying, "7 take away _____ equals _____," and find cards to match their stories. For example a student may say, "7 take away 3 equals 4," and the student collects the 3 and 4 cards from the pile.

Extend by allowing students to work with cards to 10 where the number in the first blank is 10.

▲ Under the Cup

Materials: 10 counters, 1 cup, linking cubes

Play as described in the previous lesson, providing students with 7 counters to start.

Exercise 2 • page 81

Extend

★ Super Number Bonds

Draw a number bond similar to the one shown below. Students start with 10 in the whole and then break 10 into two parts. For example, 10 can be broken into 4 and 6. Then students break 4 and 6 into two parts, and so on. Challenge students to find how many different ways they can complete the Super Number Bond.

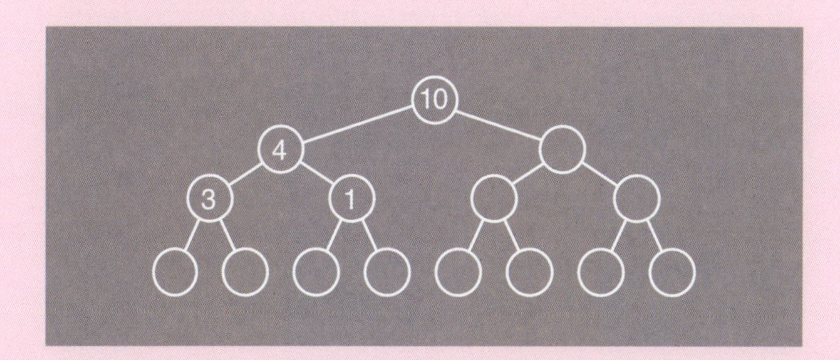

Objective

- Write subtraction sentences.

Lesson Materials

- Bear counters or other counters shaped like objects (bugs, dinosaurs, fruit, vehicles, etc.), 10 per pair of students
- Subtraction Sentences (BLM) in dry erase sleeves

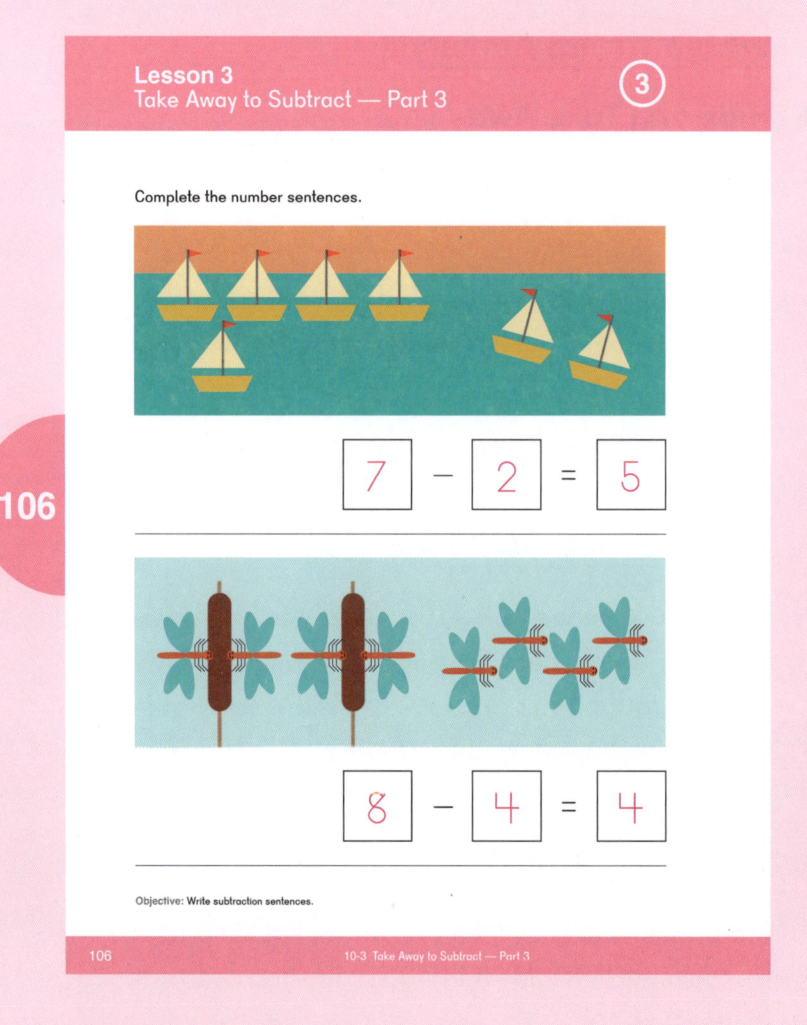

Explore

Have students tell a subtraction story. Use counters to model the story, then write the subtraction sentence on the board. Have students say the subtraction sentence using the words "minus" and "equals."

Learn

Provide pairs of students with counters. Have one student tell a story with the counters, and move some of the counters away. The partner says the subtraction sentence for the story.

For example, Partner 1 might say, "There were 6 trains waiting at the station. 2 trains chugged away with passengers. How many trains are still waiting at the station?" Partner 2 counts the remaining trains and says, "6 minus 2 is 4. There are 4 trains still at the station."

Pass out a Subtraction Sentence (BLM) and dry erase marker to each pair of students. Demonstrate a few of the stories with a number bond and counters. Have students write the numbers in the correct boxes to make subtraction sentences starting with the whole and subtracting one of the parts.

Whole Group Activity

Chant "10 Fat Sausages," asking students how many sausages are in the pan after each verse.

Ten fat sausages, sizzling in a pan, one went pop and the other went bang (clap!)
Eight fat sausages, sizzling in a pan, one went pop and the other went bang!
Six fat sausages...
Four fat sausages...
Two fat Sausages...
No fat sausages sizzling in a pan

Repeat, writing the subtraction sentences on the board.

Small Group Activities

Textbook Pages 106–107. Students should use counters as needed. Ensure that students are identifying the "whole" in each situation:

- There were 7 boats and 2 sailed away.
- There were 8 dragonflies on cattails and 4 flew away.
- There were 9 balloons and 1 drifted away.
- There were 9 rabbits and 3 hopped away.

▲ How Many are Missing?

Materials: 10 linking cubes

Player One shows a tower of 10 linking cubes. He puts the tower behind his back and snaps it into 2 parts. Keeping one part of the tower behind his back, he shows Player Two the rest of the cubes. Player Two says how many are behind Player One's back. If necessary, Player Two can use additional cubes to rebuild the tower to the designated number.

▲ Grab Bag

Materials: Bag of counters, Blank Number Bond Template (BLM), Subtraction Sentences (BLM)

Students reach into the bag and grab a handful of counters. This becomes the whole on their Blank Number Bond Template (BLM). Students then move some of the counters to one part, and determine the remaining part on their number bonds. Students record the subtraction situation on Subtraction Sentences (BLM). Students put the counters back in the bag and repeat the activity.

Exercise 3 • page 83

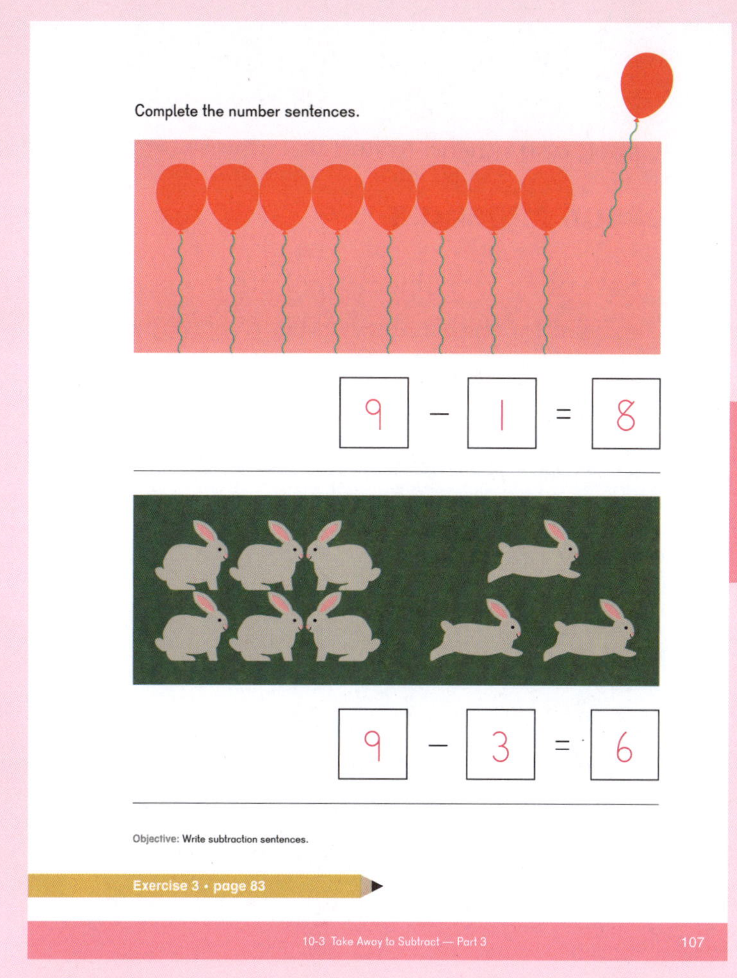

107

Extend

★ How Many Can You Make?

Materials: Number Cards (BLM) 0 to 9, Subtraction Sentences (BLM)

Students create as many correct subtraction sentences as possible using each digit only once in each sentence.

Objective

- Recognize subtraction as separating a known quantity into two parts, given one part, and having to find the other part.

Lesson Materials

- Sets of up to 5 objects that are the same except for their color or size (red and blue counting bears, red and white plates, tall and short drinking cups, etc.)
- Two-color counters, 5 per student
- Blank Number Bond Template (BLM) or partitioned plate

Explore

Show students 5 objects as described in **Lesson Materials** and have them say how they are the same and how they are different. Repeat with other objects.

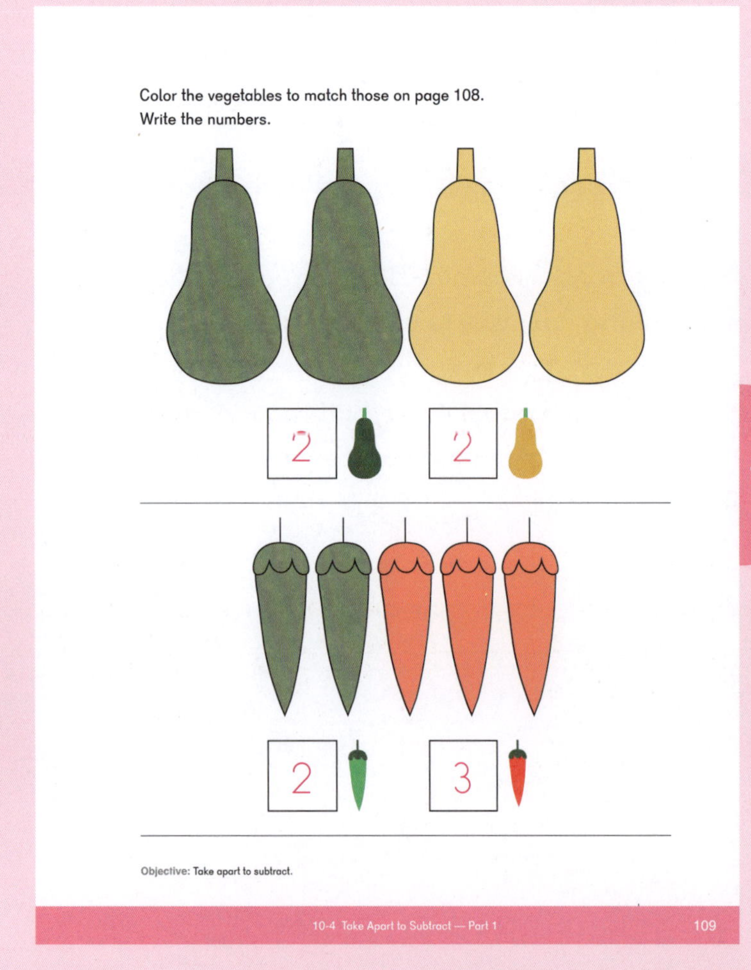

Learn

Provide students with two-color counters and the Blank Number Bond Template (BLM). Have them model stories about the objects from **Explore**.

Show students 5 bears, 1 red and 4 blue. Ask students to model the same with their counters. Ask, "How many bears altogether?" Students place all 5 counters in the whole of the number bond. Ask, "If 1 bear is red, how many are blue?" Students move 1 counter to one of the parts and the 4 remaining counters to the other part.

Write a number bond and the subtraction sentence "$5 - 1 = \square$" on the board to match the situation with the bears: 5 in the whole and 1 in one of the parts.

Repeat the activity with other objects and the gourds and peppers on page 108.

Whole Group Activity

▲ What's Different?

Materials: 3 hula hoops or painter's tape

Create a template for a number bond with hula hoops or painter's tape on the floor. Invite some students to stand in the whole. Choose an attribute that is different about the students, but don't tell them what you are thinking. Start inviting students to move into a part of the bond based on the attribute. Pause and ask students if they can tell what the different attribute is that you are thinking. If a student guesses correctly, ask her to tell a subtraction story about the students. Choose another group of students and repeat the activity. If they are incorrect, keep sorting (or moving) students into the correct part.

Small Group Activities

Textbook Pages 109–110

▲ Sort and Take Apart

Materials: A set of up to 10 objects with several attributes (such as buttons, shells, or attribute blocks), Subtraction Sentences (BLM)

Provide students with a set of up to 10 objects with several attributes. Have students put the objects into 2 groups. For example, a student may sort buttons based on size (large and small). A student may notice he has 8 buttons, 5 are small and 3 are large. Ask the student to say a subtraction sentence to match the situation. 8 minus 5 is 3. Alternatively, have students complete the Subtraction Sentences (BLM).

▲ Toss Up

Materials: 10 two-color counters per student

Students grab a handful of counters and drop them on the floor or table. Students count all, then sort them by color. Students complete a subtraction sentence to match the colors showing.

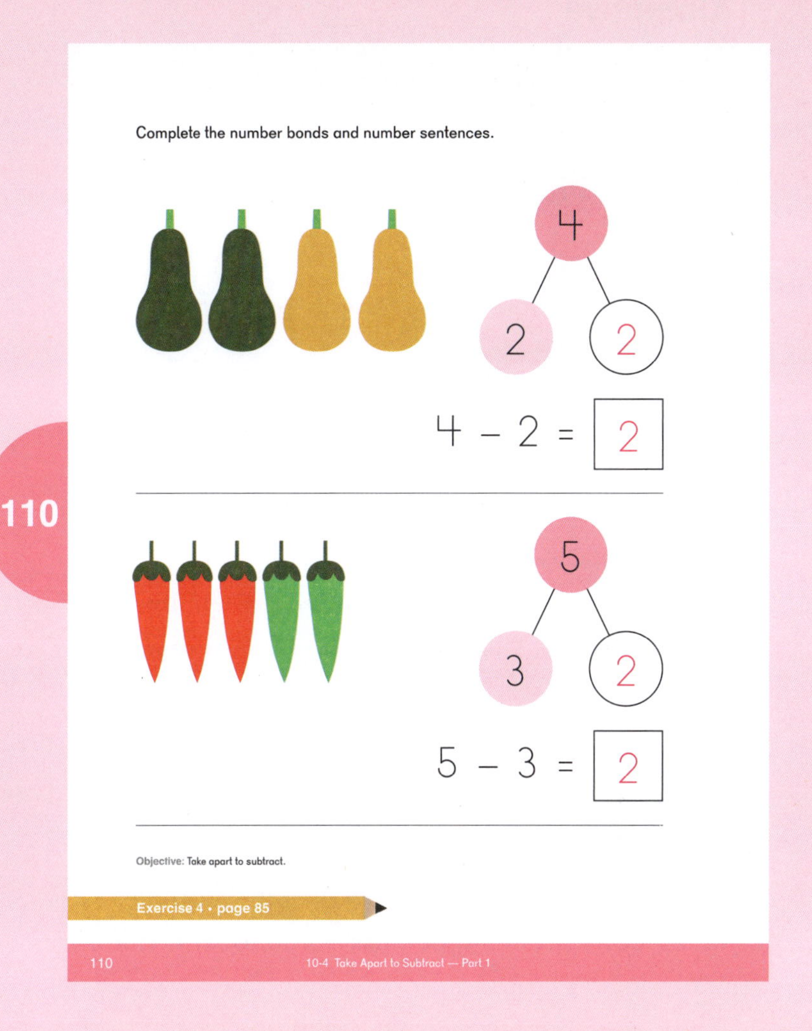

110

Complete the number bonds and number sentences.

4
2 2

$4 - 2 = 2$

5
3 2

$5 - 3 = 2$

Objective: Take apart to subtract.

Exercise 4 • page 85

110 10-4 Take Apart to Subtract — Part 1

Exercise 4 • page 85

Extend

★ Subtraction Stories

Materials: Storybooks or magazines

Use a storybook or magazine and find pictures where subtraction stories can be made. Have students tell the subtraction story to a friend or record it. Students can draw pictures of the subtraction stories and write the number sentences.

Objective

- Recognize the relationship between number bonds and subtraction sentences.

Lesson Materials

- Counters
- Blank Number Bond Template (BLM) or partitioned plates

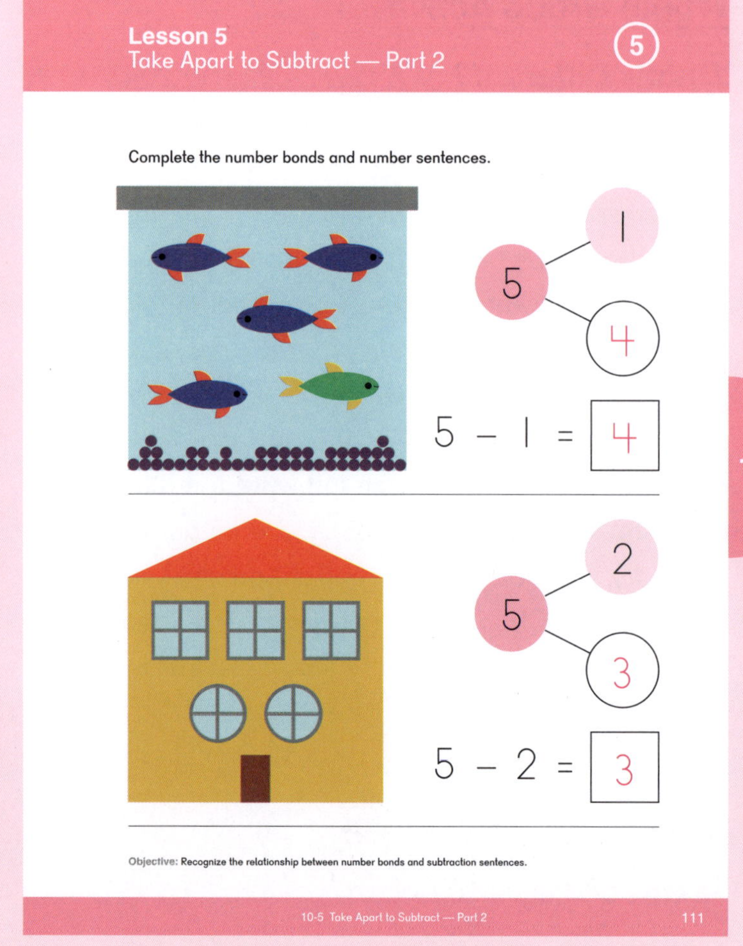

This lesson is a practice day to reinforce the concept of subtracting from a whole and that the whole is made up of parts. The relationship between the number bond and subtraction sentences is explored more here.

Have students work with partners and take turns. One student uses the counters and Blank Number Bond Template (BLM) or plate to tell a subtraction story, always beginning with the "whole." Partners write the corresponding number sentence. Partners switch roles and repeat.

Whole Group Activity

▲ Show Me the Bond

Materials: Number Cards — Large (BLM)

Hold up a Number Card — Large (BLM) and tell students that they will be finding the parts in a number bond with a whole of the number on the card. For example, if you hold up number card 4, show one part with your fingers (2 fingers), and the students will show the missing part with their fingers (2 fingers). 2 and 2 make 4.

Alternatively, students write the missing part on personal whiteboards.

Small Group Activities

Textbook Pages 111–112

Practice with games and activities from previous lessons.

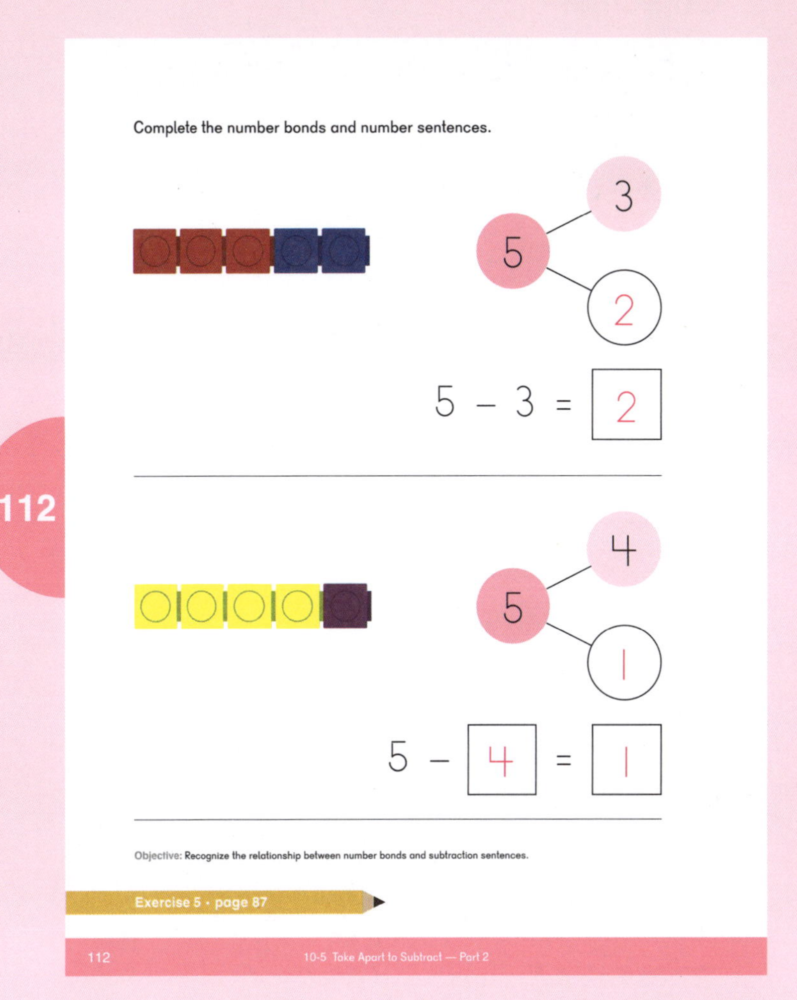
112

Complete the number bonds and number sentences.

5 – 3 = 2

5 – 4 = 1

Objective: Recognize the relationship between number bonds and subtraction sentences.

Exercise 5 • page 87

112 10-5 Take Apart to Subtract — Part 2

▲ How Many are Missing?

Materials: 10 linking cubes

Play as directed in Lesson 3.

▲ Grab Bag

Materials: Bag of counters, number bond mat, Subtraction Sentences (BLM)

Play as directed in Lesson 3.

▲ Sort and Take Apart

Materials: A set of up to 10 objects with several attributes (such as buttons, shells, or attribute blocks), Subtraction Sentences (BLM)

Play as directed in Lesson 4.

▲ Toss Up

Materials: 10 two-color counters per student

Play as directed in Lesson 4.

Exercise 5 • page 87

Extend

★ How Many Ways?

Using an extended number bond similar to the one shown, students complete the number bond starting with the whole, decompose the whole into parts and then decompose the parts into smaller parts. For a challenge, ask students to use the numbers 0 to 10 only once in each part.

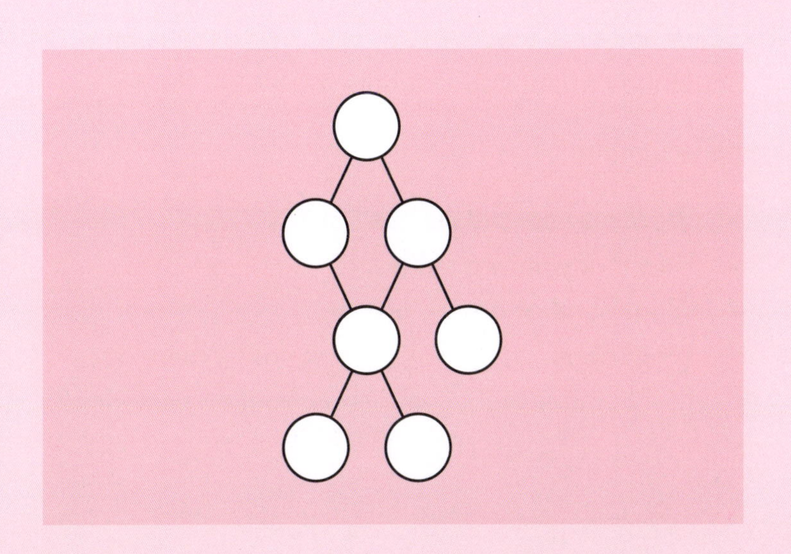

Challenge students to complete as many different extended number bonds as they can.

Objective

- Count back 1, 2, or 3 to subtract on a number path.

Lesson Materials

- Number Paths (BLM)

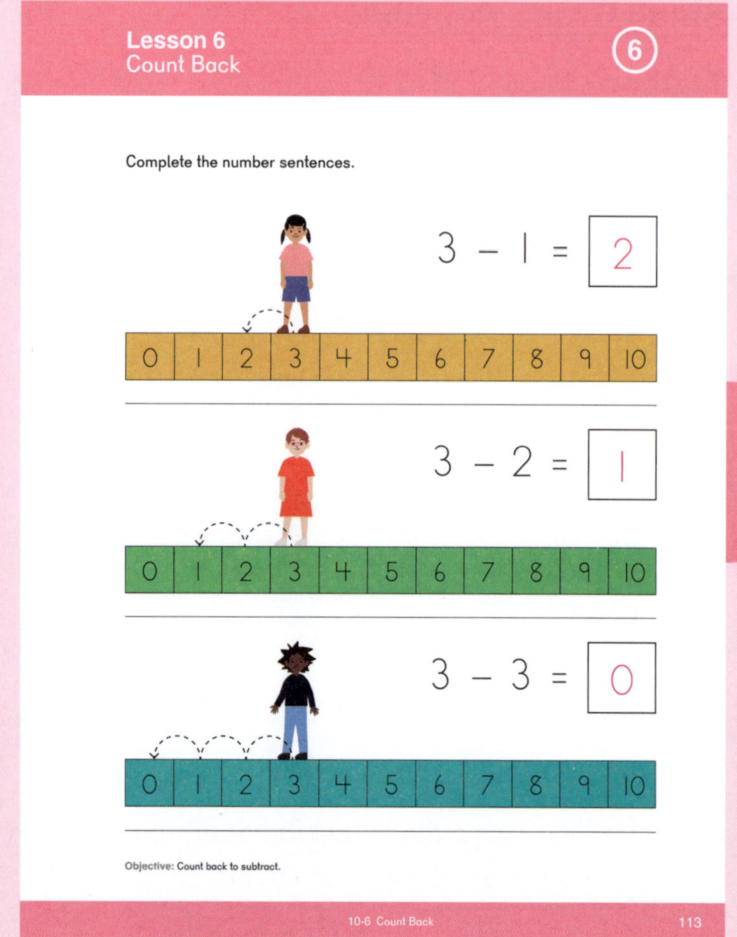
Lesson 6
Count Back
6

Complete the number sentences.

3 − 1 = 2

0 1 2 3 4 5 6 7 8 9 10

3 − 2 = 1

0 1 2 3 4 5 6 7 8 9 10

3 − 3 = 0

0 1 2 3 4 5 6 7 8 9 10

Objective: Count back to subtract.

10-6 Count Back 113

113

Explore

Sing "Ten in the Bed." Provide students with a Number Path (BLM). Have students start with their finger on the 10. Students move their fingers along the path while singing the song.

> *There were 10 in the bed and the little one said, "Roll over! Roll over!"*
>
> *So they all rolled over and 1 fell out.*
>
> *There were 9 in the bed and the little one said, "Roll over! Roll over!"*
>
> *So they all rolled over and 1 fell out.*
>
> *There were 8 in the bed...*

Continue to:

> *There was 1 in the bed and the little one said, "Alone at last!"*

Learn

Choose a number on the path. Have students count back out loud by 1, 2, or 3. Write the number sentences prompting students to provide the parts.

For example:

- What number did we start on?
- How many did we subtract?
- What number are we on now?
- What does ______ – ______ equal? Where does that number belong in the subtraction sentence?

Whole Group Activity

▲ Tap and Clap

Give students a subtraction sentence taking away 1, 2, or 3. Have them tap their laps for the starting number, and clap to count back to find the answer. For example, you say, "4 – 2. Tap 4, clap 2 times as you count back 3, 2."

Small Group Activities

Textbook Pages 113–114

▲ Roll and Subtract

Materials: Number Path (BLM), counters, die with numbers 1, 1, 2, 2, 3, 3

Place a counter on 10 to begin. On each turn, players roll the die and count the corresponding number of spaces on their number paths. The first player to get to 0 wins. The winner scores a point and the game is repeated.

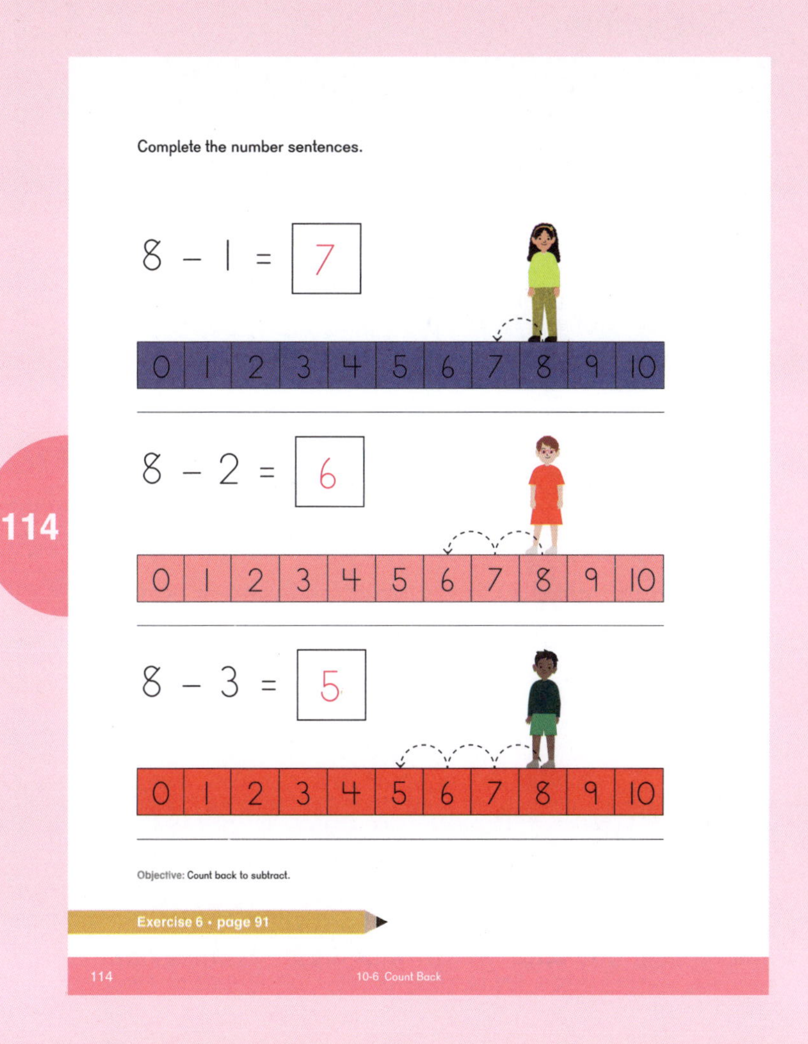

Take it Outside

▲ Next Number Hop: Subtract

Materials: Chalk, painter's tape, paper plates, Number Cards (BLM) 0 to 9, large cube-shaped box

Create large number paths outside with chalk, or inside with painter's tape. You could also use paper plates with numbers on them:

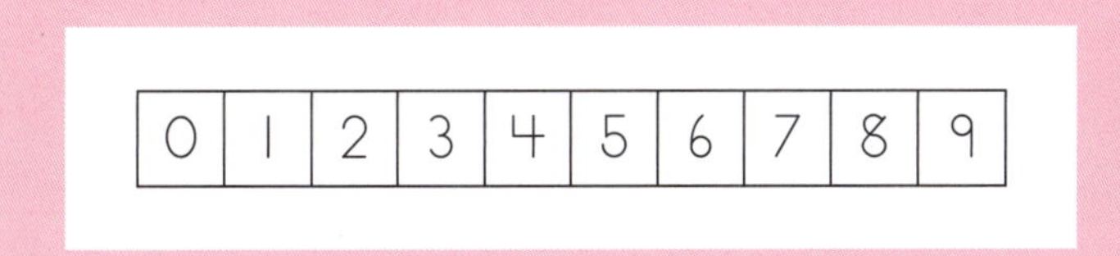

Call a number from 0 to 9 and have students stand on that number. Ask them to subtract 0, 1, or 2 from that number and walk to the new number.

To extend, make a large die out of a box with sides of 0, 0, 1, 1, 2, and 2. Have players take turns drawing from a stack of Number Cards (BLM) 0 to 9 and directing their partner to that starting spot on the number path.

The same player that drew the card then rolls the die and tells his partner how many steps to take. The partner walks that many steps back on the path and says the new number.

Exercise 6 • page 91

Extend

★ Same, 1 More, or 1 Less

Materials: 4 of each Number Card (BLM) 0 to 9

The Dealer shuffles the cards and deals 5 cards to each player. The rest of the deck is placed facedown. Play starts with the top card on the deck flipped faceup. Players take turns playing a card that is either 1 more, 1 less, or the same as the card showing.

Objective

- Subtract from 3, 4, and 5.

Lesson Materials

- 5 two-color counters
- Blank Number Bond Template (BLM)
- Number Path (BLM)

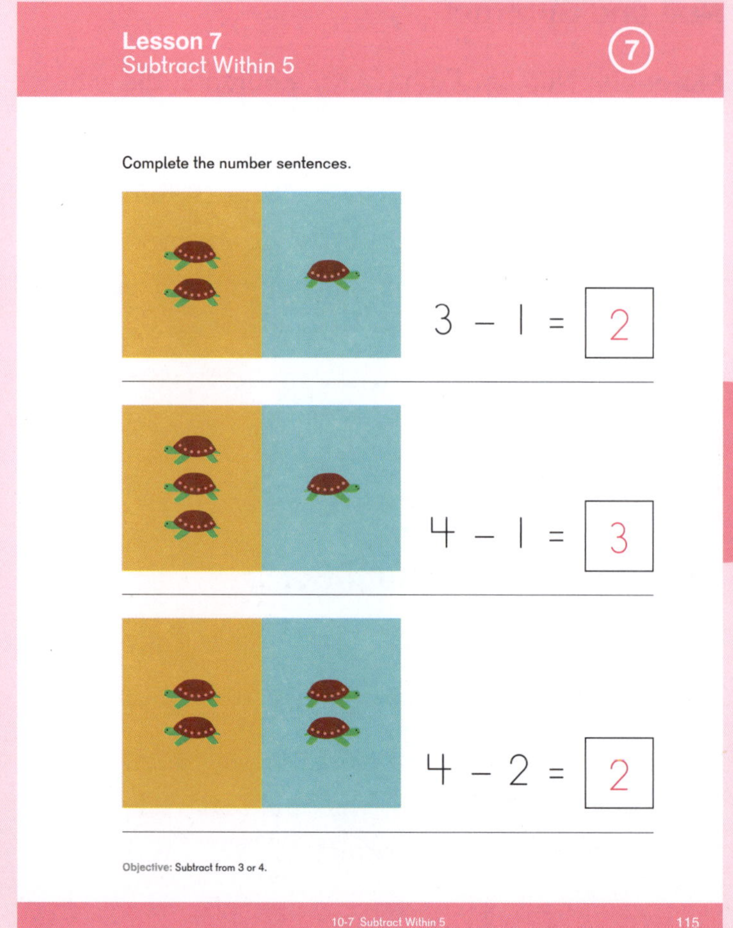

115

The following lessons are designed to help students develop fluency in subtracting from numbers up to 10.

Games and activities from previous lessons can be repeated for practice.

Explore

Provide students with counters and a Blank Number Bond Template (BLM). Have them put 3 counters in the whole of their blank number bonds and then make as many different arrangements as they can, breaking apart 3 into 2 parts.

Learn

Write all of the ways students find to create subtraction sentences for subtracting from 3. Students can use a number bond and counters or a Number Path (BLM) if needed for support.

- 3 – 0 = 3
- 3 – 1 = 2
- 3 – 2 = 1
- While students have not specifically been introduced to subtracting zero, they may find the number sentence 3 – 3 = 0.

Repeat the activity with 4 counters:

- 4 – 0 = 4
- 4 – 1 = 3
- 4 – 2 = 2
- 4 – 3 = 1
- Optional: 4 – 4 = 0

And with 5 counters:

- 5 – 0 = 5
- 5 – 1 = 4
- 5 – 2 = 3
- 5 – 3 = 2
- 5 – 4 = 1
- Optional: 5 – 5 = 0

Whole Group Activity

Still Standing

Invite up to 5 students to the front of the room. Ask, “How many students are standing up now?” Give a direction, such as, “Kneel down if you are wearing blue.” Ask students:

- How many students were standing up front?
- How many students are kneeling down?
- How many students are still standing?

Small Group Activities

Textbook Pages 115–116

▲ Subtraction Smash

Materials: Play dough or clay, dice

Partners each roll a die and make that number of play dough balls to begin. They take turns smashing 1 ball and saying how many balls remain. Partners check each other's answers.

▲ Domino Parking Lot

Materials: Dominoes, Domino Parking Lot (BLM)

Students draw a domino and subtract the lesser number from the greater and park it in the stall with the answer. For example:

Domino	Expression	Parking Lot
	"4 – 2"	Domino gets parked in the 2 stall.
	"4 – 1"	Domino gets parked in the 3 stall.

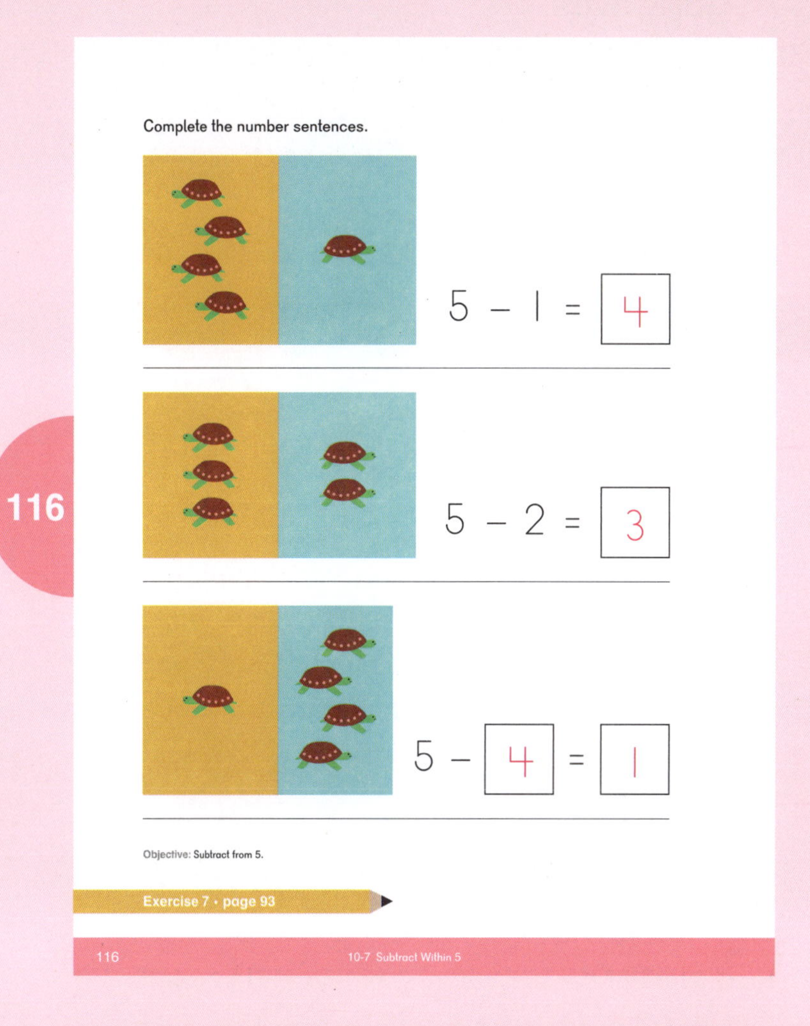

Exercise 7 • page 93

Objectives

- Subtract from 6 through 10.
- Write subtraction sentences.

Lesson Materials

- Two-color counters, 10 per pair of students
- Blank Number Bond Template (BLM)
- Number Path (BLM)

Lesson 8
Subtract Within 10 — Part 1

Complete the number sentences.

6 – 2 = 4

7 – 2 = 5

8 – 2 = 6

Objective: Subtract from 6, 7, and 8.

10-0 Subtract Within 10 — Part 1 117

117

Explore

Have students recall the activity from the previous lesson to make as many different arrangements as they can, breaking apart a number into 2 parts. Assign pairs of students a number from 6 through 10. Have the partners find all the ways to break apart their numbers.

Learn

Record all of the ways students can find to create a subtraction sentence for subtracting from each number. Students can use a Blank Number Bond Template (BLM) and counters or a Number Path (BLM), if needed for support. A list of all possible subtraction sentences is included below for reference only. Subtraction with zero has not been formally taught, however, it is included as students may come up with it.

6 subtraction sentences

- 6 – 0 = 6
- 6 – 1 = 5
- 6 – 2 = 4
- 6 – 3 = 3
- 6 – 4 = 2
- 6 – 5 = 1
- 6 – 6 = 0

7 subtraction sentences

- 7 – 0 = 7
- 7 – 1 = 6
- 7 – 2 = 5
- 7 – 3 = 4
- 7 – 4 = 3
- 7 – 5 = 2
- 7 – 6 = 1
- 7 – 7 = 0

8 subtraction sentences

- 8 – 0 = 8
- 8 – 1 = 7
- 8 – 2 = 6
- 8 – 3 = 5
- 8 – 4 = 4
- 8 – 5 = 3
- 8 – 6 = 2
- 8 – 7 = 1
- 8 – 8 = 0

9 subtraction sentences

- 9 – 0 = 9
- 9 – 1 = 8
- 9 – 2 = 7
- 9 – 3 = 6
- 9 – 4 = 5
- 9 – 5 = 4
- 9 – 6 = 3
- 9 – 7 = 2
- 9 – 8 = 1
- 9 – 9 = 0

10 subtraction sentences

- 10 – 0 = 10
- 10 – 1 = 9
- 10 – 2 = 8
- 10 – 3 = 7
- 10 – 4 = 6
- 10 – 5 = 5
- 10 – 6 = 4
- 10 – 7 = 3
- 10 – 8 = 2
- 10 – 9 = 1
- 10 – 10 = 0

Small Group Activities

Textbook Pages 117–118

▲ Subtraction Bowling

Materials: 10 pins (small paper cups, counting bears, dominoes, or markers), small ball, Bowling Recording Sheet (BLM)

Players take turns bowling. They record the number of pins that were knocked down from 10 and subtract to find how many are still standing. Students can record each turn on the Bowling Recording Sheet (BLM).

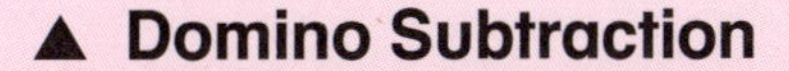

▲ Domino Subtraction

Materials: Dominoes

Students draw a domino from the pile and write the subtraction sentence, beginning with all of the pips, or the whole, and subtracting the pips on one half of the domino. For example:

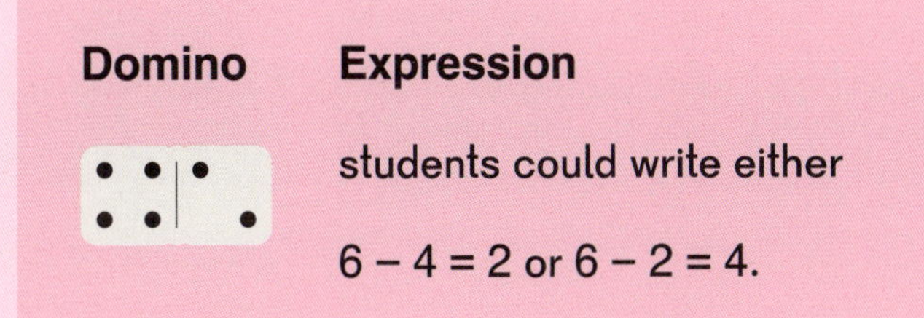

Domino	Expression
(domino with 4 and 2 pips)	students could write either $6 - 4 = 2$ or $6 - 2 = 4$.

118

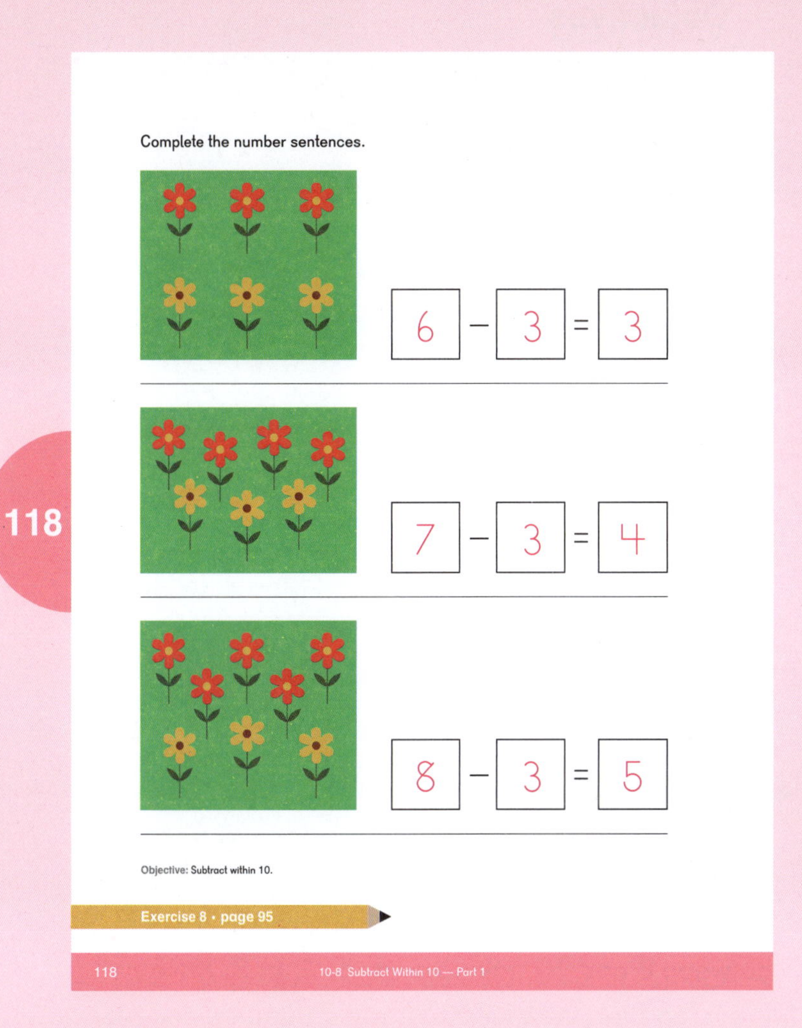

Complete the number sentences.

$6 - 3 = 3$

$7 - 3 = 4$

$8 - 3 = 5$

Objective: Subtract within 10.

Exercise 8 • page 95

118 10-8 Subtract Within 10 — Part 1

Exercise 8 • page 95

Extend

★ Subtraction Patterns

Materials: Subtraction Fact Cards (BLM), using only the cards for "a number – 2" (2 – 2, 3 – 2, up to 10 – 2)

Have students put the cards in order with the smallest whole (2 – 2) first, and the greatest whole (10 – 2) last.

Ask them to describe the pattern in the answers based on the order of the cards. Repeat the activity subtracting other numbers.

Objectives

- Subtract from 6 through 10.
- Write subtraction sentences.

Lesson Materials

- Picture book(s)

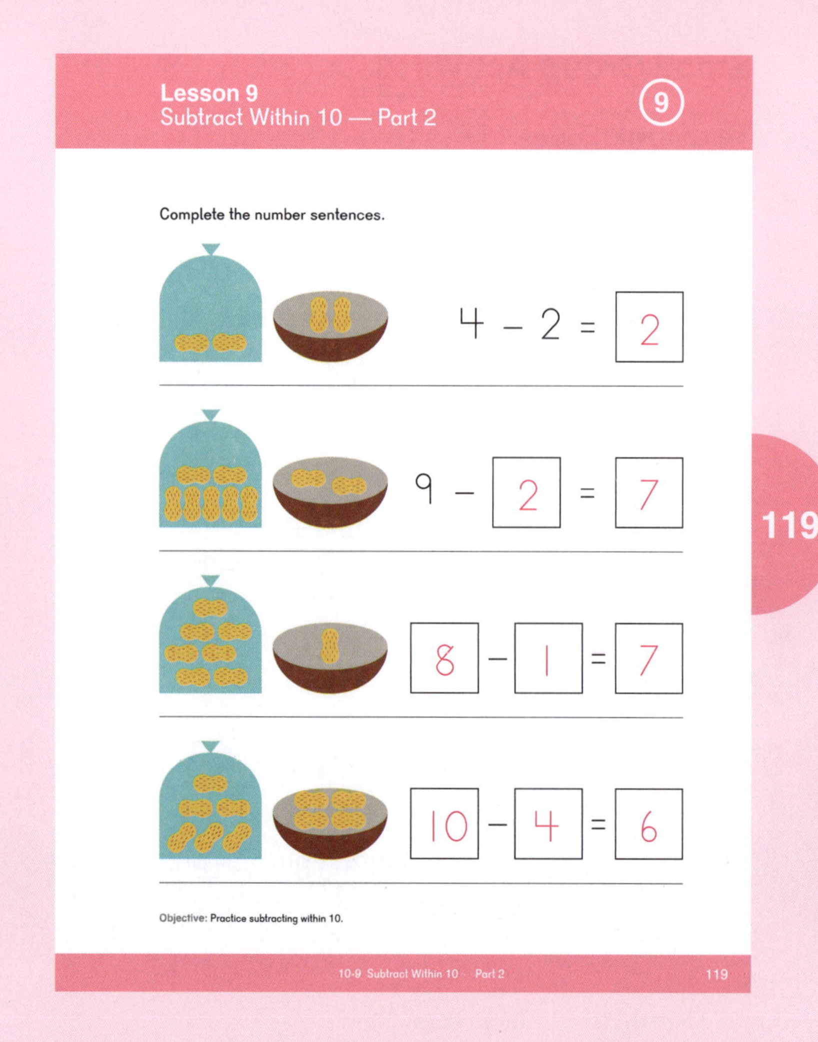

119

This lesson is designed to help students develop fluency with subtracting from numbers up to 10.

You may choose to start the lesson by reading a story from the list of recommended stories found at the beginning of the chapter. Pause and ask students to tell subtraction stories as you read.

Whole Group Activity

▲ Tap and Clap

Give students a subtraction sentence with a whole between 6 and 10, taking away 1, 2, or 3. Have them tap their laps for the starting number, and clap to count back to find the answer. For example, you could say, "4 – 2. Tap 4, clap 2 times as you count back 3, 2."

Small Group Activities

Textbook Pages 119–120. Ask students what pattern they notice in the subtraction from 10 number sentences.

▲ Clear the Board

Materials: Blank Ten-frames (BLM), 10 counters, die modified with sides 1, 1, 2, 2, 3, 3

Players begin with a Blank Ten-frame (BLM) with counters on all the spaces. They take turns rolling the die and removing that number of counters while saying or recording the subtraction sentence.

The first player to clear his board is the winner.

$10 - 2 = 8$

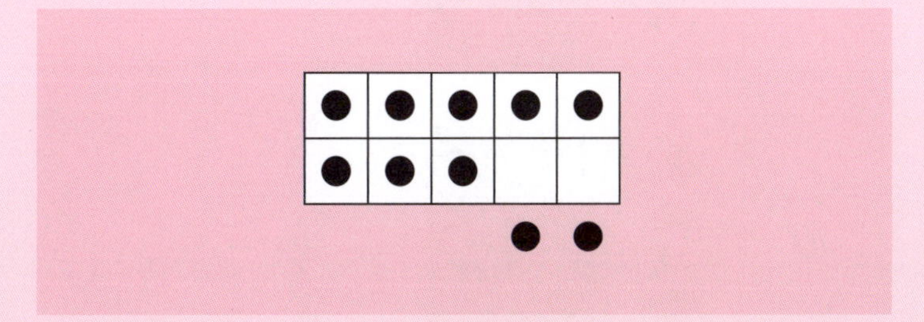

120

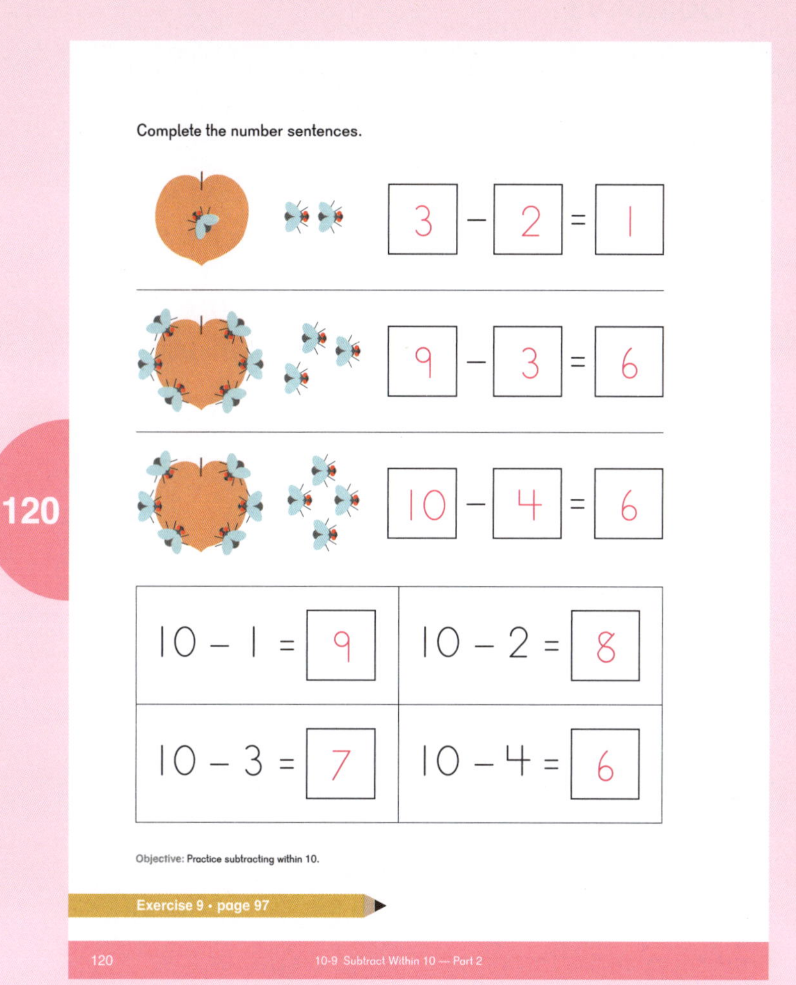

Exercise 9 • page 97

Extend

★ How Many Ways?

Materials: Number Cards (BLM) 0 to 10, Subtraction Template (BLM)

Students create as many correct number sentences as possible, using each number only once in each sentence.

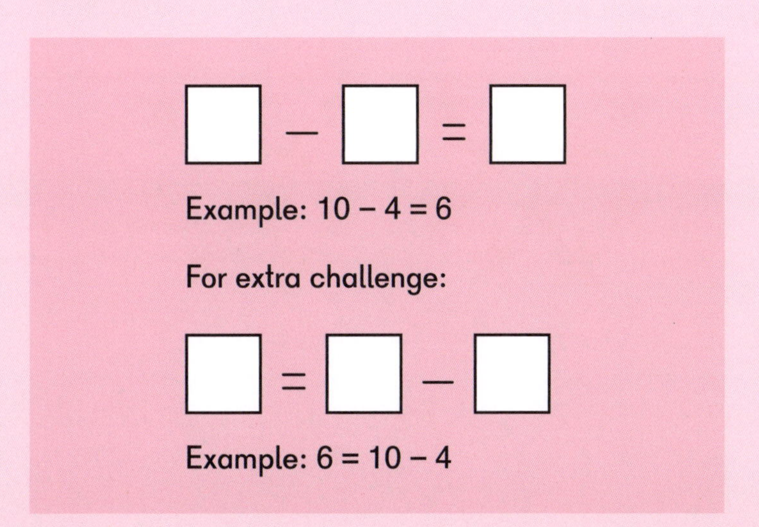

Example: $10 - 4 = 6$

For extra challenge:

Example: $6 = 10 - 4$

Objective

- Practice subtraction within 10.

Practice lessons are designed for further practice and assessment as needed.

Students can complete the pages and workbook pages as practice and/or as assessment.

Use activities and extensions from the chapter for additional review and practice.

Small Group Activity

▲ Draw and Write — Subtraction

Materials: Number Bond Cards (BLM)

Students draw a Number Bond Card (BLM) and write the corresponding subtraction equation on their whiteboards. Students can also illustrate a picture that represents the number bond on paper.

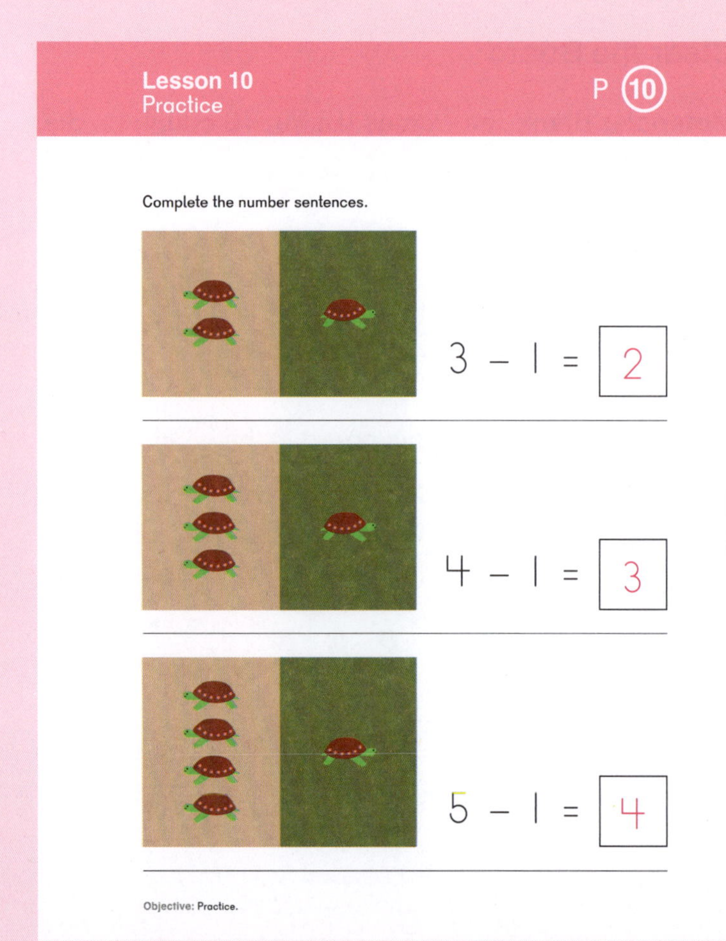

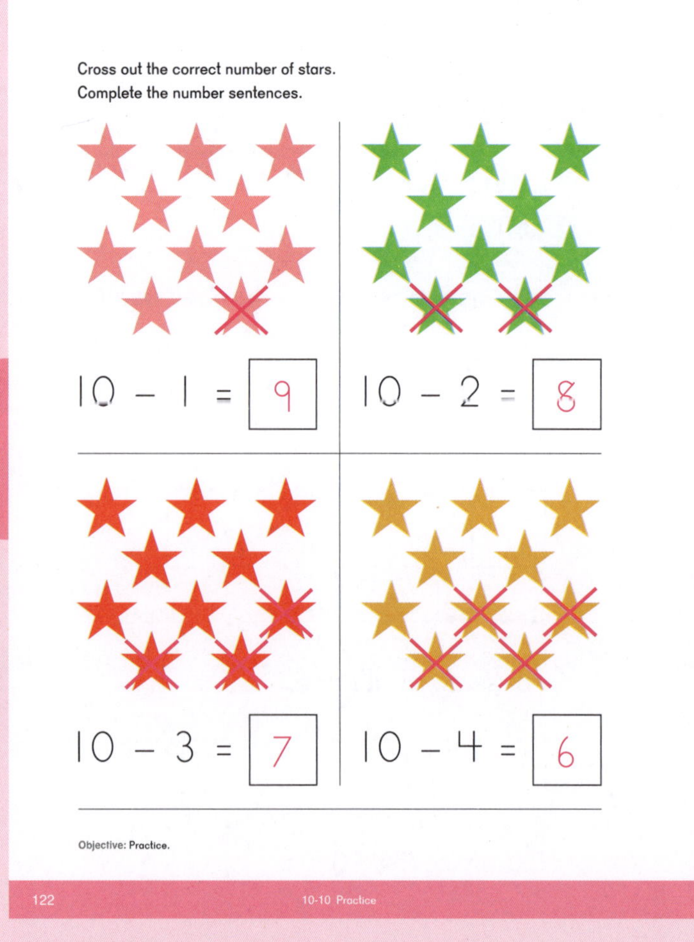

Exercise 10 • page 99

Extend

★ Subtraction Sort

Materials: Subtraction Facts Cards (BLM), Number Cards (BLM) 1 to 9

Students can work independently or with partners. They begin by drawing two Number Cards (BLM). These cards are the cards that the Subtraction Facts Cards (BLM) will be sorted under.

Students then draw a Subtraction Facts Card (BLM) and find the answer. If the difference is one of the Number Cards (BLM), they put it under that Number Card (BLM). If the difference does not match a number card, the fact card is discarded.

For a greater challenge, play head-to-head. The winner is the first player to put 5 facts under each number card.

6	3
7 – 1	5 – 2
10 – 4	3 – 0

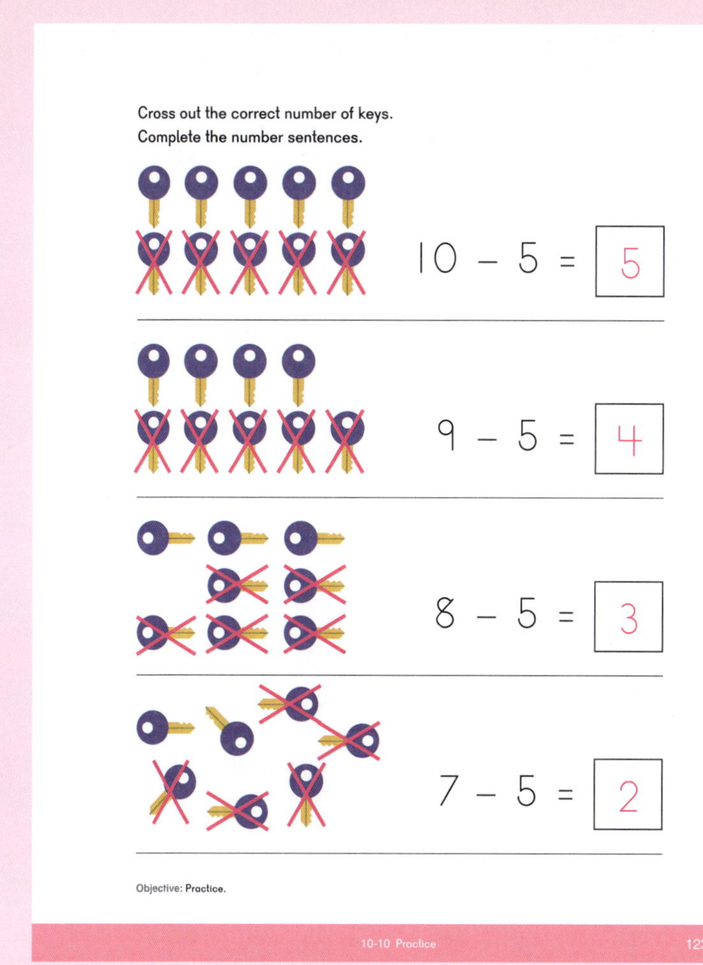

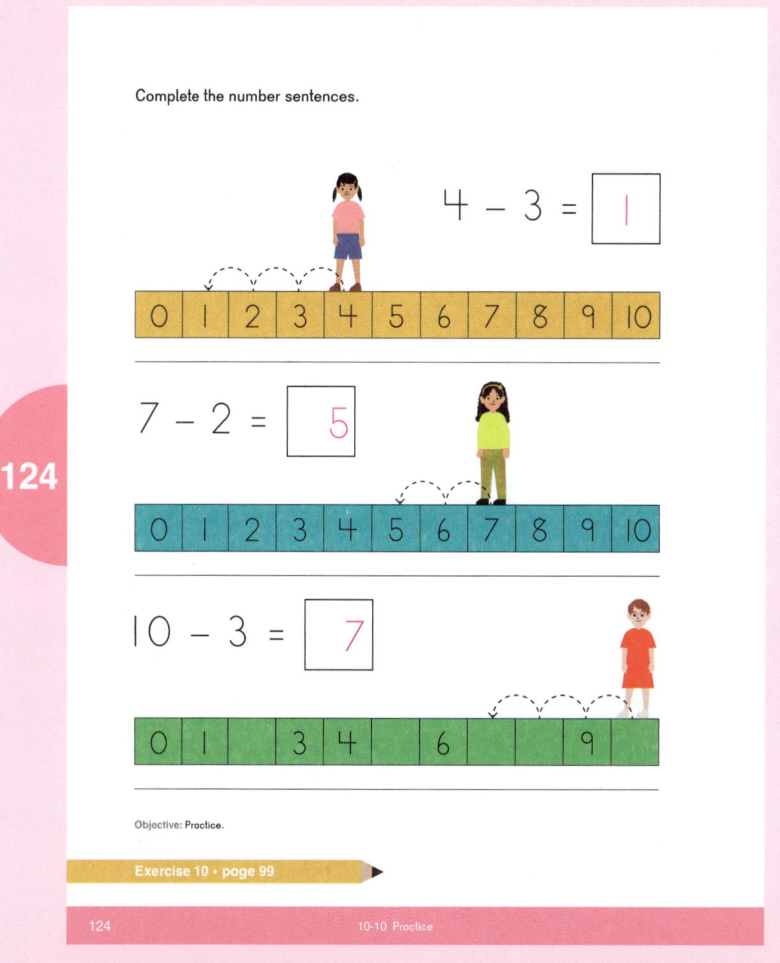

Exercise 1 • pages 79–80

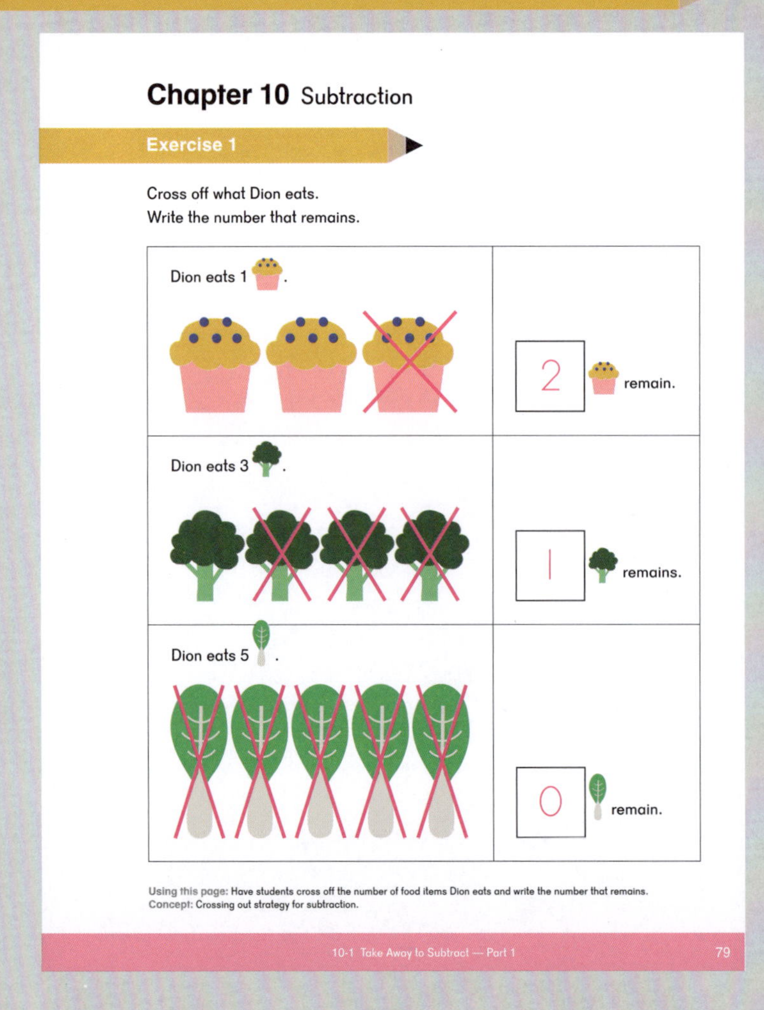

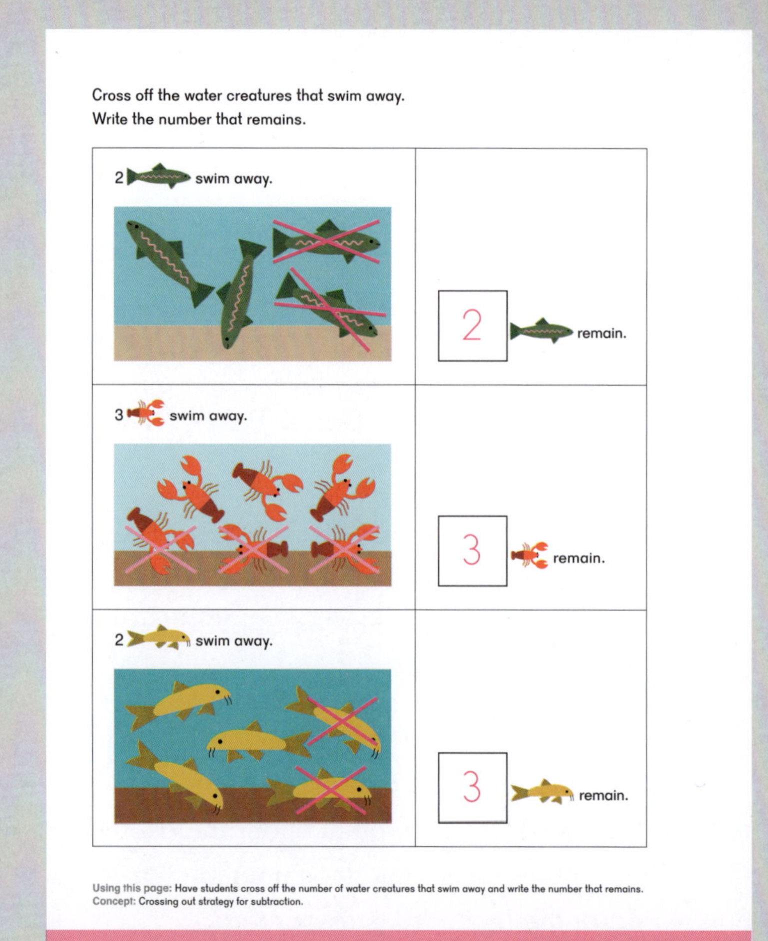

Exercise 2 • pages 81–82

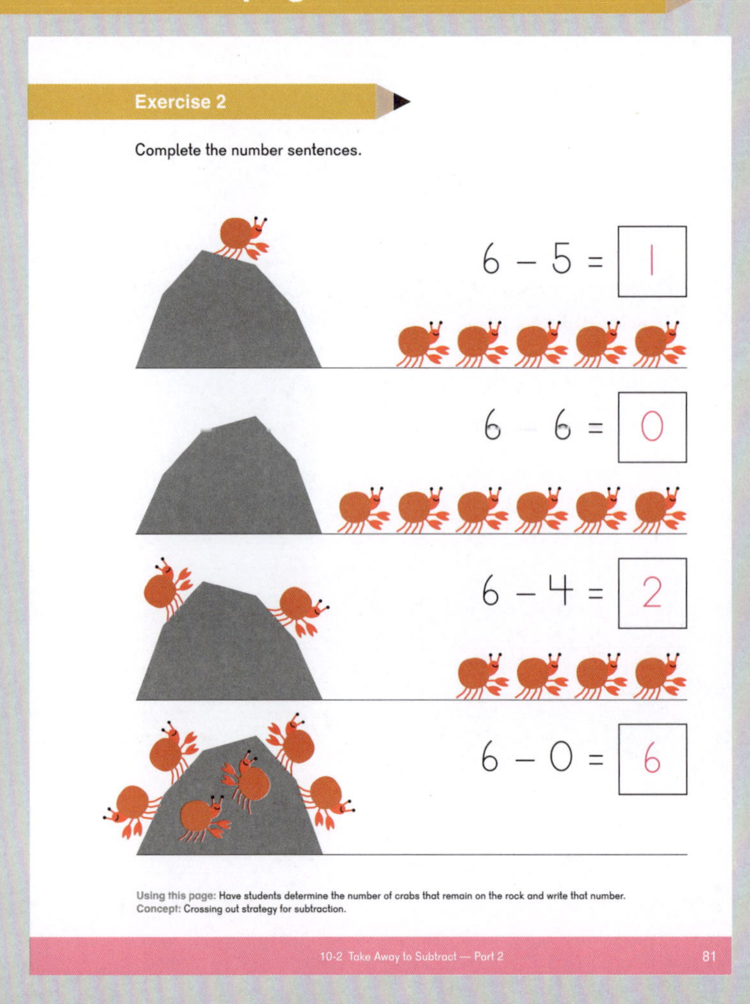

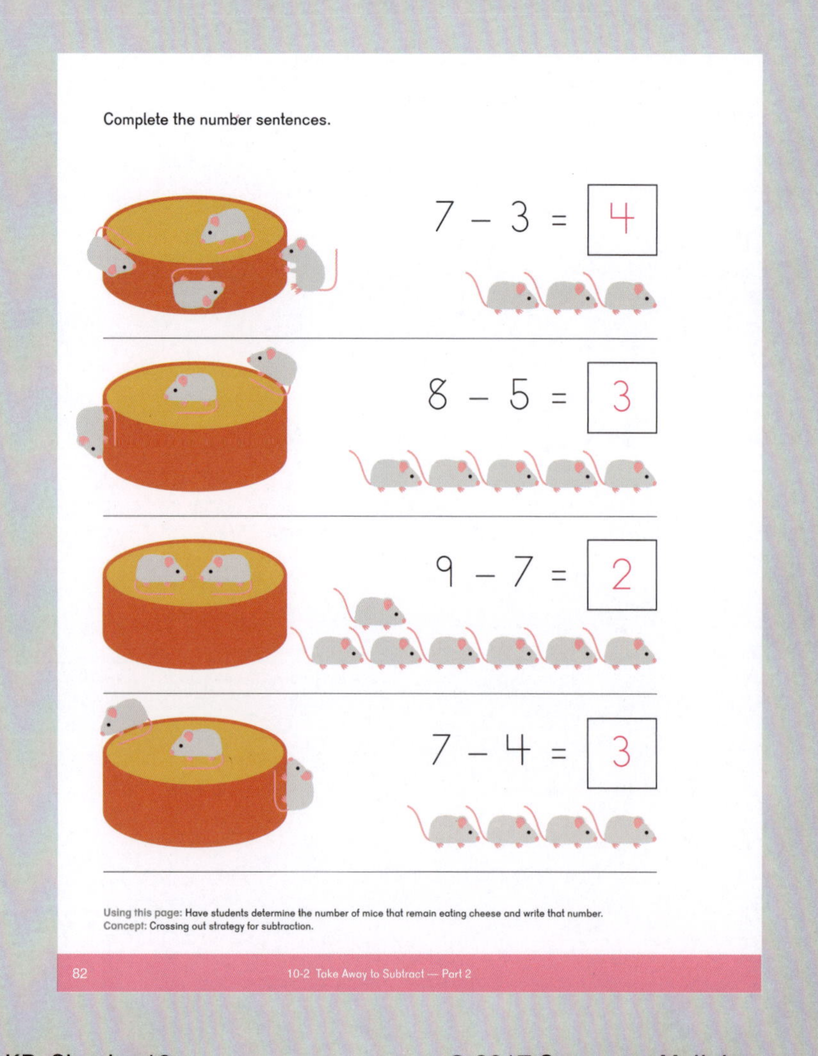

Exercise 3 • pages 83–84

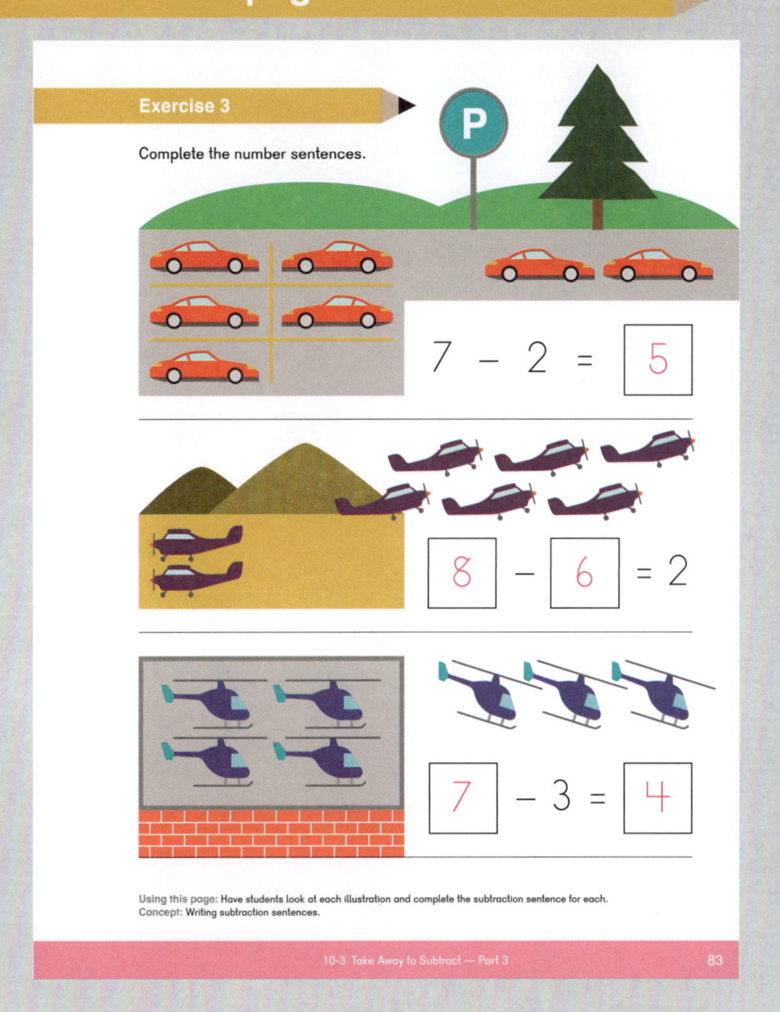

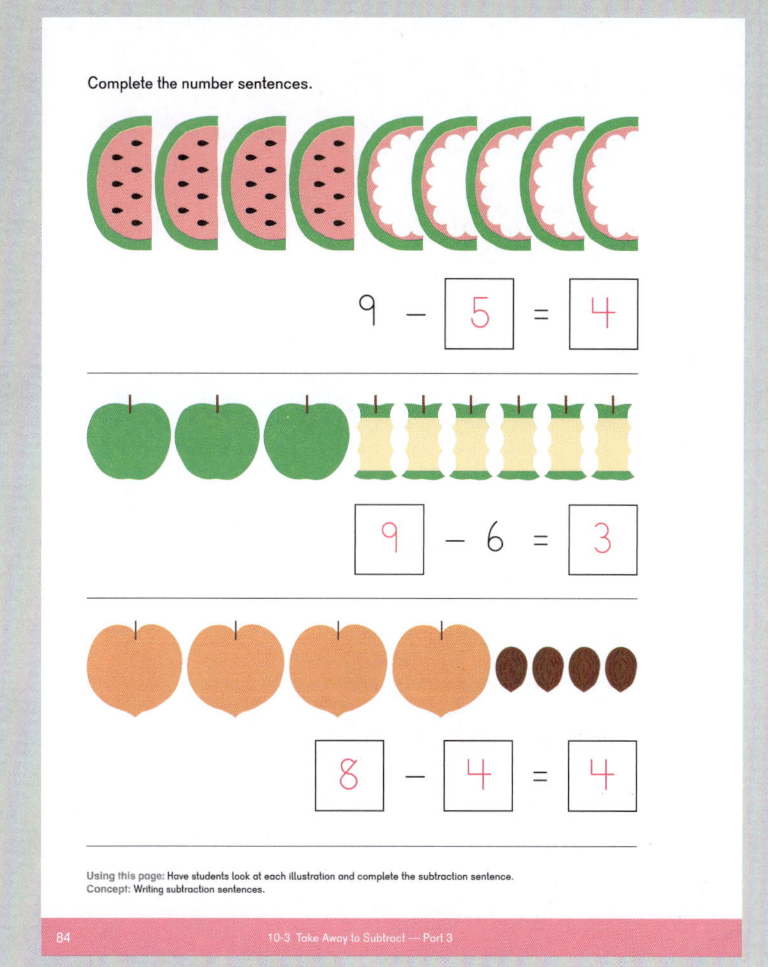

Exercise 4 • pages 85–86

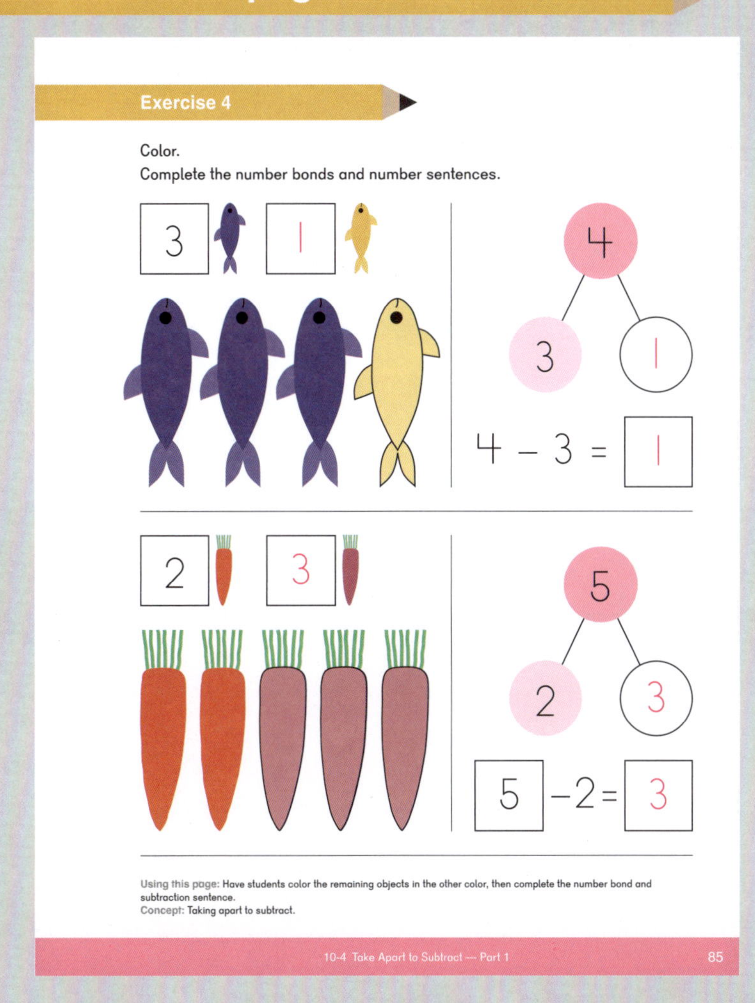

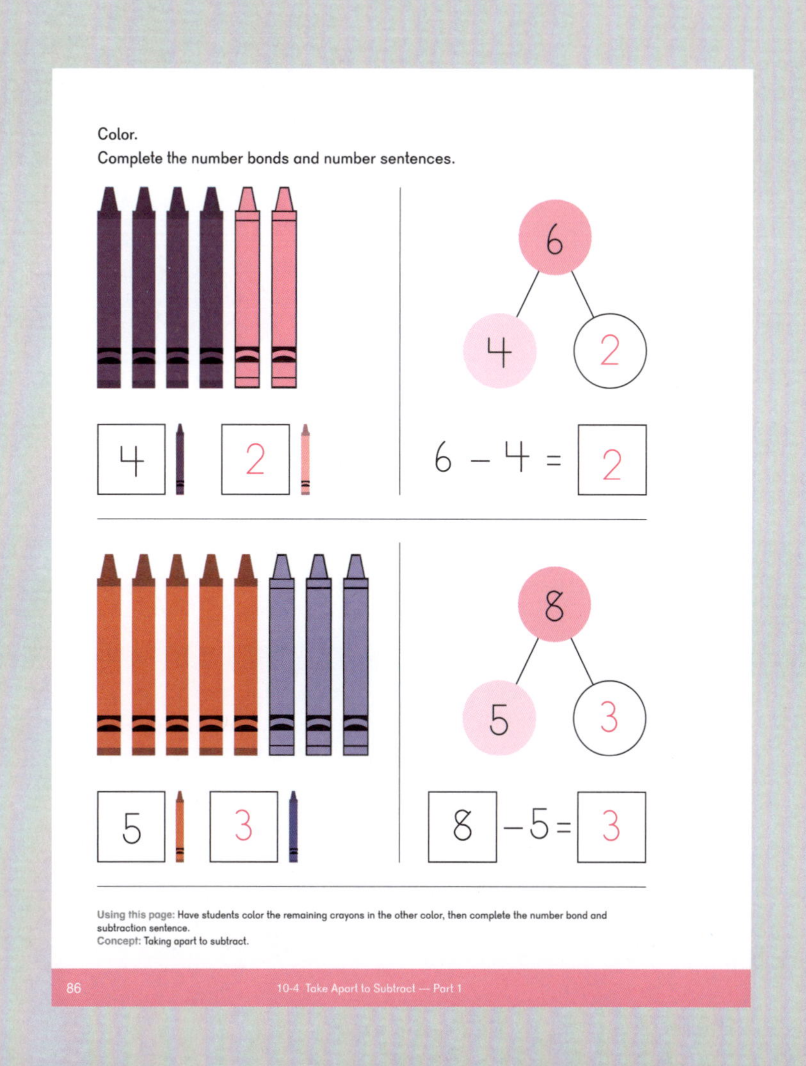

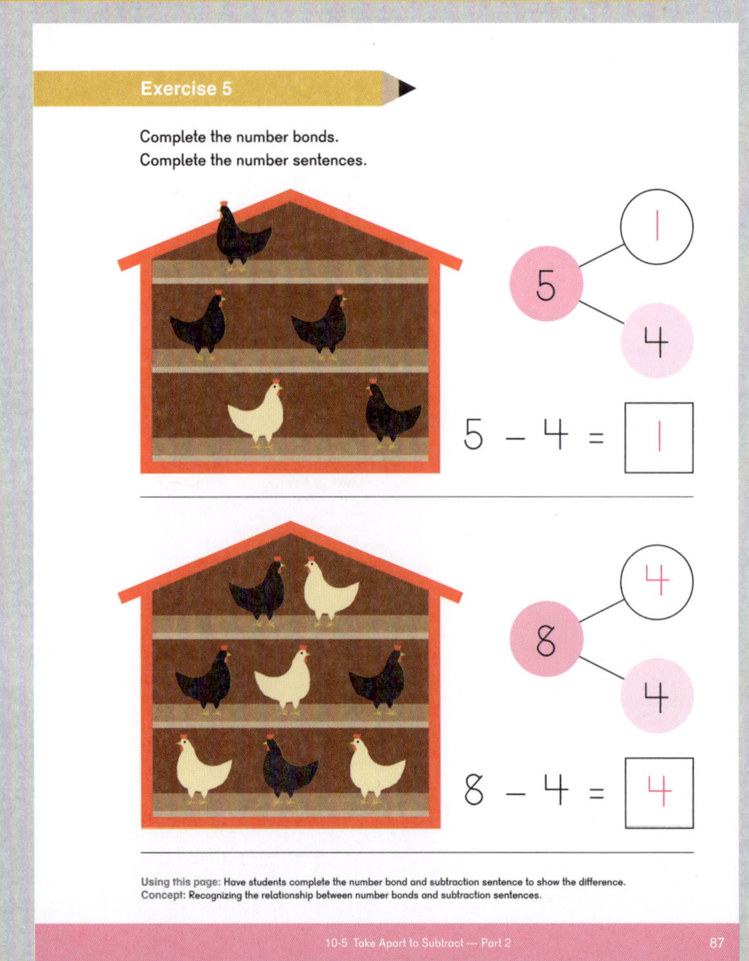
Exercise 5
Complete the number bonds.
Complete the number sentences.
5
1
4
5 – 4 = 1
8
4
4
8 – 4 = 4
Using this page: Have students complete the number bond and subtraction sentence to show the difference.
Concept: Recognizing the relationship between number bonds and subtraction sentences.
10-5 Take Apart to Subtract — Part 2
87

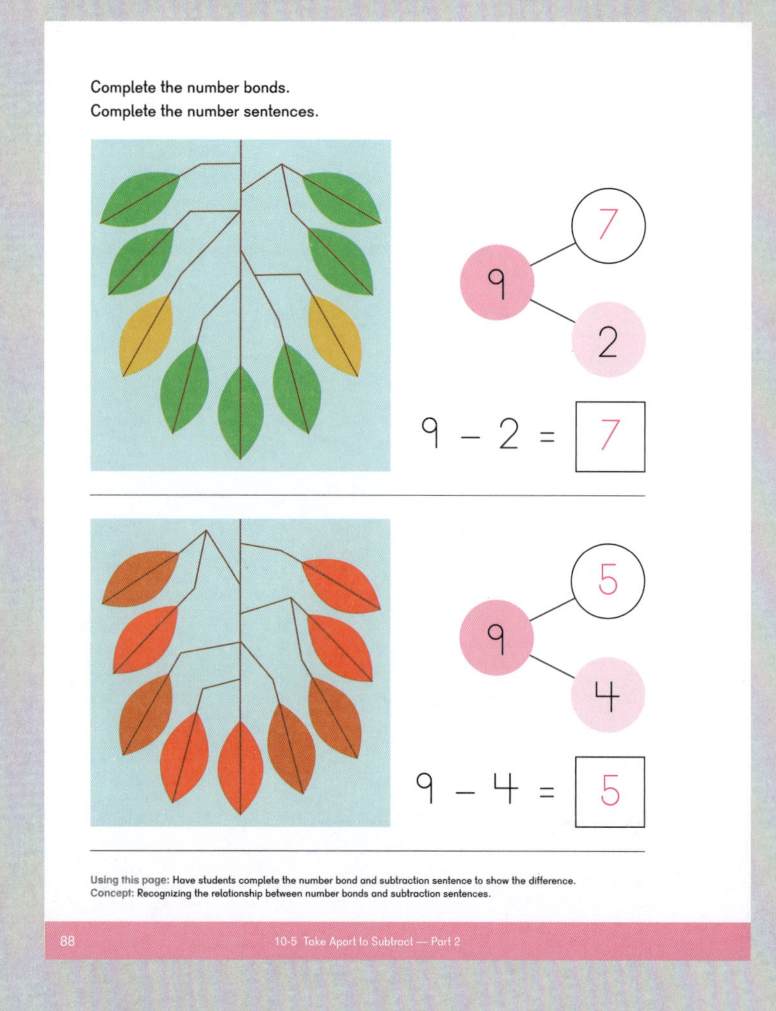
Complete the number bonds.
Complete the number sentences.
9
7
2
9 – 2 = 7
9
5
4
9 – 4 = 5
Using this page: Have students complete the number bond and subtraction sentence to show the difference.
Concept: Recognizing the relationship between number bonds and subtraction sentences.
88
10-5 Take Apart to Subtract — Part 2

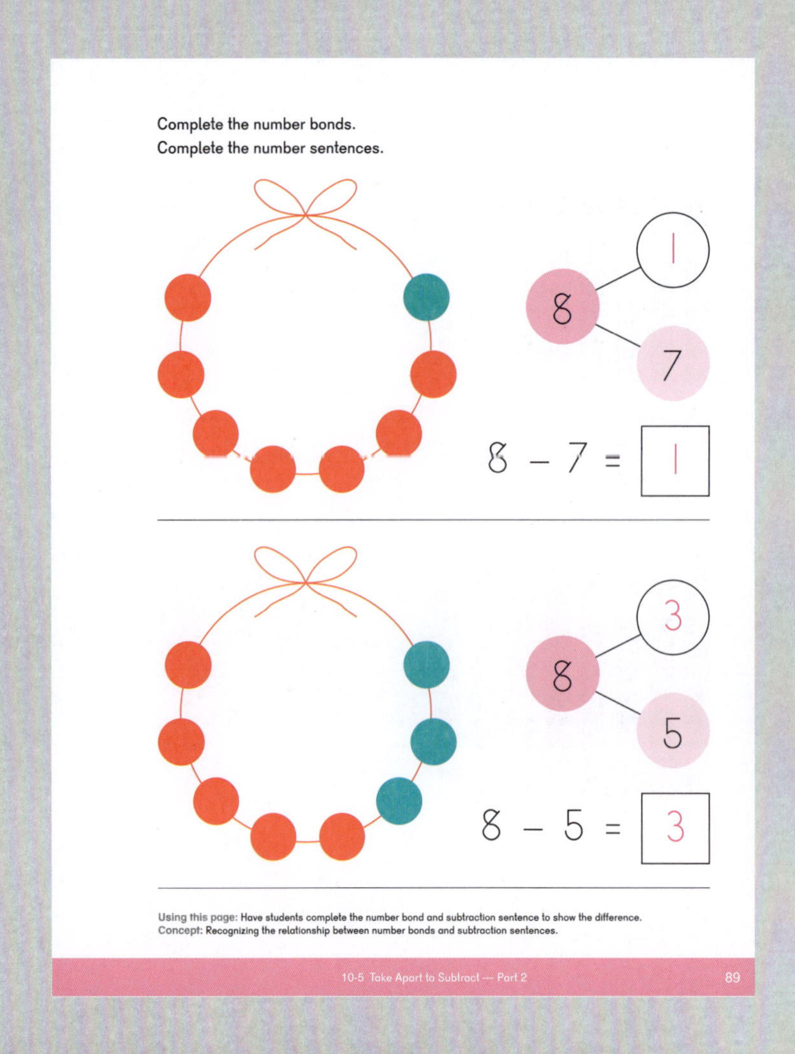
Complete the number bonds.
Complete the number sentences.
8
1
7
8 – 7 = 1
8
3
5
8 – 5 = 3
Using this page: Have students complete the number bond and subtraction sentence to show the difference.
Concept: Recognizing the relationship between number bonds and subtraction sentences.
10-5 Take Apart to Subtract — Part 2
89

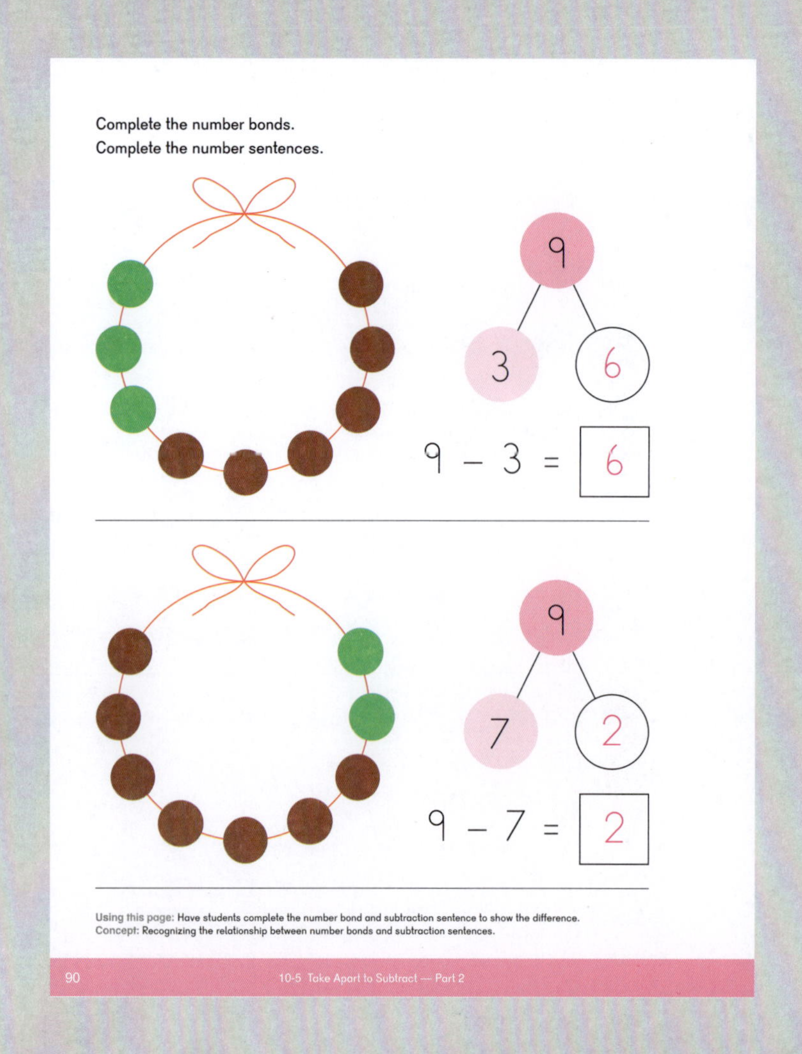
Complete the number bonds.
Complete the number sentences.
9
3
6
9 – 3 = 6
9
7
2
9 – 7 = 2
Using this page: Have students complete the number bond and subtraction sentence to show the difference.
Concept: Recognizing the relationship between number bonds and subtraction sentences.
90
10-5 Take Apart to Subtract — Part 2

Exercise 6 • pages 91–92

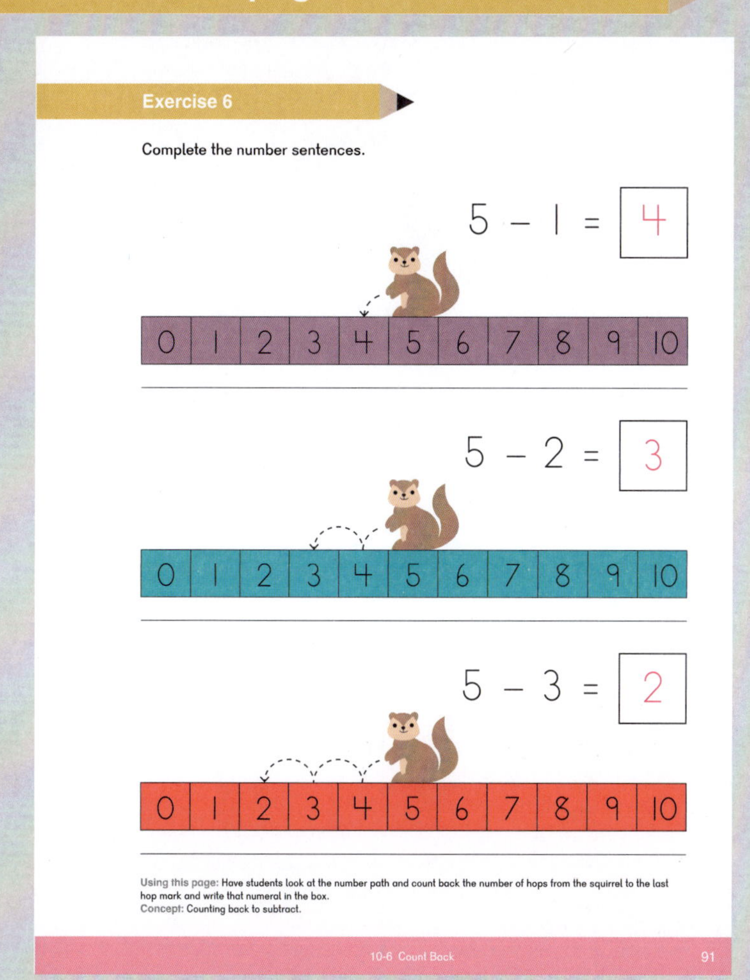

Exercise 6

Complete the number sentences.

5 − 1 = 4

0 1 2 3 4 5 6 7 8 9 10

5 − 2 = 3

0 1 2 3 4 5 6 7 8 9 10

5 − 3 = 2

0 1 2 3 4 5 6 7 8 9 10

Using this page: Have students look at the number path and count back the number of hops from the squirrel to the last hop mark and write that numeral in the box.
Concept: Counting back to subtract.

10-6 Count Back 91

Complete the number sentences.

9 − 3 = 6

0 1 2 3 4 5 6 7 8 9 10

8 − 3 = 5

0 1 2 3 4 5 6 7 8 9 10

7 − 3 = 4

0 1 2 3 4 5 6 7 8 9 10

6 − 3 = 3

0 1 2 3 4 5 6 7 8 9 10

Using this page: Have students look at the number path and count back three hops from the ant to the last hop mark and write that numeral in the box.
Concept: Counting back to subtract.

92 10-6 Count Back

Exercise 7 • pages 93–94

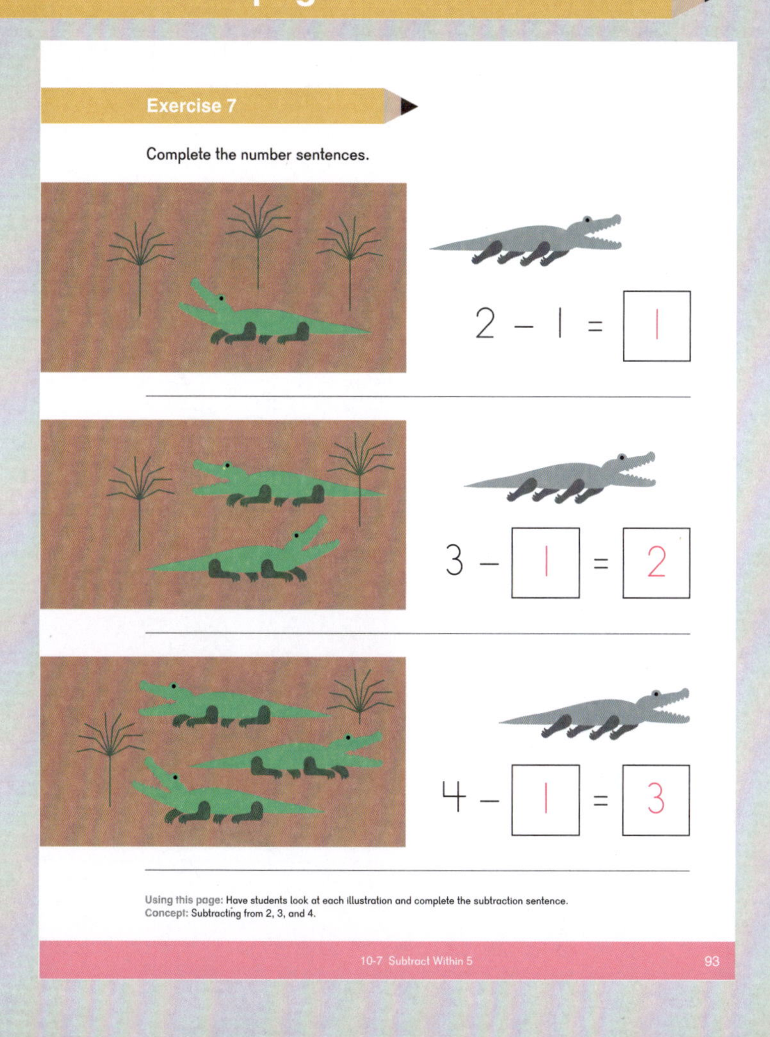

Exercise 7

Complete the number sentences.

2 − 1 = 1

3 − 1 = 2

4 − 1 = 3

Using this page: Have students look at each illustration and complete the subtraction sentence.
Concept: Subtracting from 2, 3, and 4.

10-7 Subtract Within 5 93

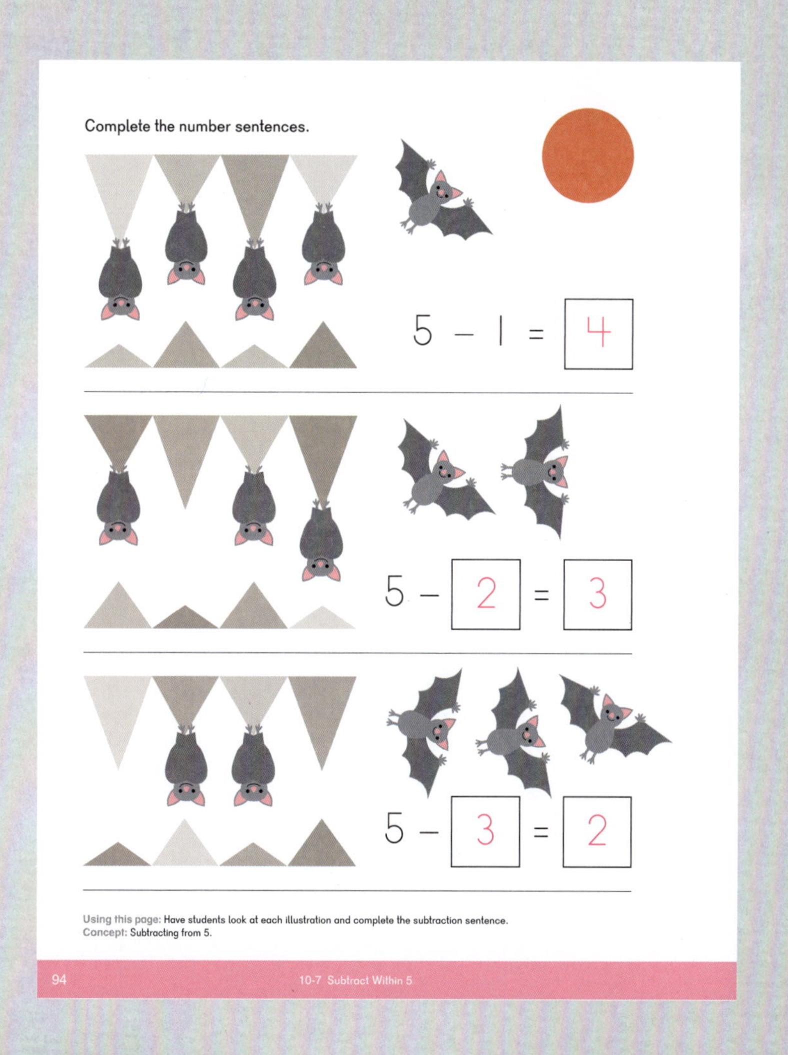

Complete the number sentences.

5 − 1 = 4

5 − 2 = 3

5 − 3 = 2

Using this page: Have students look at each illustration and complete the subtraction sentence.
Concept: Subtracting from 5.

94 10-7 Subtract Within 5

Exercise 8 • pages 95–96

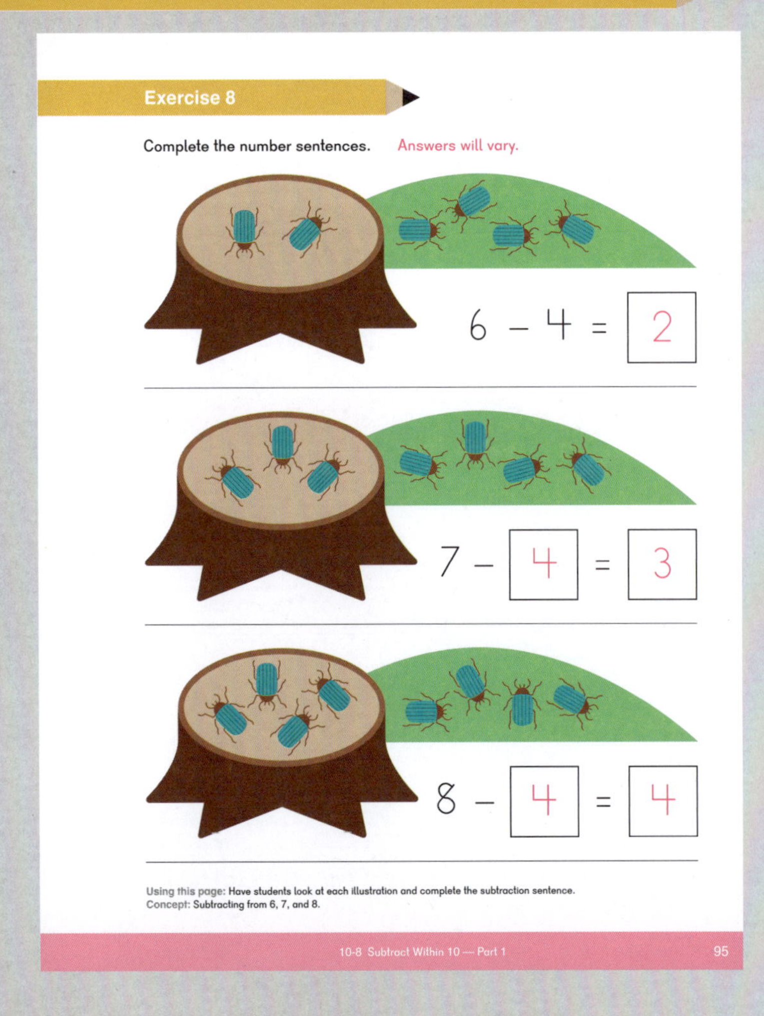

Exercise 8

Complete the number sentences. Answers will vary.

6 – 4 = 2

7 – 4 = 3

8 – 4 = 4

Using this page: Have students look at each illustration and complete the subtraction sentence.
Concept: Subtracting from 6, 7, and 8.

10-8 Subtract Within 10 — Part 1 95

Complete the number sentences. Answers will vary.

9 – 3 = 6

9 – 4 = 5

9 – 5 = 4

Using this page: Have students look at each illustration and complete the subtraction sentence.
Concept: Subtracting from 9.

96 10-8 Subtract Within 10 — Part 1

Exercise 9 • pages 97–98

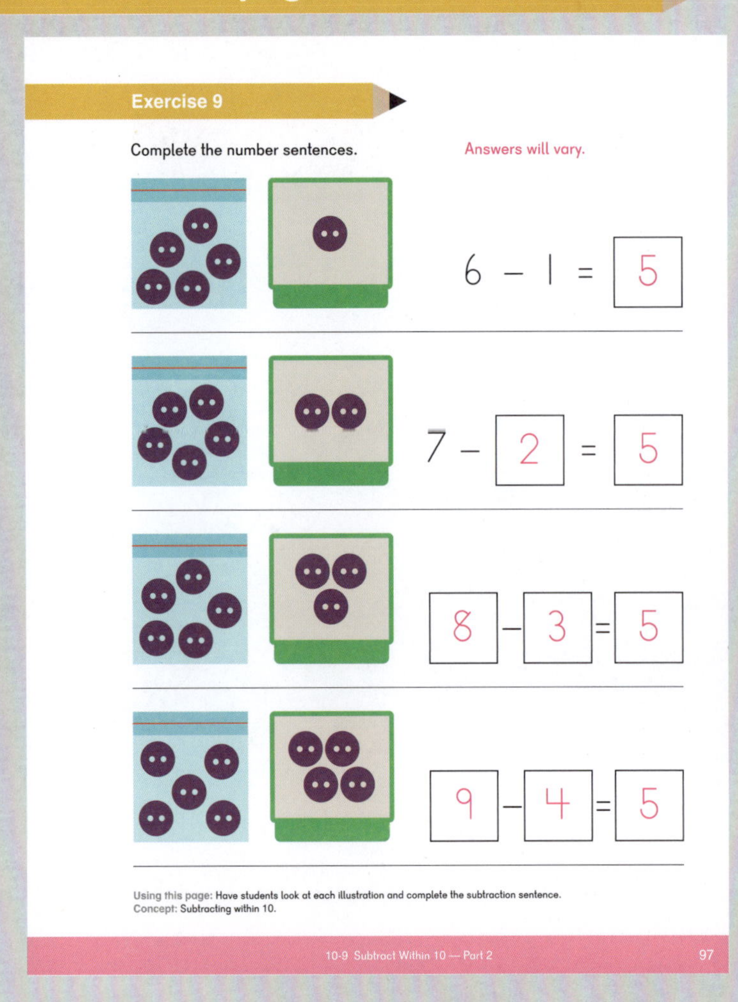

Exercise 9

Complete the number sentences. Answers will vary.

6 – 1 = 5

7 – 2 = 5

8 – 3 = 5

9 – 4 = 5

Using this page: Have students look at each illustration and complete the subtraction sentence.
Concept: Subtracting within 10.

10-9 Subtract Within 10 — Part 2 97

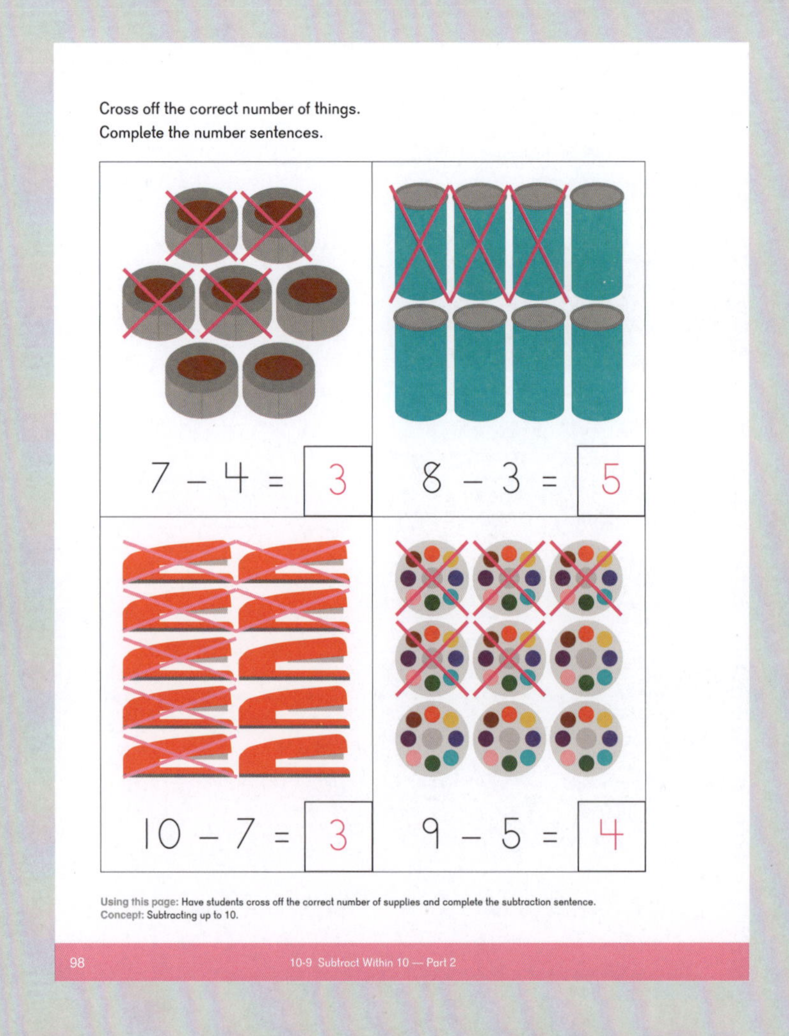

Cross off the correct number of things.
Complete the number sentences.

7 – 4 = 3

8 – 3 = 5

10 – 7 = 3

9 – 5 = 4

Using this page: Have students cross off the correct number of supplies and complete the subtraction sentence.
Concept: Subtracting up to 10.

98 10-9 Subtract Within 10 — Part 2

Exercise 10 • pages 99–102

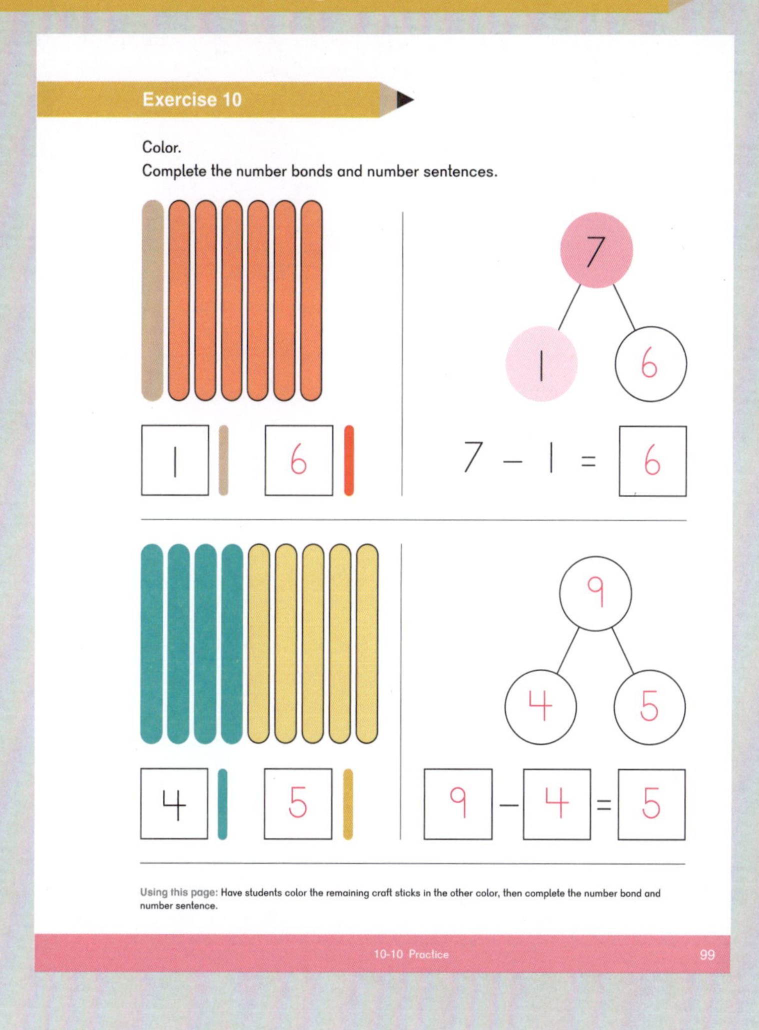

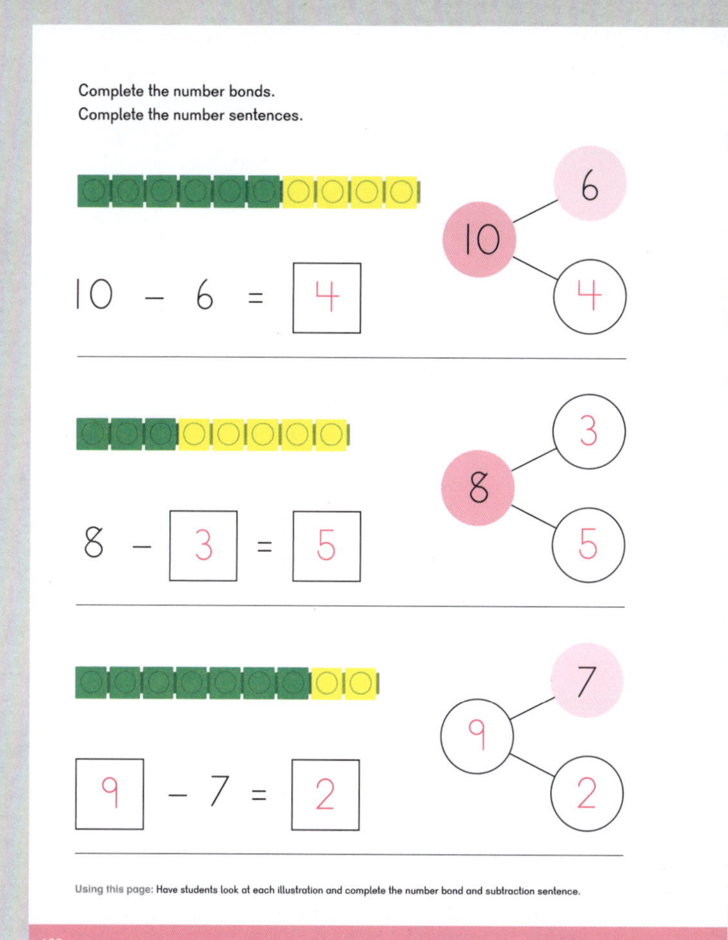

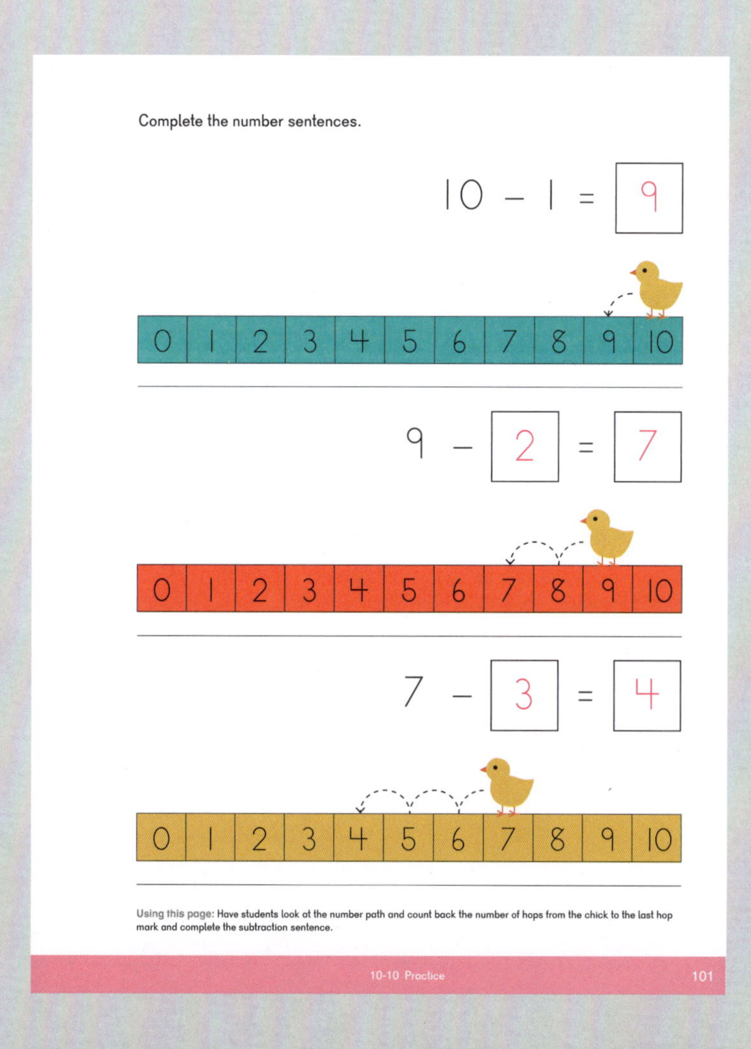

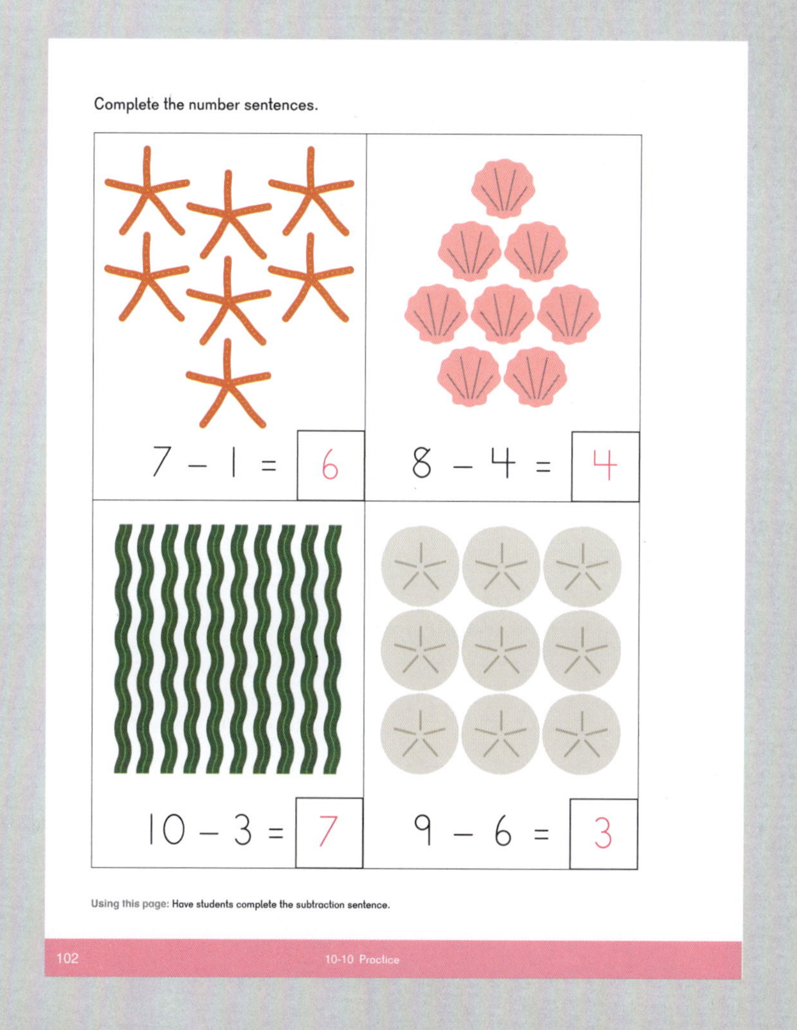

Notes

Suggested number of class periods: 7–8

	Lesson	Page	Resources	Objectives
	Chapter Opener	p. 157	TB: p. 125	
1	Add and Subtract	p. 158	TB: p. 126 WB: p. 103	Compare situations that involve addition to situations that involve subtraction and understand the difference. Recognize the relationship between four addition and subtraction number sentences that can be written for a given picture that depicts a part-whole situation.
2	Practice Addition and Subtraction	p. 160	TB: p. 128 WB: p. 105	Recognize the four related addition and subtraction sentences that can be written for a part-whole situation.
3	Part-Whole Addition and Subtraction	p. 161	TB: p. 129 WB: p. 107	Distinguish between addition and subtraction part-whole situations.
4	Add to or Take Away	p. 163	TB: p. 131 WB: p. 109	Distinguish between addition and subtraction situations involving adding on or taking away.
5	Put Together or Take Apart	p. 165	TB: p. 133 WB: p. 111	Distinguish between addition and subtraction part-whole situations.
6	Practice	p. 166	TB: p. 134 WB: p. 113	Practice skills from the chapter.
	Workbook Solutions	p. 168		

Notes

Students previously learned concretely, using objects, that two quantities (parts) can combine to make a new quantity (the whole). They have also learned that a quantity, or whole, can be split into two quantities (parts). The two parts and the whole form a number bond. Students first represented this idea pictorially and then with numerals. In the previous two chapters, they extended this concept to the more abstract representation of an equation with symbols for addition, subtraction, and equality. Additionally, students learned to relate situations involving adding on and taking away to a part-whole concept.

In this chapter, they will learn about the relationship between addition and subtraction, and be able to represent this relationship as the abstract written equation.

Students will work further with all of these concepts again in Grade 1, but those who can grasp this material in Kindergarten will be well prepared to recall and solidify their understanding in Grade 1. Any games and activities from the previous chapters in **Dimensions Math® KB** can also be played throughout this chapter.

It is assumed that all students will have access to recording tools. When a lesson refers to a whiteboard, any writing materials can be used.

Materials

- 10-sided die
- 20 Chart
- 6-sided die
- Counters
- Die with modified sides: +, +, +, −, −, −
- Dot painters
- Drawing tools
- Recording sheet
- Set of dominoes up to fives
- Stickers
- Two-color counters
- Whiteboards

Blackline Masters

- Blank Number Bond Template
- Game Board to 20
- Number Bond Fact Cards
- Number Bonds for 10
- Number Cards

Letters Home

- Chapter 11 Letter

Notes

Lesson Materials

- Counters
- Blank Number Bond Template (BLM)

Chapter 11

Addition and Subtraction

125

Ask students to look at the two pictures on page 125. Discuss what is different and which each represents – addition or subtraction.

In the top figure, by moving to the right on the number path, Sofia is "counting on" to add a quantity.

In the bottom figure, by moving to the left on the number path to a lesser number, Alex is "counting back" to subtract a quantity.

Ask students to write a number sentence for both situations, then have them represent the situations with counters on a Blank Number Bond Template (BLM). In each case, ask what number is the whole and which are the parts. In Sofia's case, 3 is the whole, and in Alex's case, 3 is the part.

Have students tell addition stories within a sum of 10 and have them write addition sentences for each.

Then repeat the exercise with subtraction stories within 10 and tell how to write subtraction sentences for each.

Extend

★ How Many Ways?

Materials: Number Bonds for 10 (BLM)

Using Number Bonds for 10 (BLM) , have students complete number bonds for all the combinations with sums of 10. Ask students to use the number bonds to make addition and subtraction sentences to match each bond.

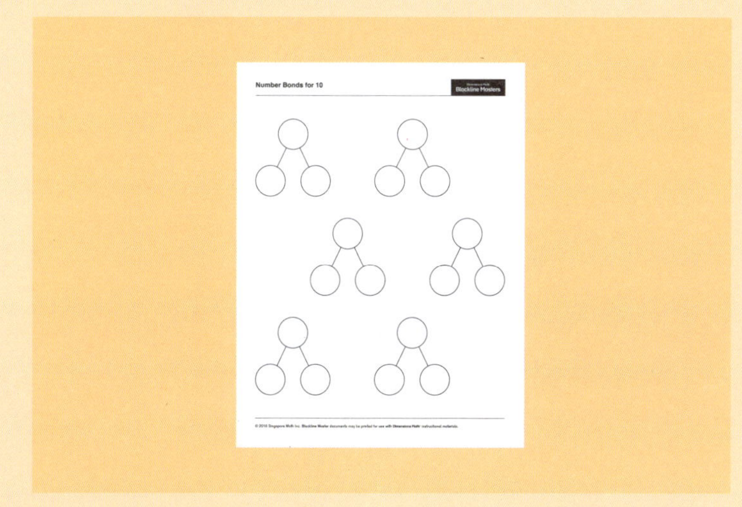

Objectives

- Compare situations that involve addition to situations that involve subtraction and understand the difference.
- Recognize the relationship between four addition and subtraction number sentences that can be written for a given picture that depicts a part-whole situation.

Lesson Materials

- Counters, 8 per student or pair

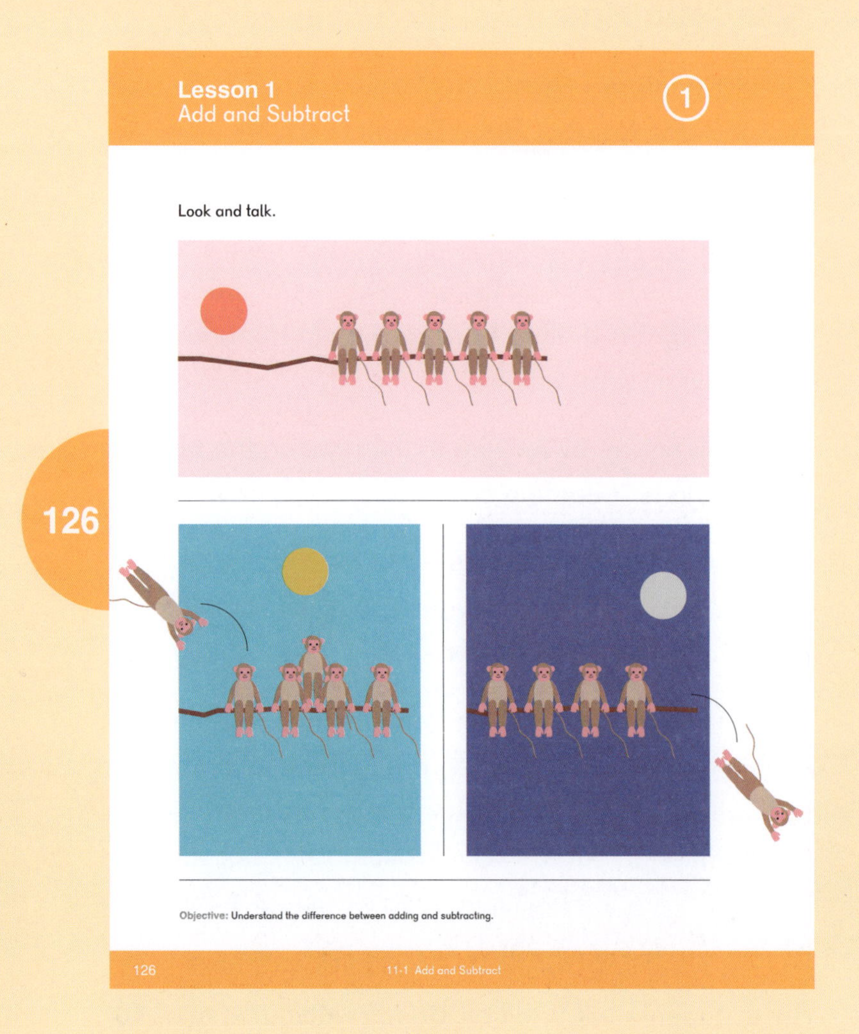

Explore

Provide students or partners with 8 counters and ask them to tell and model an addition story. Ask how they know they are adding. Possible responses:

- I started with 5 in my group and I put 3 more in.
- I started with 5, then at the end I had 8, so I added 3.

Ask students to begin with 8 counters again, but this time model a subtraction story. Ask how they know they are subtracting. You might hear, "I started with 8 in my group and I took some away."

Learn

Discuss page 126. Have students identify which picture shows addition and which shows subtraction. Ask students to tell a story for each picture.

Have them model 1 monkey joining the 5 monkeys with counters and say the addition sentence, "There were 5 monkeys on a branch. 1 monkey joins them. There are 6 monkeys in all. $5 + 1 = 6$."

Repeat for the monkey going away. Say, "There were 5 monkeys on a branch. 1 monkey went away. There are 4 monkeys still on the branch. $5 - 1 = 4$."

For each situation, ask them to look at the answer or final number. Is it greater or less than the number they started with?

Whole Group Activity

▲ Addition or Subtraction

Have students take turns telling number stories involving either addition or subtraction. After hearing a story, ask students to make either the plus sign or the minus sign with their hands to show if the story was an addition story or a subtraction story.

Small Group Activities

Textbook Page 127. Have students tell stories about the picture that relate to the number sentences. For example:

- There are 3 rabbits on top and 4 on the bottom. How many rabbits are there in all?
- There are 4 rabbits with carrots and 3 rabbits with no carrots. How many rabbits are there in all?

- There are 7 rabbits. 4 have carrots. How many don't have carrots?
- There are 7 rabbits. 3 are on top. How many are on the bottom?
- 4 rabbits have carrots. There are 7 rabbits in all. How many rabbits do not have carrots?

▲ Draw the Number Bond

Materials: Number Bond Fact Cards (BLM), and stickers, dot painters, or drawing tools

Have students draw a Number Bond Fact Card (BLM) and illustrate a number story using the numbers on the card. Have students write the addition or subtraction sentence that matches their illustration.

To extend, have students illustrate both an addition and subtraction story to match the number bond card and write the number sentences to match.

▲ Off the Board

Materials: Game Board to 20 (BLM), 6-sided die, die with modified sides labeled: +, +, +, –, –, –

Students place a marker on 10 on the Game Board to 20 (BLM). Students take turns rolling the number die and the +/– die and moving either higher or lower on the number chart based on the roll. The first player to land off the board, either lower than 1 or higher than 20, is the winner.

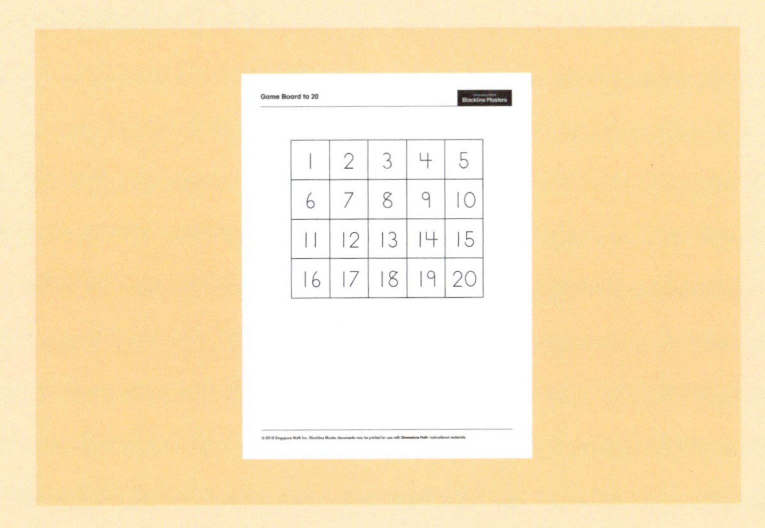

1	2	3	4	5
6	7	8	9	10
11	12	13	14	15
16	17	18	19	20

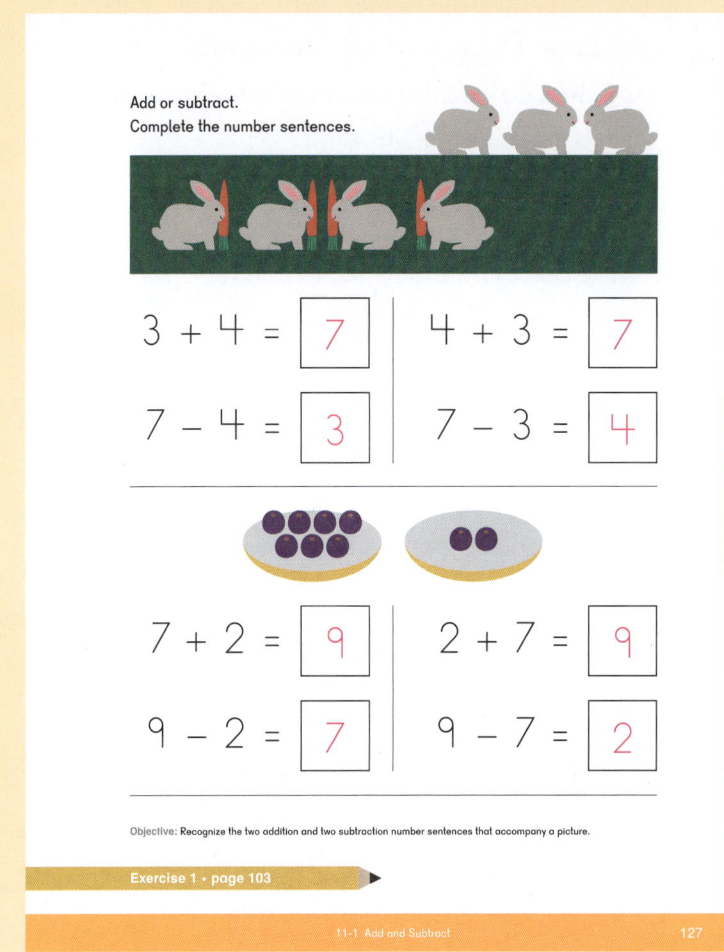

Exercise 1 • page 103

Extend

★ Race to 20

Materials: 10-sided die, die with modified sides: +, +, +, –, –, –

Students start with the number 10. Students take turns rolling the dice and writing an addition or subtraction sentence. For example, on the first roll a student writes 10 + 6 = 16. On the next roll, a student writes 16 – 4 = 12. Students continue to roll and write number sentences until a player reaches 20.

Alternatively, players could race to 0.

Objective

- Recognize the four related addition and subtraction sentences that can be written for a part-whole situation.

Lesson Materials

- Two-colored counters, 6 per student

128

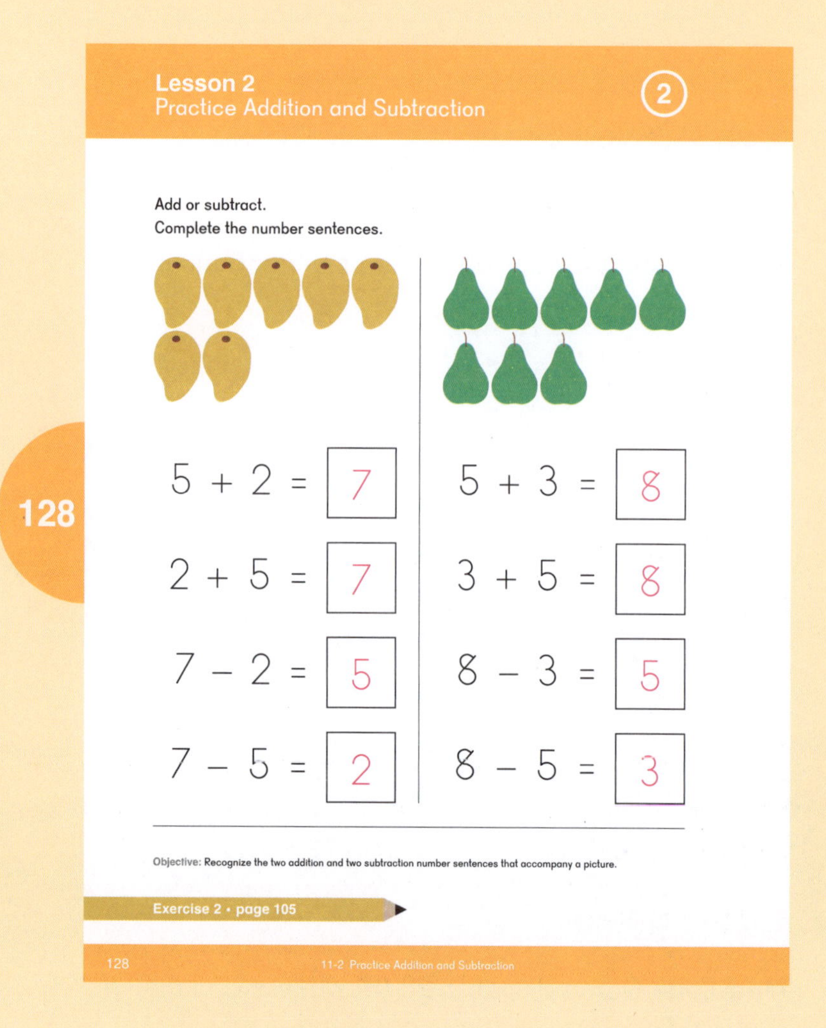

Lesson 2
Practice Addition and Subtraction 2

Add or subtract.
Complete the number sentences.

5 + 2 = 7
2 + 5 = 7
7 − 2 = 5
7 − 5 = 2

5 + 3 = 8
3 + 5 = 8
8 − 3 = 5
8 − 5 = 3

Objective: Recognize the two addition and two subtraction number sentences that accompany a picture.

Exercise 2 • page 105

128 11-2 Practice Addition and Subtraction

Explore

Provide each student with 6 two-color counters. Ask students to drop the counters and tell an addition and a subtraction story based on how they land. For example, a student has 2 red counters and 4 white counters showing. She might say:

- 2 red counters and 4 white counters are 6 counters altogether.
- There were 6 chips and 2 are red, so 4 counters are white.

Learn

As students share their stories, write the number bonds to match their stories. Have students notice that the number bond is the same for both the addition and subtraction story because the whole or total is the same. Looking at the number bond, write two addition and two subtraction sentences to match.

Whole Group Activity

▲ Number Strings

Materials: Two-colored counters

Give students three related addition or subtraction situations. Have students use counters or calculate mentally to determine the answer.

For example, you might say:

- There were 4 dogs at the park. 2 dogs left. 3 more dogs came. How many dogs were at the park at the end?

Small Group Activities

Textbook Page 128

▲ Writing Number Sentences

Materials: Two-colored counters

Students grab a small handful of counters. Students drop the counters and write a number bond and two addition and two subtraction sentences to match the drop.

Exercise 2 • page 105

Extend

★ Writing Number Strings

Have students write number strings similar to the Whole Group Activity for future use.

Lesson 3 Part-Whole Addition and Subtraction

Objective

- Distinguish between addition and subtraction part-whole situations.

Lesson Materials

- Two-color counters
- Blank Number Bond Template (BLM)

Explore

Tell students part-whole number stories.

Examples:

- Dion has 4 cars. Emma has 2 cars. How many cars do they have in all?
- Dion has 5 cars. 3 cars are red and the rest are blue. How many blue cars does Dion have?

Allow students time to develop a visual image of the story being told. Ask students if they know the whole and a part, or both parts. Have students model the story on a Blank Number Bond Template (BLM) with two-color counters. Have students share how they knew whether to add or subtract.

Emphasize that if the number story asks us to find a whole, we add.

If the number story gives a whole and a part, and asks us to find a part, we subtract.

Learn

Look at the basketballs on page 129. Ask students to tell addition and subtraction stories about the balls. With each story, ask students to model with counters and a Blank Number Bond Template (BLM). For an extension, ask students to write the addition or subtraction sentences to match each story. Repeat with the balloons.

129

Whole Group Activity

▲ Add or Subtract?

Tell addition or subtraction stories and ask students to write the addition or subtraction sign on individual whiteboards to show whether they need to add or subtract to find the answer.

Small Group Activities

Textbook Page 130

▲ Draw the Number Sentence

Materials: 10-sided die or Number Cards (BLM) 1 to 10, die with modified sides: +, +, +, –, –, –

Have students roll the dice, or draw 2 Number Cards (BLM), and the die modified with the addition and subtraction sign. Using the roll, have students create and illustrate the number sentence with pictures and/or words.

Students can share with a partner, or the pages can be saved and put into a classroom book.

▲ Off the Board

Materials: Game Board to 20 (BLM), 6-sided die, die with modified sides labeled: +, +, +, –, –, –

Students place a marker on 10 on the Game Board to 20 (BLM). Students take turns rolling the number die and the +/– die and moving either higher or lower on the number chart based on the roll. The first player to land off the board, either lower than 1 or higher than 20, is the winner.

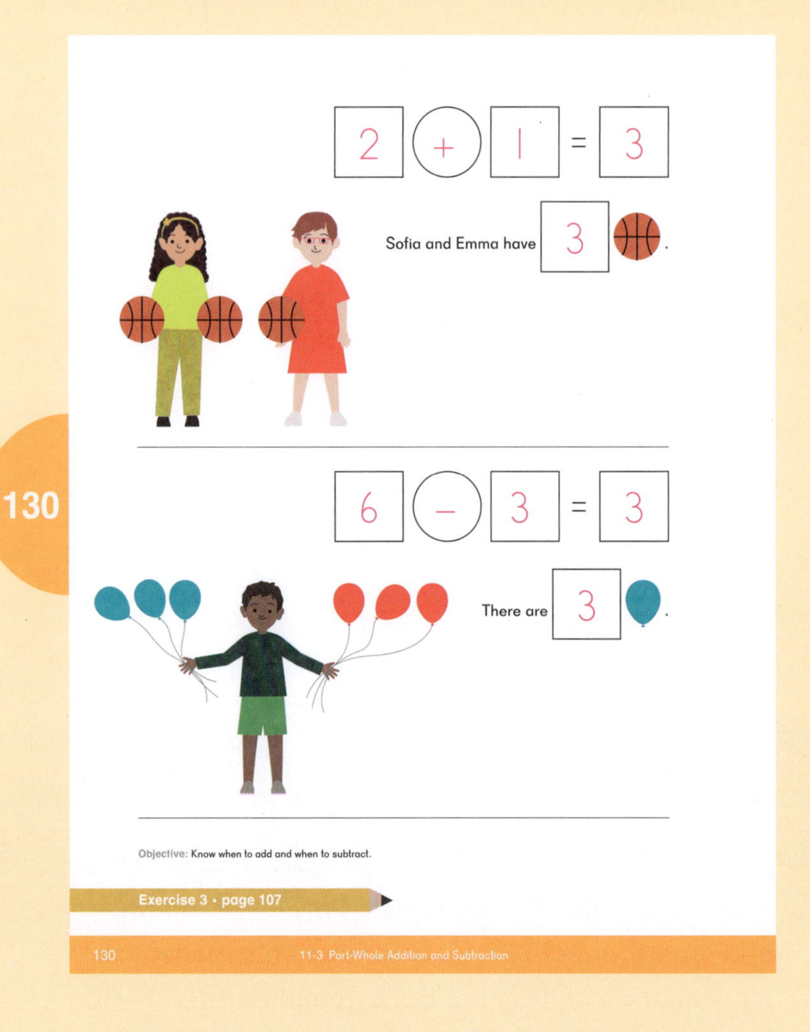

Exercise 3 • page 107

Extend

★ Salute!

Materials: 4 sets of Number Cards (BLM) 0 to 5

Salute! is played with 3 students. Players shuffle and split the cards between two of the players. The third player is the Caller.

When the Caller says, "Salute!" the players place the top card from their piles on their foreheads to salute each other. Players can't see their own cards.

The Caller tells the players the total or "whole" of the two numbers on the cards. (Think of the three players as a number bond.)

The players hear the whole and subtract the other's number to find their own.

The player who says her missing part first is the winner. Winners can collect the two cards or players can play through their piles or take turns being the Caller.

Objective

- Distinguish between addition and subtraction situations involving adding on or taking away.

Lesson Materials

- Counters

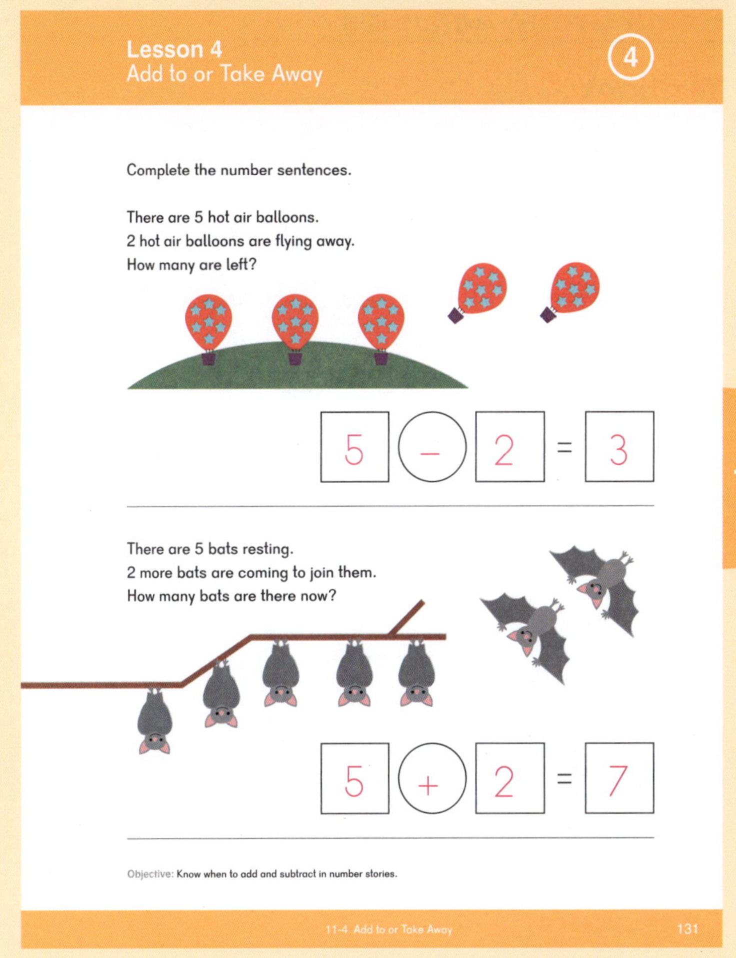
Lesson 4
Add to or Take Away
4

Complete the number sentences.

There are 5 hot air balloons.
2 hot air balloons are flying away.
How many are left?

5 – 2 = 3

There are 5 bats resting.
2 more bats are coming to join them.
How many bats are there now?

5 + 2 = 7

Objective: Know when to add and subtract in number stories.

11-4 Add to or Take Away 131

131

Explore

Tell students a number story where the initial quantity changes. Have them act out the story with counters.

- Alex has 1 book and Mei gives him 2 books. How many books does Alex have in all?
- Mei had 5 library books. She returned some of them to the library. She kept 2 books to read. How many books did Mei return to the library?

Learn

Have students share how they knew whether to add or subtract in the stories from **Explore**. Discuss that in these types of stories, something is changing: There were some books and my friend gave me more books. I had some books and I gave some away.

Have students write number bonds and number sentences on a whiteboard for the stories in **Explore**. Ensure that they are writing the correct signs for addition and subtraction.

Discuss additional problems with students. Guide students to think about the parts and wholes in these problems. If one part is going away, the whole is the initial quantity. Similarly, if a part is joining another part, the whole is the initial quantity and the amount that joined.

- I have 10 dogs. 3 leave. Some are still there. Is the whole the 10 dogs, the 3 that leave, or the dogs that are still there?
- I have 5 pens. Clara brings 3 more. What's the whole? Is it Clara's part of 3 pens, or my part 5 pens? Or something else?

Whole Group Activity

▲ Student Number Bonds

Call a group of 8 students to the front of the class. Tell either an addition story with a sum of 8 or a subtraction story that starts with 8. Have students arrange themselves into either the two parts or the whole and one part of a number bond to match the story. Ask the remaining students to hold up fingers to show the missing part.

Small Group Activities

Textbook Pages 131–132

▲ How Did it Change?

Materials: Between 1 and 10 counters per student

Students work in pairs. Player 1 shows a number of counters between 1 and 10. Player 2 counts the counters, records the number and then closes his eyes. Player 1 removes or adds to the counters. Player 2 opens his eyes and determines how the group changed. Did Player 1 add to or take away from the group? How many were added or taken away? Player 2 completes the number sentence determining if the initial quantity was added to or subtracted from.

▲ Make it the Same

Materials: 4 sets of Number Cards (BLM) 0 to 10, counters

Provide students with 4 sets of Number Cards (BLM) 0 to 10. Students shuffle the cards and place them in a pile between them. Each player chooses a card and flips it faceup. Students compare cards and determine what change needs to happen so that they will have the same. For example, students flip a 4 and a 7. The player with the 4 says, "I need to add 3 more to make 7," and the player with the 7 needs to say, "I need to take away 4 to make 3." Students could use counters for assistance.

Exercise 4 • page 109

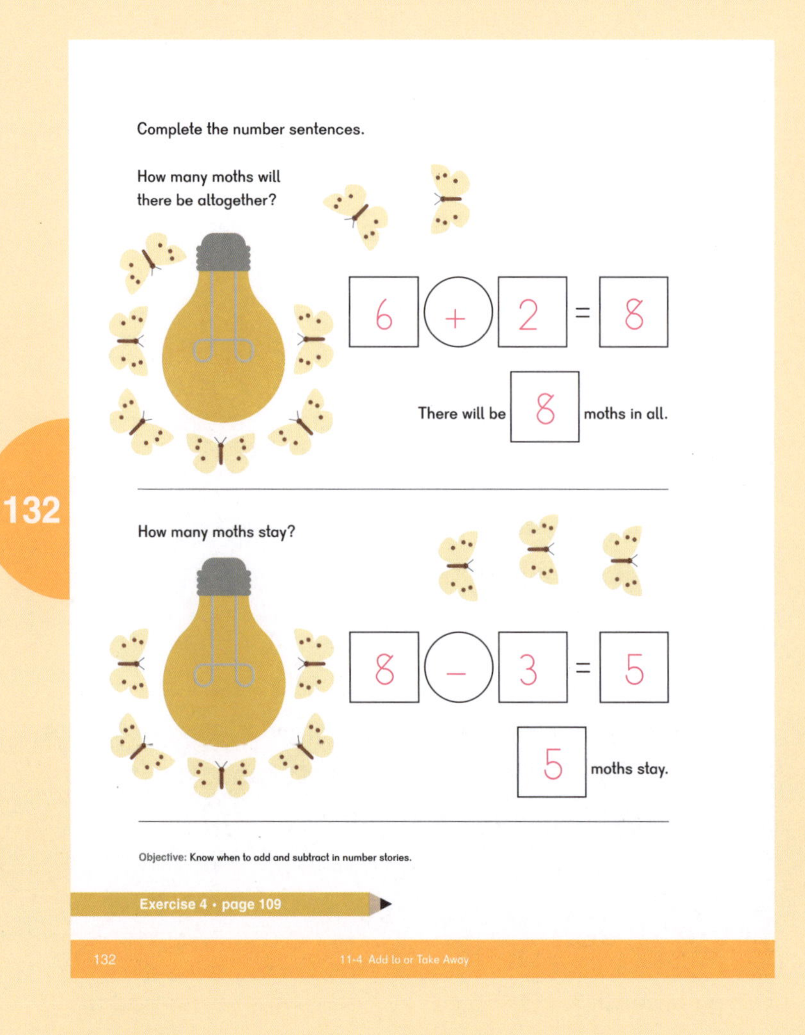

Extend

★ Choice Day

Students can choose to do any of the extend activities from the chapter.

- **Race to 20** (page 159 of this Teacher's Guide)
- **Number Strings** (page 160 of this Teacher's Guide)
- **Salute!** (page 162 of this Teacher's Guide)

Objective

- Distinguish between addition and subtraction part-whole situations.

Lesson Materials

- Counters

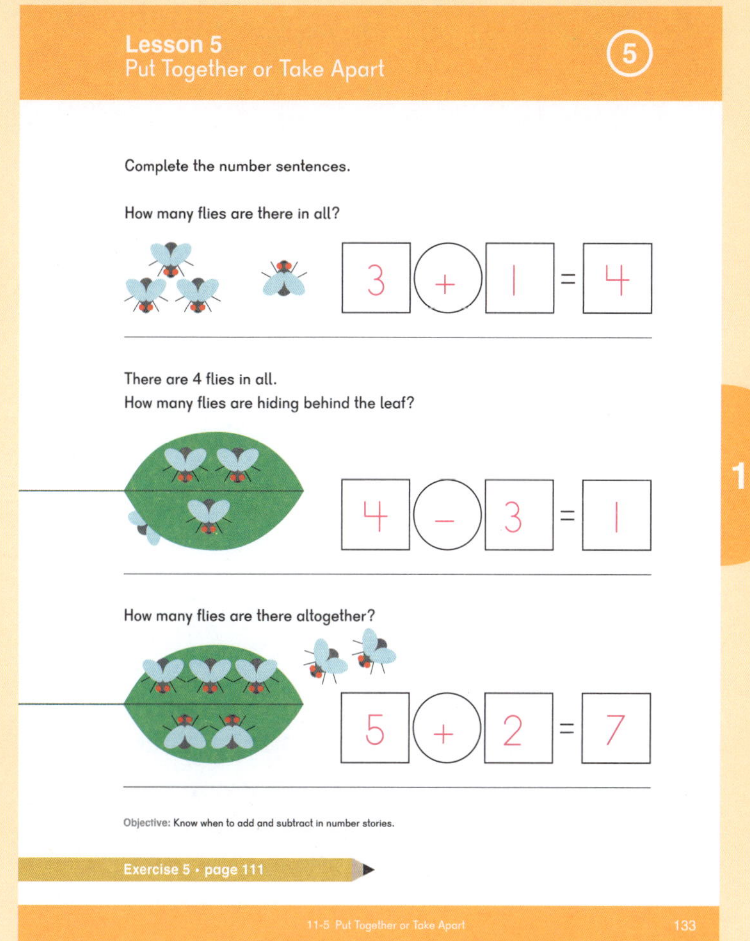

Explore

Tell students put-together and take-apart number stories. Ask students to model the stories with counters and have them share how they knew whether to add or subtract. Examples:

- Sofia has 1 storybook and 2 math books. How many books does she have in all?
- Emma has 5 books. 3 are math books and the rest are storybooks. How many storybooks does Emma have?

Learn

Discuss that the total number of items in the stories from **Explore** always stay the same and they are grouped in different ways. They could be put into 2 groups by size, color, type of object, etc., but nothing is coming or going in the story.

Repeat the number stories used in **Explore**, and have students write number sentences on a whiteboard. Ensure they are writing the correct sign for addition and subtraction.

Whole Group Activity

▲ Group Up Stories

Ask students to group themselves by any number between 2 and 10. Once in groups, students notice things about them that are the same or different, and tell a number story about their group. For example, a group might say:

- There are 3 girls and 4 boys. How many in all?
- There are 7 in all and 3 are girls. How many boys are there?

Students who aren't in the number story group can write the equations on a whiteboard.

Small Group Activities

Textbook Page 133

▲ Greatest Sum

Materials: 4 sets of Number Cards (BLM) 0 to 5

One player shuffles and deals a deck of 4 sets of Number Cards (BLM) from 0 to 5 facedown. Players flip the top two cards from their pile and find the sum. The player with the greatest sum in each round wins the cards.

Exercise 5 • page 111

Lesson 6 Practice

Objective

- Practice skills from the chapter.

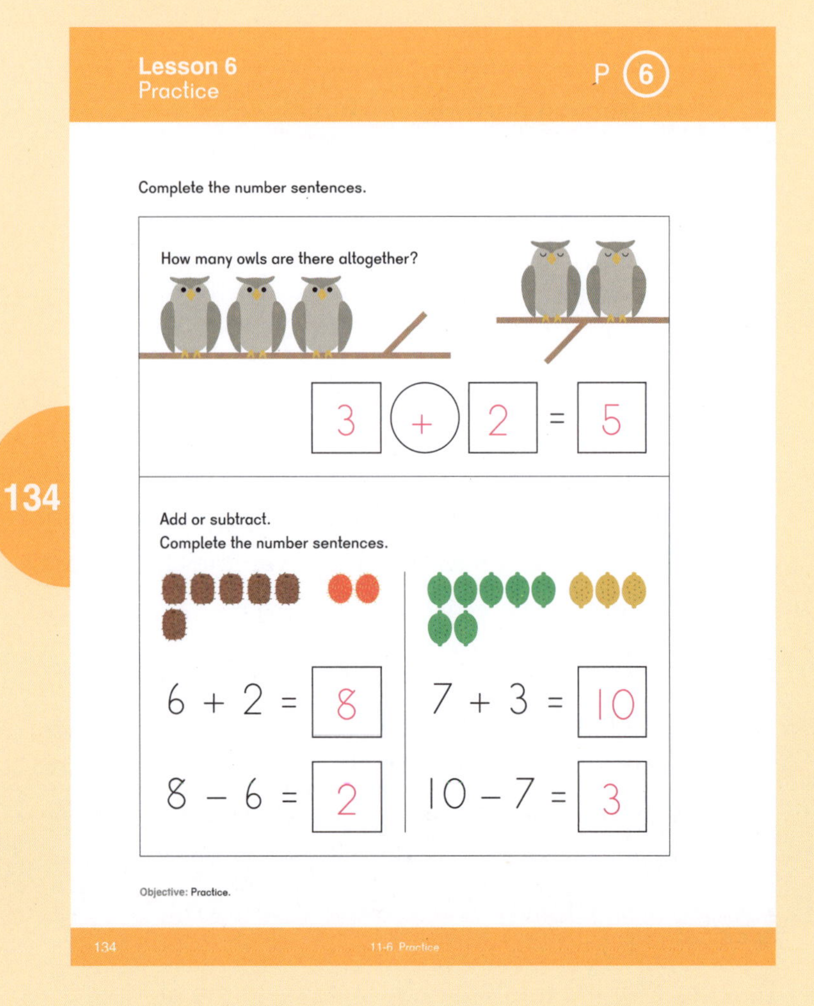

Practice lessons are designed for further practice and assessment as needed.

Students can complete the textbook pages and workbook pages as practice and/or as assessment. Use activities and extensions from the chapter for additional review and practice.

Whole Group Activity

▲ Group Up

Call out a number. Students have 10 seconds to get themselves into groups of that size. It might be impossible for everyone to get in a group every time, but each new number gives everyone another chance.

To begin the activity, say, "Put yourselves into groups of 3." Once students get the idea, call out addition or subtraction problems such as, "Groups of 7 – 4."

Small Group Activity

▲ Domino Equations

Materials: Set of dominoes up to double fives, recording sheet

This activity works well with both partners and larger groups.

Players spread out a set of dominoes facedown, then select one domino each. Players write an equation for the numbers on the two halves of the domino to find the total. The player with the greatest total wins the dominoes.

After five rounds, the player with the most dominoes wins the game.

To extend, have students draw two dominoes and add the total pips on each.

Exercise 6 • page 113

Extend

★ Nutty Number Sentences

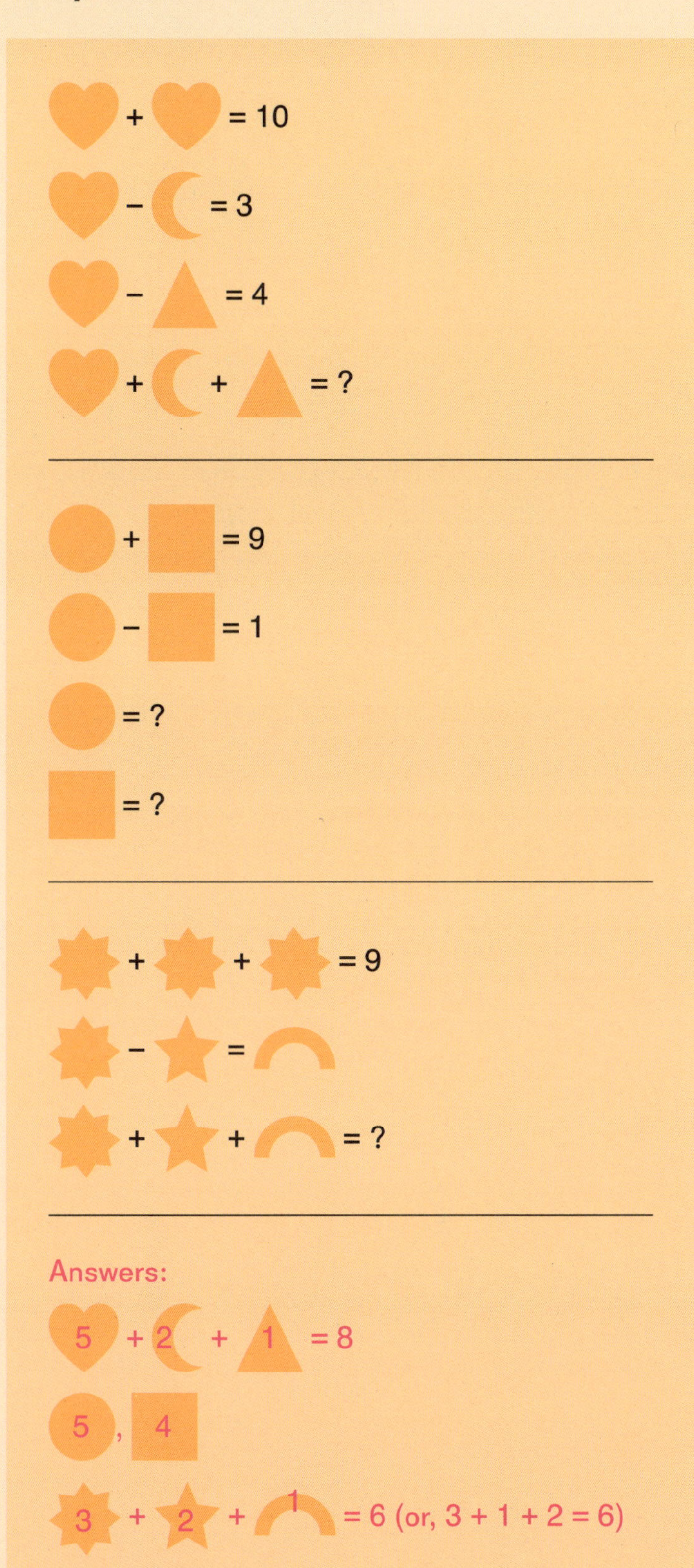

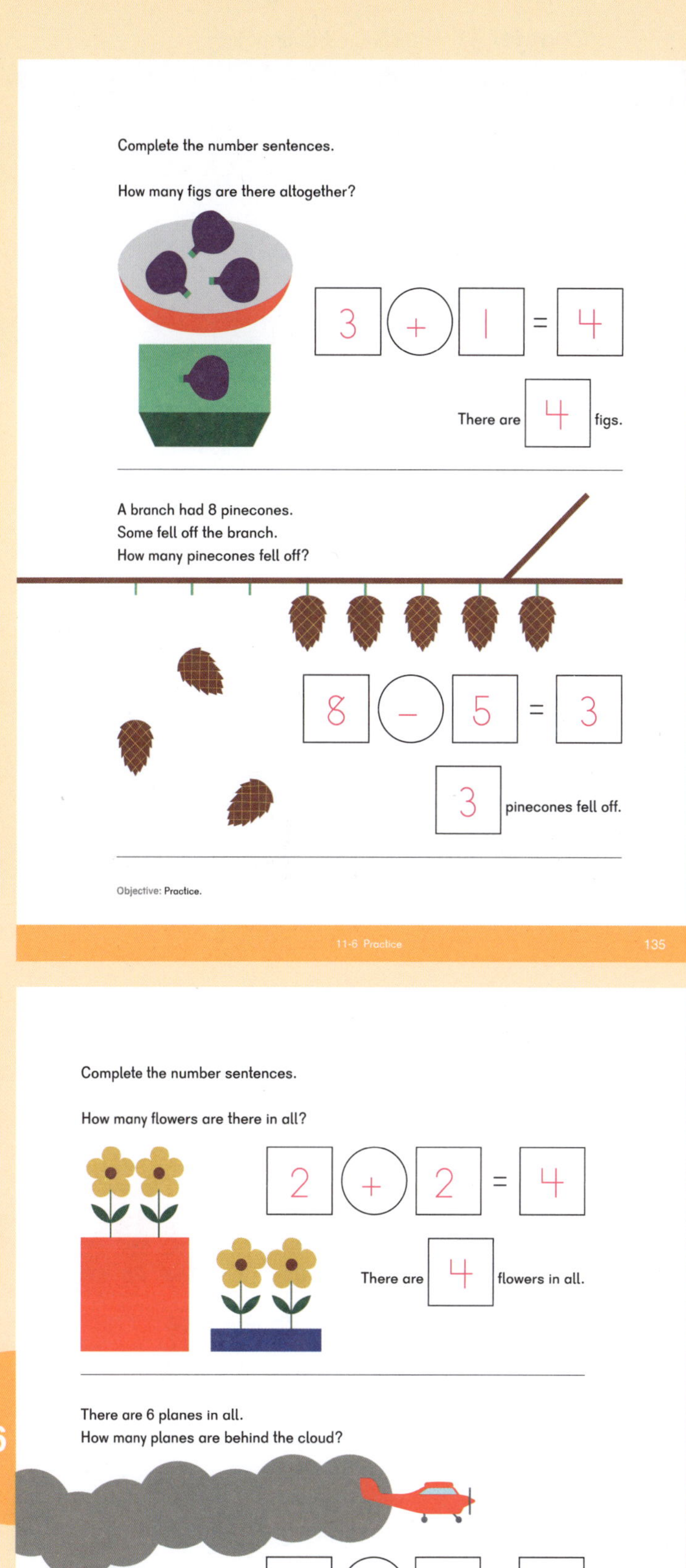

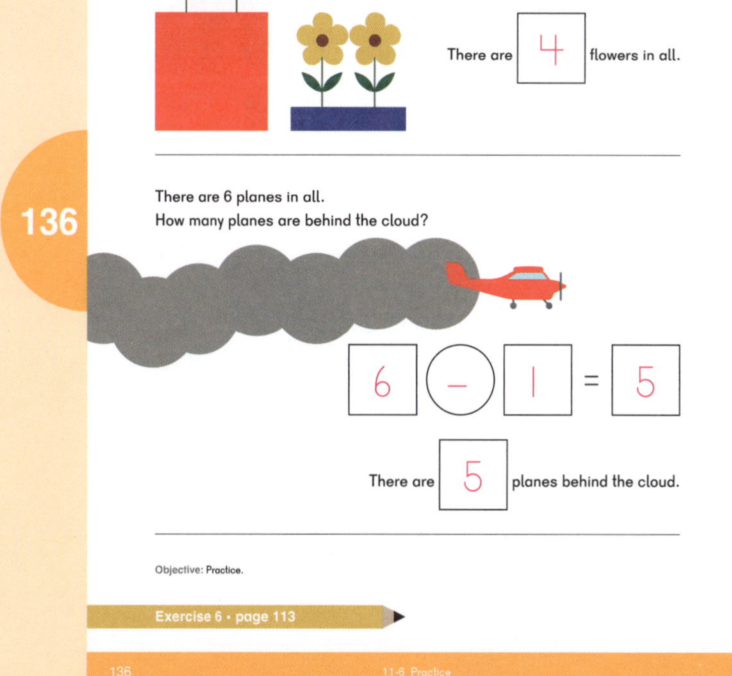

Exercise 1 • pages 103–104

Chapter 11 Addition and Subtraction

Exercise 1

Add or subtract.
Complete the number sentences.

5 + 4 = 9 | 4 + 5 = 9
9 − 4 = 5 | 9 − 5 = 4

6 + 3 = 9 | 3 + 6 = 9
9 − 3 = 6 | 9 − 6 = 3

Using this page: Have students look at the illustration and complete the addition and subtraction sentences.
Concept: Recognizing the two addition and two subtraction number sentences that accompany a picture.

Add or subtract.
Complete the number sentences.

8 + 1 = 9 | 1 + 8 = 9
9 − 1 = 8 | 9 − 8 = 1

7 + 3 = 10 | 3 + 7 = 10
10 − 3 = 7 | 10 − 7 = 3

Using this page: Have students look at the illustration and complete the addition and subtraction sentences.
Concept: Recognizing the two addition and two subtraction number sentences that accompany a picture.

Exercise 2 • pages 105–106

Exercise 2

Add or subtract.
Complete the number sentences.

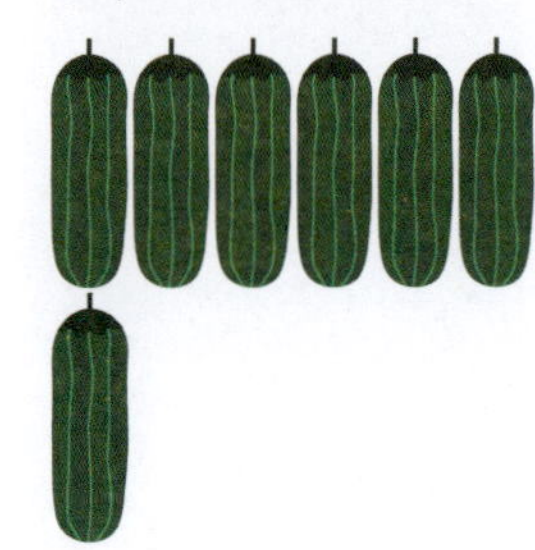

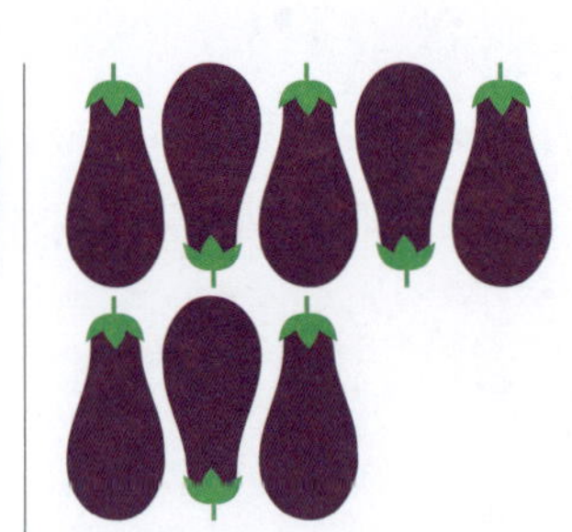

6 + 1 = 7 | 5 + 3 = 8
1 + 6 = 7 | 3 + 5 = 8
7 − 1 = 6 | 8 − 3 = 5
7 − 6 = 1 | 8 − 5 = 3

Using this page: Have students look at the illustration and complete the addition and subtraction sentences.
Concept: Recognizing the two addition and two subtraction number sentences that accompany a picture.

Add or subtract.
Complete the number sentences.

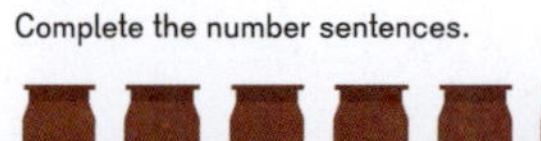

6 + 4 = 10 | 4 + 6 = 10
10 − 4 = 6 | 10 − 6 = 4

9 + 1 = 10 | 1 + 9 = 10
10 − 1 = 9 | 10 − 9 = 1

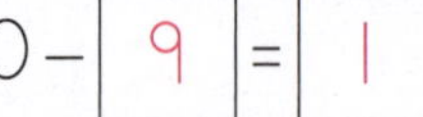

Using this page: Have students look at the illustration and complete the addition and subtraction sentences.
Concept: Recognizing the two addition and two subtraction number sentences that accompany a picture.

Exercise 3 • pages 107–108

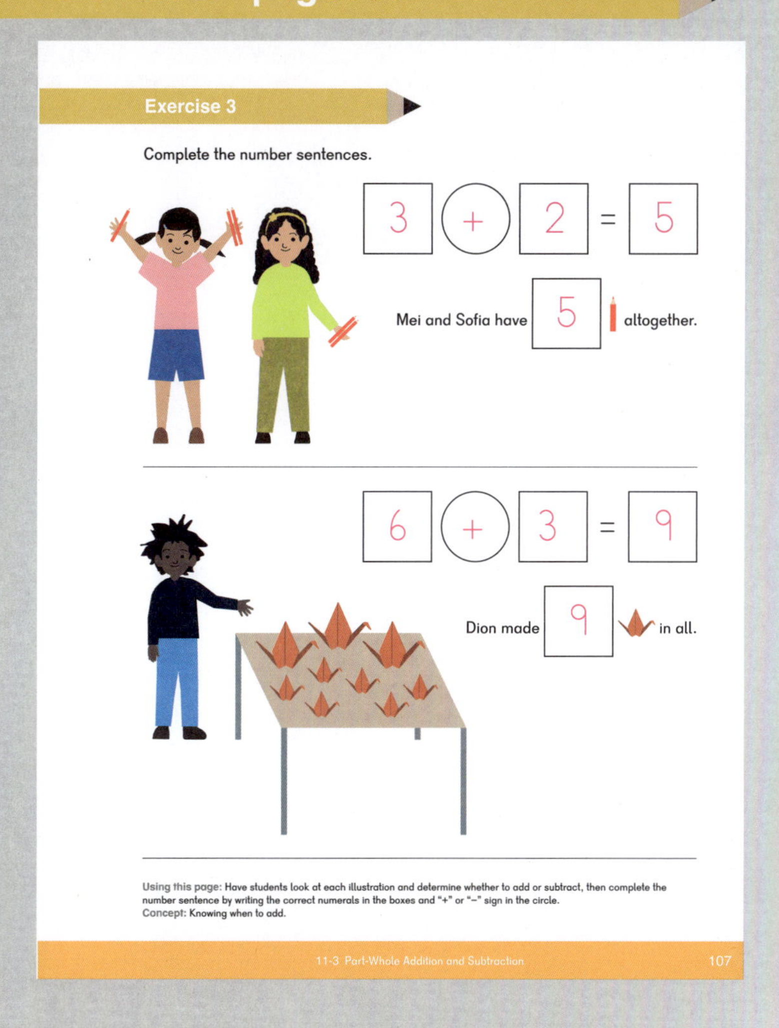

Exercise 3

Complete the number sentences.

3 + 2 = 5

Mei and Sofia have 5 altogether.

6 + 3 = 9

Dion made 9 in all.

Using this page: Have students look at each illustration and determine whether to add or subtract, then complete the number sentence by writing the correct numerals in the boxes and "+" or "–" sign in the circle.
Concept: Knowing when to add.

11-3 Part-Whole Addition and Subtraction 107

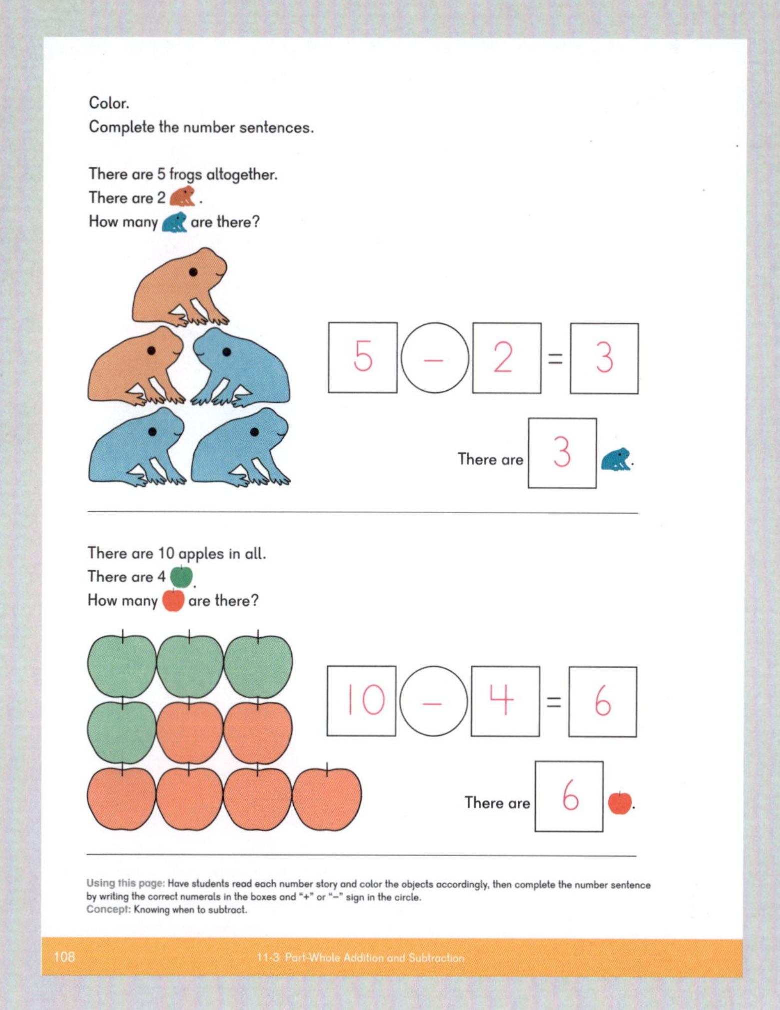

Color.
Complete the number sentences.

There are 5 frogs altogether.
There are 2 .
How many are there?

5 – 2 = 3

There are 3 .

There are 10 apples in all.
There are 4 .
How many are there?

10 – 4 = 6

There are 6 .

Using this page: Have students read each number story and color the objects accordingly, then complete the number sentence by writing the correct numerals in the boxes and "+" or "–" sign in the circle.
Concept: Knowing when to subtract.

108 11-3 Part-Whole Addition and Subtraction

Exercise 4 • pages 109–110

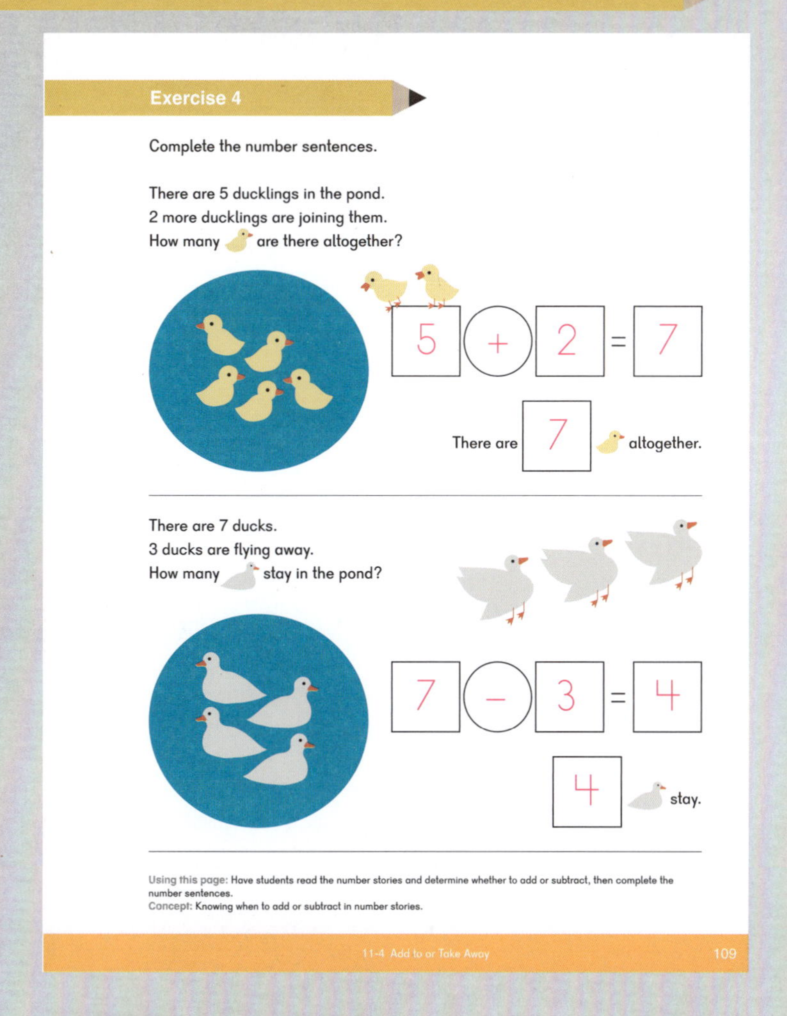

Exercise 4

Complete the number sentences.

There are 5 ducklings in the pond.
2 more ducklings are joining them.
How many are there altogether?

5 + 2 = 7

There are 7 altogether.

There are 7 ducks.
3 ducks are flying away.
How many stay in the pond?

7 – 3 = 4

4 stay.

Using this page: Have students read the number stories and determine whether to add or subtract, then complete the number sentences.
Concept: Knowing when to add or subtract in number stories.

11-4 Add to or Take Away 109

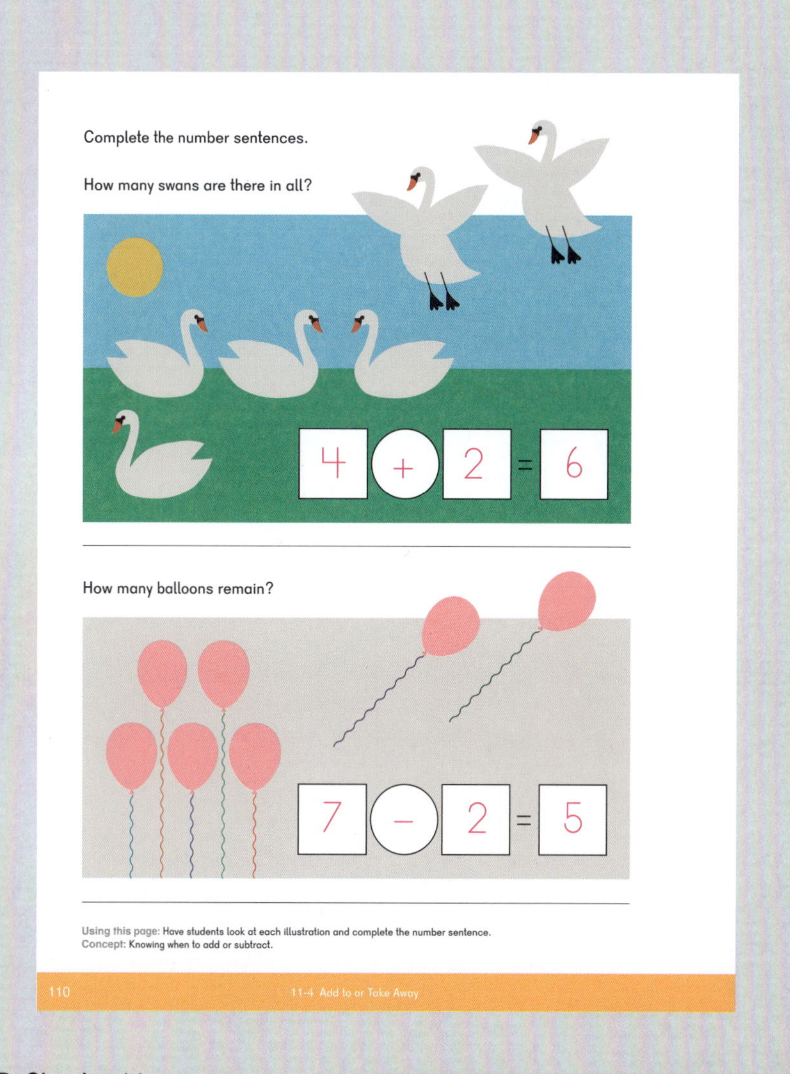

Complete the number sentences.

How many swans are there in all?

4 + 2 = 6

How many balloons remain?

7 – 2 = 5

Using this page: Have students look at each illustration and complete the number sentence.
Concept: Knowing when to add or subtract.

110 11-4 Add to or Take Away

Exercise 5 • pages 111–112

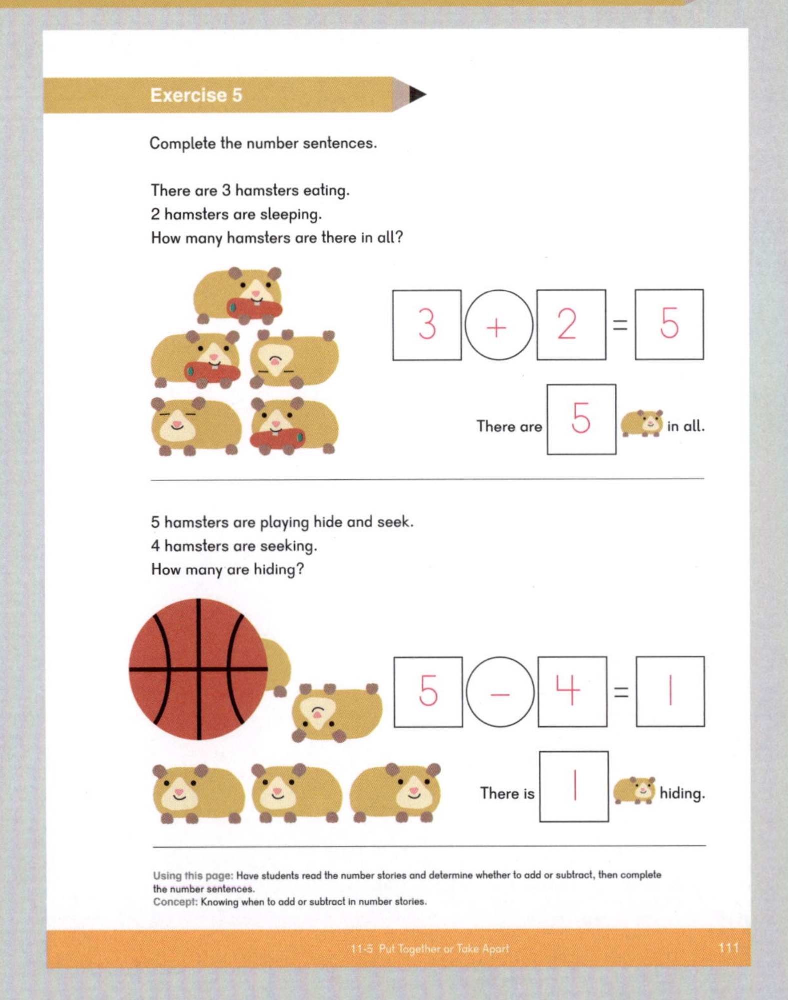

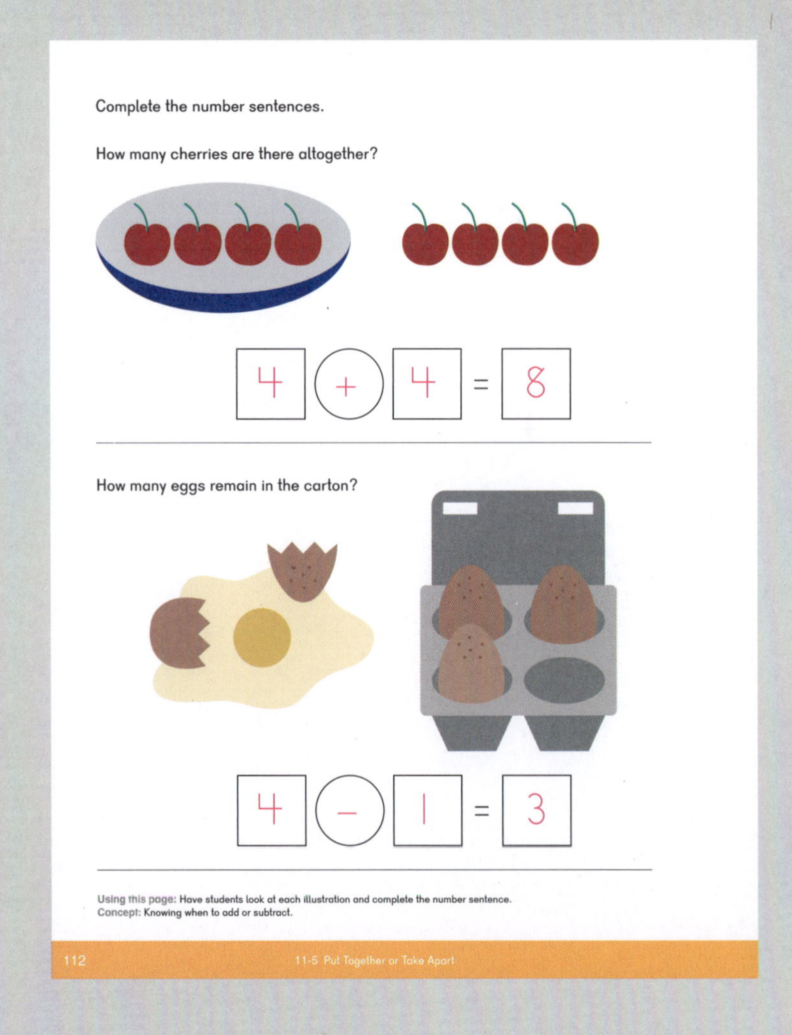

Exercise 6 • pages 113–116

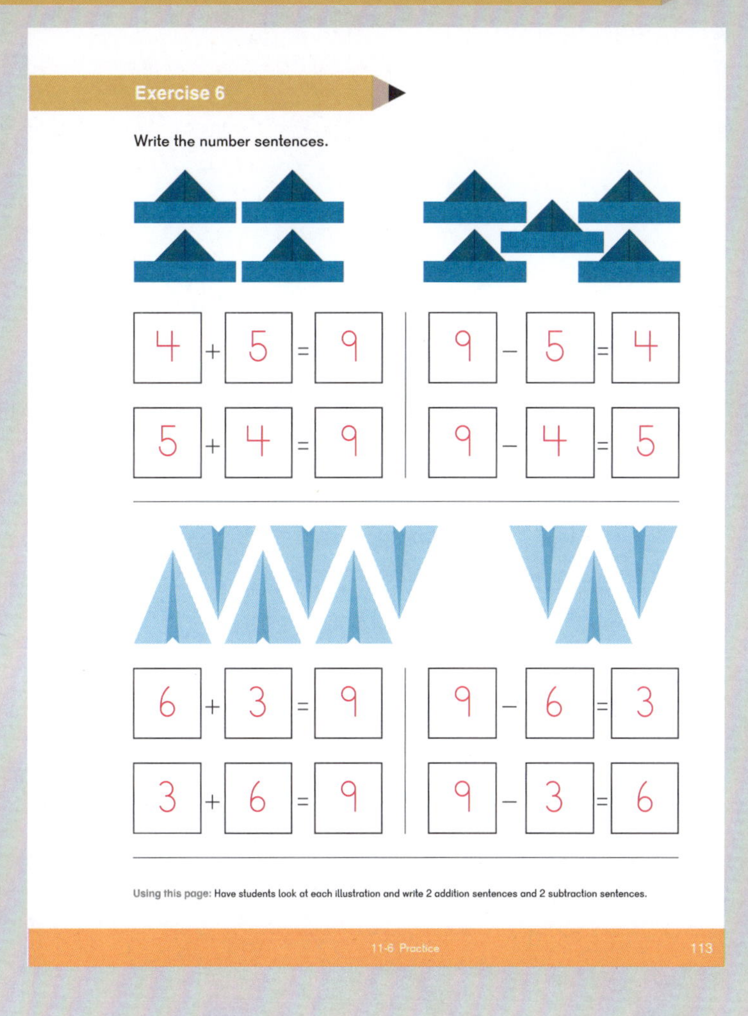

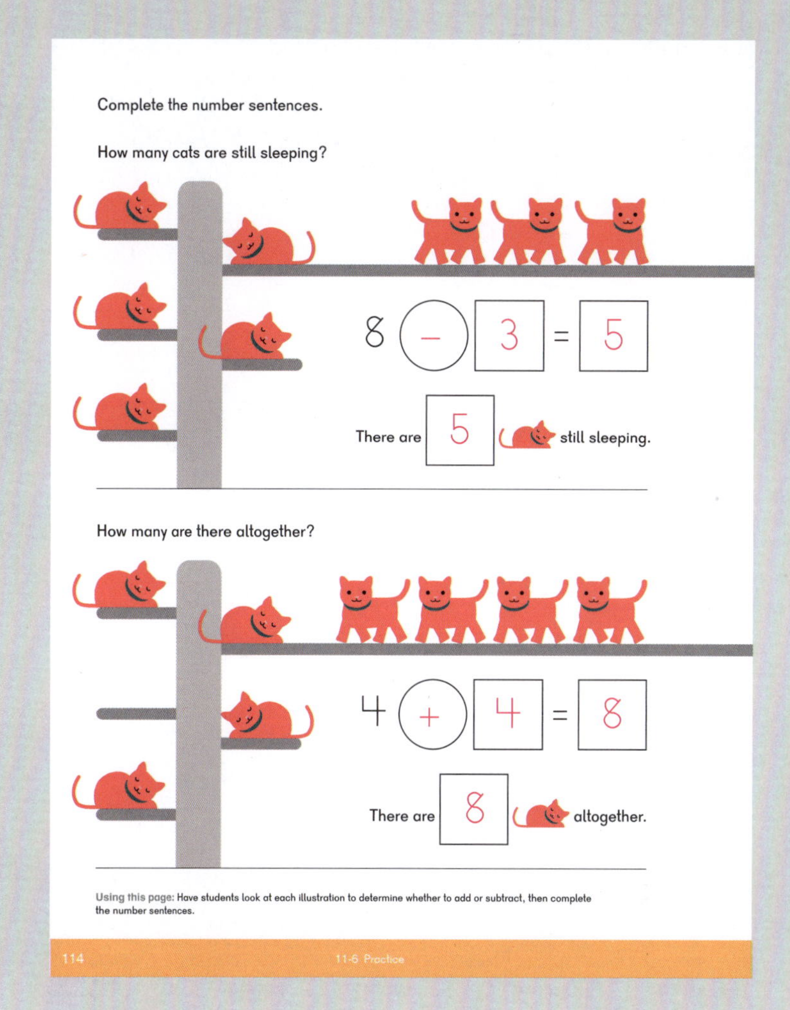

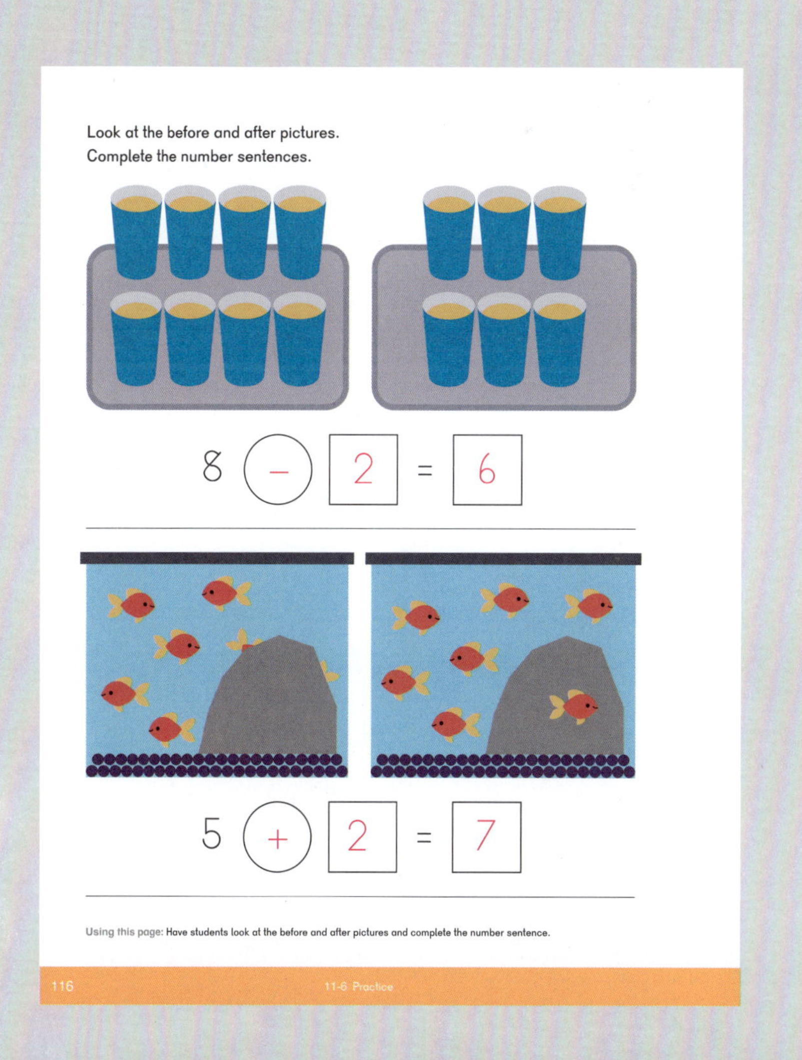

Notes

Suggested number of class periods: 11–12

	Lesson	Page	Resources	Objectives
	Chapter Opener	p. 177	TB: p. 137	
1	Count by Tens — Part 1	p. 178	TB: p. 138 WB: p. 117	Count by tens from 10 to 50. Recognize numbers and number words for tens to 50.
2	Count by Tens — Part 2	p. 180	TB: p. 140 WB: p. 119	Count by tens to 100. Recognize number words for tens, from 6 tens to 10 tens.
3	Numbers to 30	p. 182	TB: p. 142 WB: p. 121	Count sets of 20 to 30 objects. Understand which digit represents tens and which digit represents ones in two-digit numbers to 30.
4	Numbers to 40	p. 184	TB: p. 144 WB: p. 123	Count sets of 30 to 40 objects. Understand which digit represents tens and which digit represents ones in two-digit numbers to 40.
5	Numbers to 50	p. 187	TB: p. 147 WB: p. 125	Count a set of 40 to 50 objects. Understand which digit represents tens and which digit represents ones in two-digit numbers to 50.
6	Numbers to 80	p. 189	TB: p. 149 WB: p. 127	Count a set of 50 to 80 objects. Understand which digit represents tens and which digit represents ones in two-digit numbers to 80.
7	Numbers to 100 — Part 1	p. 191	TB: p. 151 WB: p. 131	Count a set of up to 100 objects. Understand which digit represents tens and which digit represents ones in two-digit numbers to 99.
8	Numbers to 100 — Part 2	p. 193	TB: p. 153 WB: p. 133	Sequence numbers 1 to 100.
9	Count by Fives — Part 1	p. 195	TB: p. 155 WB: p. 135	Count by fives to 50.
10	Count by Fives — Part 2	p. 197	TB: p. 157 WB: p. 137	Count by fives to 50.
11	Practice	p. 199	TB: p. 159 WB: p. 139	Practice skills from the chapter.
	Workbook Solutions	p. 202		

Notes

In **Dimensions Math® KB** Chapter 7: Numbers to 20, students learned to count and write numbers to 20 and to understand the structure of numbers between 10 and 20 as "ten and some more." With numbers greater than 20, students first encounter the idea of counting tens as units: 2 tens make 20, 3 tens make 30. The idea of 10 as a countable unit is challenging to young students, and this chapter is the first step in helping students develop that concept.

The lessons begin with rote counting by tens and identifying tens number words (thirty through ninety). The term "hundred" is used, as is the term "10 tens." Students will count and interpret numbers to 30 in terms of tens and ones. Later lessons in this chapter extend this understanding to numbers within 40, 50, 80, and finally, 100. The meaning of the digits in the tens and ones place will be stressed.

While most of the chapter emphasizes tens, Lesson 9: Count by Fives — Part 1 and Lesson 10: Count by Fives — Part 2 introduce students to counting by fives to 50. Many students will naturally extend this pattern to 100 using a hundred chart.

Counting by fives and tens will be reinforced in Chapter 14: Money. In that chapter, after students recognize coins and learn their values they will count by fives to find the total value of a set of nickels.

Place value charts are included in the textbook to help students see that a number within 100 is made up of tens and ones.

When a large hundred chart is suggested, you can project the chart on page 137 or page 152 of the textbook.

It is assumed that all students will have access to recording tools. When a lesson refers to a whiteboard, any writing materials can be used.

Materials

- 10-sided die
- 3 slightly different sized containers
- Art paper
- Beads
- Blank index cards
- Chalk
- Clear container or plastic bags
- Counters
- Craft sticks
- Crayons or markers
- Desk dividers (box lids or privacy folders)
- Dice with modified sides: 0, 1, 2, 3, 4, 5
- Dice with modified sides: 1, 1, 2, 2, 3, 3
- Dried beans
- Game markers
- Glue
- Linking cubes
- Paint
- Pipe cleaners
- Small drinking cups
- Small objects
- Soft ball or bean bag
- Strips of construction paper in a variety of colors

Blackline Masters

- Blank Ten-frame
- Hundred Chart
- Number Cards
- Numbers to 40 Chart
- Numbers to 50 Chart
- Ten-frame Cards
- Tens Number Cards
- Tens Number Words Cards
- Tens Ten-frame Cards

Storybooks

- *Toasty Toes: Counting by Tens* by Michael Dahl
- *Ants at the Picnic: Counting by Tens* by Michael Dahl
- *Let's Count to 100* by Masayuki Sebe
- *One Watermelon Seed* by Celia Barker Lottridge
- *The Chicken Problem* by Jennifer Oxley
- *Ninety-Three in My Family* by Erica S. Perl
- *One is a Snail, Ten is a Crab* by April Pulley Sayre & Jeff Sayre
- *Richard Scarry's Best Counting Book Ever* by Richard Scarry
- *One Hundred Hungry Ants* by Elinor J. Pinczes
- *Lots of Ladybugs* by Michael Dahl

Letters Home

- Chapter 12 Letter

Notes

Lesson Materials

- Linking cubes
- Large hundred chart

Have students count single linking cubes and put them into rods of ten. If cubes are already organized into rods of ten, students can link together 10 paper clips.

Students can get into groups and combine their rods to lay out 10 rods of 10:

Using one group's set of rods, count the rods with students by counting the tens and showing students the number patterns on the hundred chart on page 137.

Count:

- 1 ten is 10
- 2 tens is 20
- 3 tens is 30
- 4 tens is 40
- Up to: 10 tens is 100

Keep linking cubes linked into rods of ten for the next lesson.

Using a large hundred chart, have students count from 1 to 20. Then point to and count the tens from 20 to 100.

Chapter 12

Numbers to 100

1	2	3	4	5	6	7	8	9	10
11	12	13	14	15	16	17	18	19	20
21	22	23	24	25	26	27	28	29	30
31	32	33	34	35	36	37	38	39	40
41	42	43	44	45	46	47	48	49	50
51	52	53	54	55	56	57	58	59	60
61	62	63	64	65	66	67	68	69	70
71	72	73	74	75	76	77	78	79	80
81	82	83	84	85	86	87	88	89	90
91	92	93	94	95	96	97	98	99	100

137

137

Extend

★ Measuring with Cubes

Students can use the rods of 10 linking cubes to find approximate lengths of long objects. Students use as many rods of 10 as possible and single cubes to get an accurate measurement.

Students could measure:

- Classroom rug
- Length of a wall
- Length of the cubbies
- Length of the whiteboard

Lesson 1 Count by Tens — Part 1

Objectives

- Count by tens from 10 to 50.
- Recognize numbers and number words for tens to 50.

138

Lesson 1
Count by Tens — Part 1

1

Look and talk.

10 ten
20 twenty
30 thirty
40 forty
50 fifty

Objective: Count by tens, 10 to 50.

138 12-1 Count by Tens — Part 1

Explore

Ask students how many fingers they have on one hand and then two hands. Have them put their two hands in the air. Call five students to the front of the room and count their fingers by tens. Ask students to gather in groups of 30 fingers, 50 fingers, etc.

Learn

Look at textbook page 138. Count fingers again, and discuss the numbers and number words for 10, 20, 30, 40, and 50.

Whole Group Activities

▲ Magic Thumb

Pointing your thumb up or down, have students chorally count up and down within 50 by tens.

For example, start out by saying, "Let's count by tens starting at 10. First number?" The class responds, "10." Then, point your thumb up, and the class responds, "20." Then point your thumb down, and the class responds, "10." Point down again, and the class responds, "0," and so on.

▲ Find Your Match

Materials: Tens Number Cards (BLM) and Tens Ten-frame Cards (BLM) for 10 to 50

Hand out cards so that each student has one card. Have students find a partner with a card that has a corresponding quantity.

Small Group Activities

Textbook Page 139

▲ Bean or Bead Counters

Materials: Craft sticks, dried beans, glue; or beads in various colors, pipe cleaners

Provide each student with craft sticks and dried beans. Have them glue 10 beans onto each stick to make 5 sticks of 10 beans so they can skip count by tens to 50.

Save the tens sticks after the glue dries. They make good place value manipulatives and can be stored for future use.

As they are working, have students count their progress, such as, "I have 10, 20, 30, 40 beans glued already!"

Alternatively, students can string beads on pipe cleaners in groups of 10 of the same color, alternating colors with each group of 10.

▲ Match and Sort

Materials: Ten-frame Cards (BLM), Tens Number Words Cards (BLM), Tens Number Cards (BLM) for 10, 20, 30, 40, and 50

Students work in small groups to match and sort the Ten-frame Cards (BLM), Tens Number Cards (BLM), and Tens Number Words Cards (BLM). Once matched, arrange them in order from least to greatest.

Match.

40 forty

10 ten

30 thirty

20 twenty

50 fifty

Objective: Match objects to numbers when counting by tens from 10 to 50.

Exercise 1 • page 117

12-1 Count by Tens — Part 1 139

139

Exercise 1 • page 117

Extend

★ Ten-frame Fill-up

Materials: 10-sided die, Blank Ten-frame (BLM) for each player, linking cubes

Players take turns rolling the die and adding the corresponding number of cubes to their Blank Ten-frame (BLM). When a player fills the ten-frame, she removes the cubes and links them together as a 10. Play continues until a player has 5 tens or 50 cubes.

Lesson 2 Count by Tens — Part 2

Objectives

- Count by tens to 100.
- Recognize number words for tens, from 6 tens to 10 tens.

Lesson Materials

- Linking cubes in rods of 10 from the **Chapter Opener**, or Tens Ten-frame Cards (BLM)
- Tens Number Word Cards (BLM) for 60 to 100

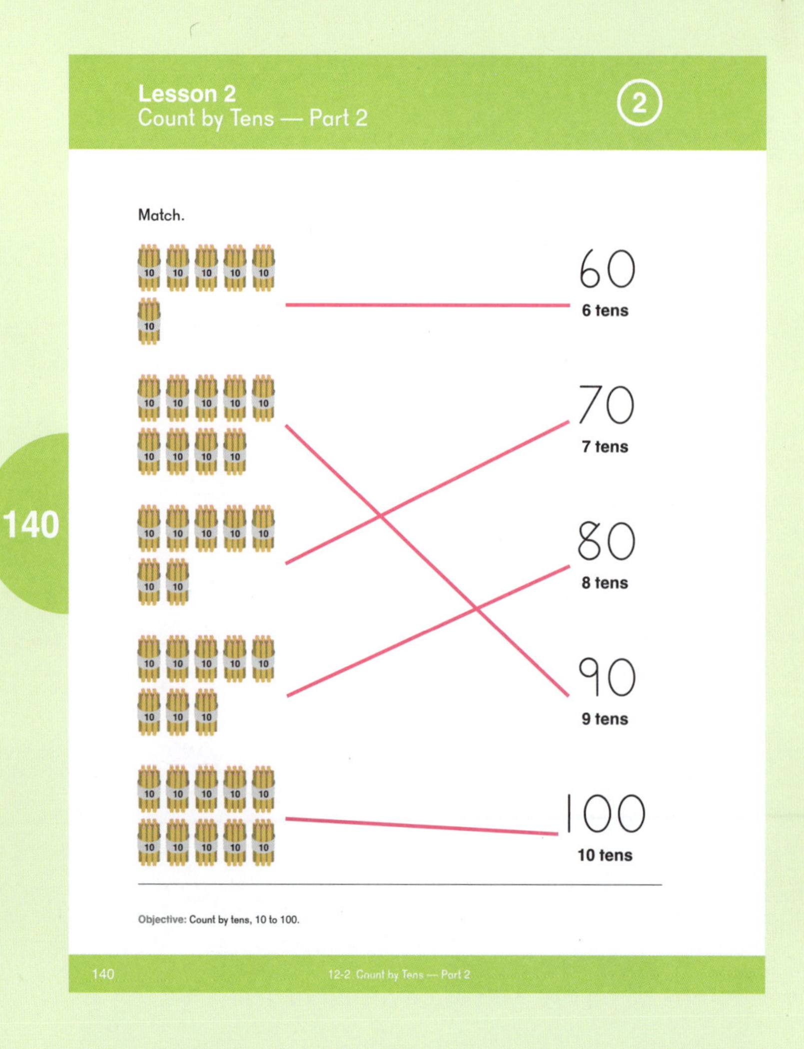

Explore

Have 5 students come in front of the class and review counting fingers by 10 to 50. Add a student and count the fingers to 6 tens or 60. Continue up to 10 tens or one hundred.

Learn

Pass out Tens Ten-frame Cards (BLM) or rods of ten linking cubes to the students from **Explore** as a replacement for their fingers. Count again to 60 with students holding up rods of ten as they are counted. Pass out Tens Number Word Cards (BLM) or Tens Ten-frame Cards (BLM) to students and count again. Have students hold up their cards as each ten is counted.

Whole Group Activities

▲ Magic Thumb

As described in the previous lesson, use your thumb to point up or down and have students chorally count on and back to 100 by tens.

▲ Count and Sit

Choose a target number between 60 and 100 that is a multiple of 10. Have students stand in a circle. The first student starts the count at 10. The count continues around the circle by tens. The student that says the target number sits. The count starts with the next student standing in the circle back at 10 and play continues. The last student standing is the winner.

Small Group Activities

Textbook Pages 140–141

▲ Bean or Bead Counters

Materials: Craft sticks, dried beans, glue; or beads in various colors, pipe cleaners

Have students make 5 craft sticks with 10 beans on each stick, or add another 50 beads to their pipe cleaners so they now have a total of 100 beans or beads.

▲ Match and Sort

Materials: Ten-frame Cards (BLM), Tens Number Words Cards (BLM), Tens Number Cards (BLM) for 10 to 100

Students work in small groups to match and sort the Ten-frame Cards (BLM), Tens Number Cards (BLM), and Tens Number Words Cards (BLM). Once matched, arrange them in order from least to greatest.

▲ Roll, Count, and Color

Materials: Die, Hundred Chart (BLM) for each player, beads or small counters

Players play in pairs and take turns rolling the die. On each turn, players collect as many counters as shown on the die. When a player has 10 counters, she colors in the first row on her Hundred Chart (BLM). Play continues until one player has filled in all 10 rows on her Hundred Chart (BLM).

Exercise 2 • page 119

Extend

★ Super Ten-frame Fill-up

Materials: 10-sided die, Blank Ten-frame (BLM) for each player, linking cubes

Players, playing in groups of 2–4, take turns rolling the die and adding the corresponding number of cubes to their Blank Ten-frames (BLM).

Rolls can be split up to fill a ten-frame. For example, if a player's ten-frame has 8 blank spaces and he rolls a 5, he can fill the ten-frame using 2 of the cubes and use the remaining 3 cubes to start filling his ten-frame again.

When a player fills the ten-frame, he removes the cubes and links them together as a 10. Play continues until a player has 10 tens or 100 cubes.

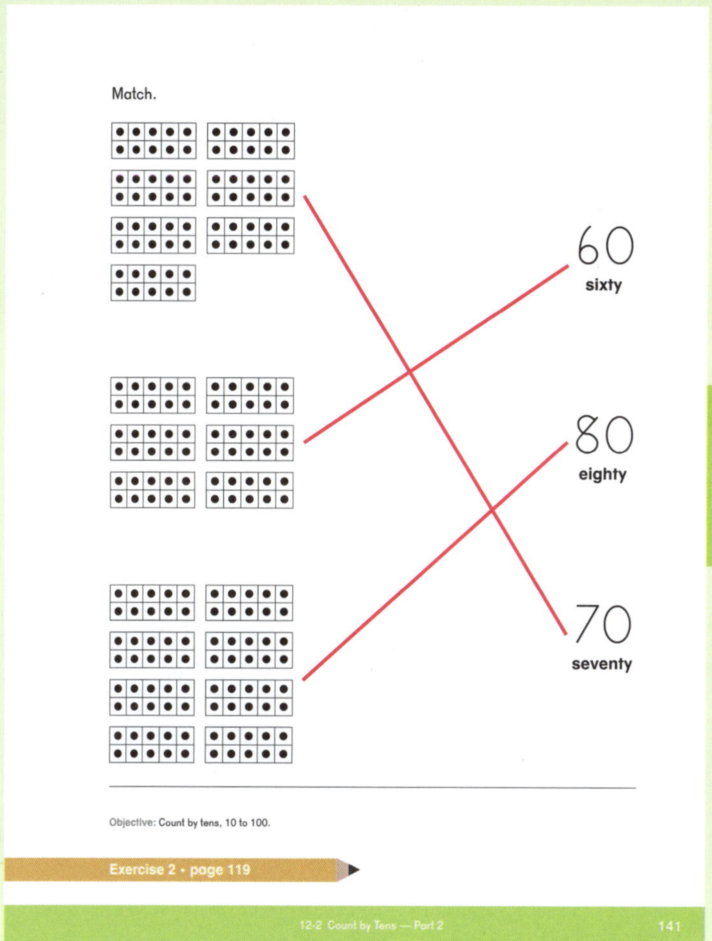

141

Lesson 3 Numbers to 30

Objectives

- Count sets of 20 to 30 objects.
- Understand which digit represents tens and which digit represents ones in two-digit numbers to 30.

Lesson Materials

- Blank Ten-frames (BLM)
- Counters, 30 per student

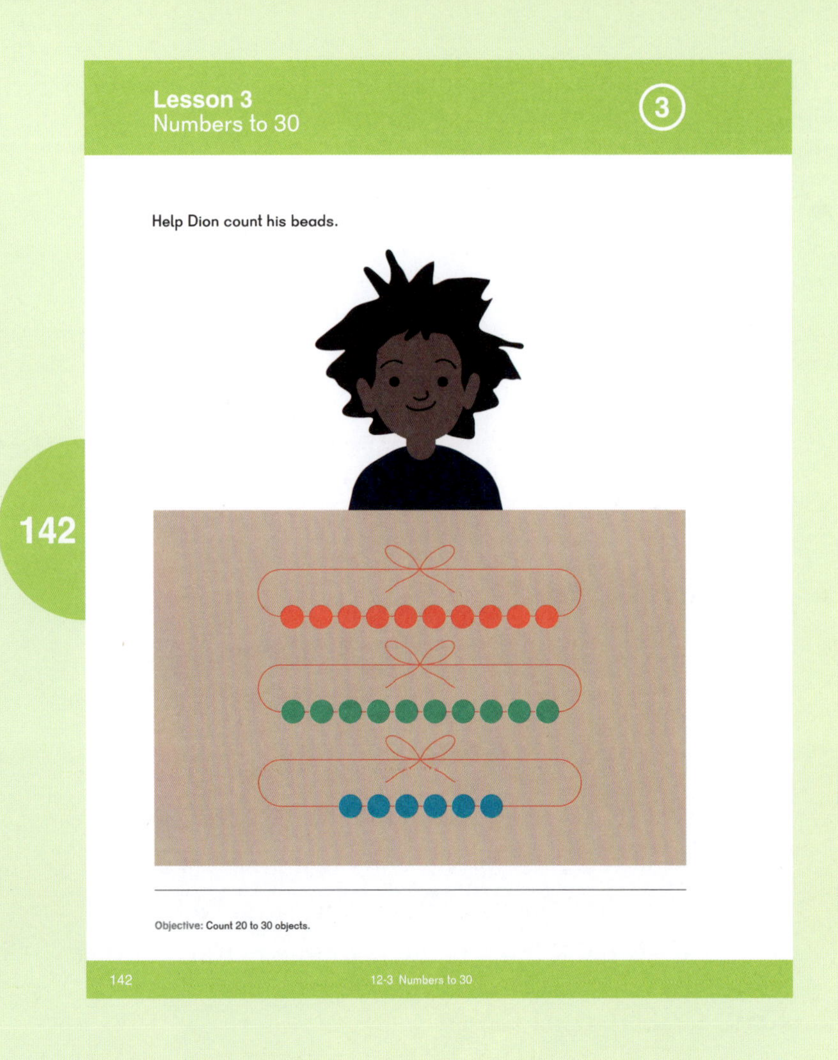

Explore

Provide pairs of students with between 20 and 30 counters and have them count how many they have in all. Discuss how the students keep track of the number of counters. Check if they are counting them one by one or making groups of 10 first.

Give each pair of students three Blank Ten-frames (BLM) and have them organize their counters.

Learn

Have each pair of students share how many counters they have in all. Ask them how many tens and how many more. Encourage students to count the tens first, as in, "2 tens and 4 more makes 24."

Write the number on the board as students say it. Point out to students that the 4 is written in the ones place and 2 tens is written in the tens place. Discuss Dion's beads on textbook page 142.

Whole Group Activities

▲ Circle Count

With students in a circle, begin by having one student say, "One." The student to the left counts, "Two." Have students continue counting around the circle to 30, in order, as quickly as they can. After a few rounds, start at 30 and count down to 0.

Alternatively, students can clap on every ten.

▲ Magic Thumb

As described in previous lessons, use your thumb to point up or down and have students chorally count on and back to 30 by ones.

Small Group Activities

Textbook Page 143

▲ Stack the Cups

Materials: 30 small drinking cups

Have students stack the cups to make a tower using 30 cups. Challenge them to count and see how many cups they can stack before the tower falls over.

▲ Guess and Count

Materials: Around 30 objects in a clear container (clear jars or plastic bags)

Have students guess how many objects are in the container. Students can share the strategies they used to guess number of objects. Have them open the container and count the objects and see how close their guesses were to the correct total.

▲ Draw and Make

Materials: Number Cards (BLM) 20 to 30, linking cubes, Bean or Bead Counters from Lessons 1 and 2

Shuffle the cards and place them facedown. Students draw a card and make the number on the card with linking cubes by showing rods of ten and some more.

Alternatively, students can build the number drawn with Bean or Bead Counters.

Exercise 3 • page 121

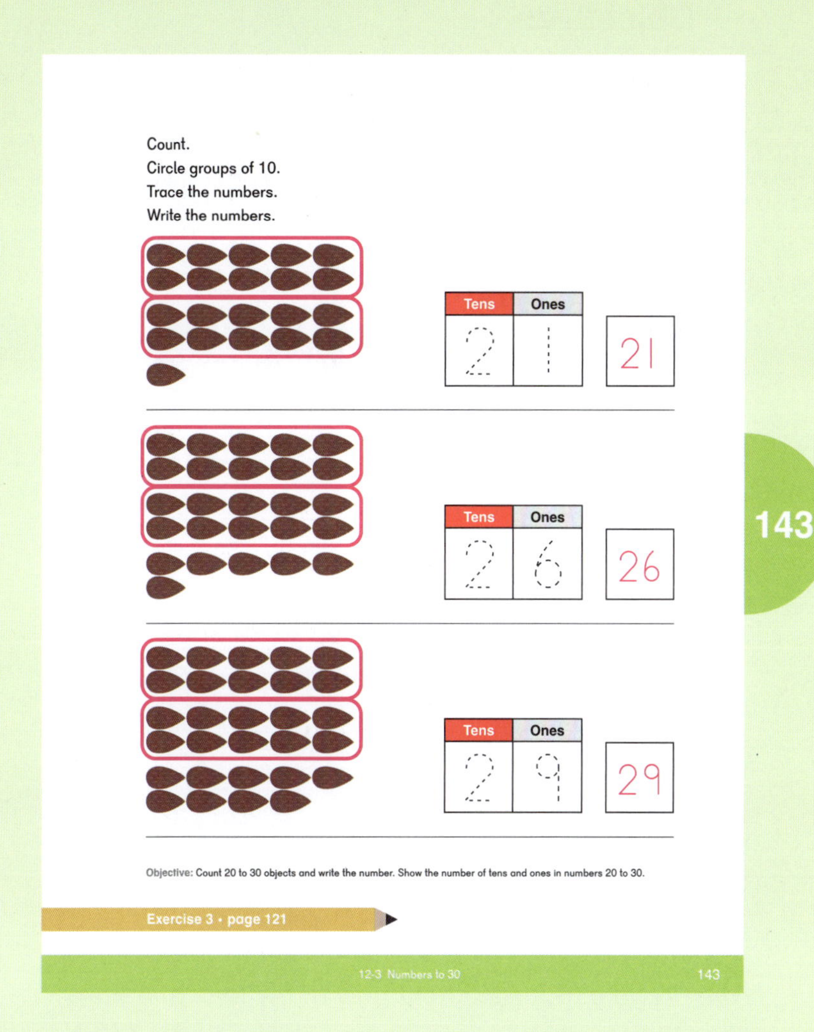

Count.
Circle groups of 10.
Trace the numbers.
Write the numbers.

Tens	Ones
2	1

21

Tens	Ones
2	6

26

Tens	Ones
2	9

29

Objective: Count 20 to 30 objects and write the number. Show the number of tens and ones in numbers 20 to 30.

Exercise 3 • page 121

12-3 Numbers to 30 143

143

Extend

★ Cup Pyramids

Materials: 30 small drinking cups

Challenge students to build the tallest pyramid-shaped tower with 30 small drinking cups. Then challenge students to build a pyramid using 20 small drinking cups, and then 10 small drinking cups.

Objectives

- Count sets of 30 to 40 objects.
- Understand which digit represents tens and which digit represents ones in two-digit numbers to 40.

Lesson Materials

- Sets of small objects, up to 40 per pair of students

Explore

Provide pairs of students with up to 40 small objects and have them count how many they have in all. Discuss how the students keep track of the number of objects.

Learn

Have each pair of students share how many objects they have in all. Ask them how many tens and how many more. Encourage students to count the tens first, as in, "3 tens and 7 more makes 37." Write the number on the board as students share.

Discuss textbook page 144. Remind students to count the number of fingers on two hands by tens. Some may also know how to count by fives. Discuss which is the fastest way to count. Write the number of fingers in each row on the board.

Whole Group Activities

▲ Show and Say

Materials: Number Cards (BLM) 1 to 9, Tens Number Cards (BLM) 10 to 40

Show students a number. Have students say how many tens and how many ones are in the number.

▲ Circle Count

With students in a circle, begin by having one student say, "One." The student to the left counts, "Two." Have students continue counting around the circle to 40, in order, as quickly as they can. After a few rounds, start at 40 and count down to 0.

Alternatively, students can clap on every ten.

▲ Magic Thumb

As described in previous lessons, use your thumb to point up or down and have students chorally count on and back to 40 by ones.

Small Group Activities

Textbook Pages 145–146

▲ Draw and Make

Materials: Number Cards (BLM) 20 to 40, linking cubes, Bean or Bead Counters from Lessons 1 and 2

Shuffle the cards and place them facedown. Students draw a card and make the number on the card with linking cubes by showing rods of ten and some more.

Alternatively, students can build the number drawn with Bean or Bead Counters.

▲ Race to 40

Materials: Completed Numbers to 40 Chart (BLM), game marker for each player, die with modified sides: 1, 1, 2, 2, 3, 3

Players start with their markers on “0” of the Completed Numbers to 40 Chart (BLM). Players take turns rolling the die and moving their markers forward the corresponding number of spaces. The first player to cross 40 wins.

Variation A: Start at 40 and move the marker back the number of spaces on the die. The first player to 0 is the winner.

Variation B: Start at 20 and roll the die and another modified die with +, +, +, −, −, −. The first student to cross either 0 or 40 wins.

▲ Guess and Count

Materials: Around 40 objects in a clear container (clear jars or plastic bags)

Have students guess how many objects are in the container. Students can share the strategies they used to guess the number of objects. Have them dump out objects and count them to see how close their guesses were to the correct total.

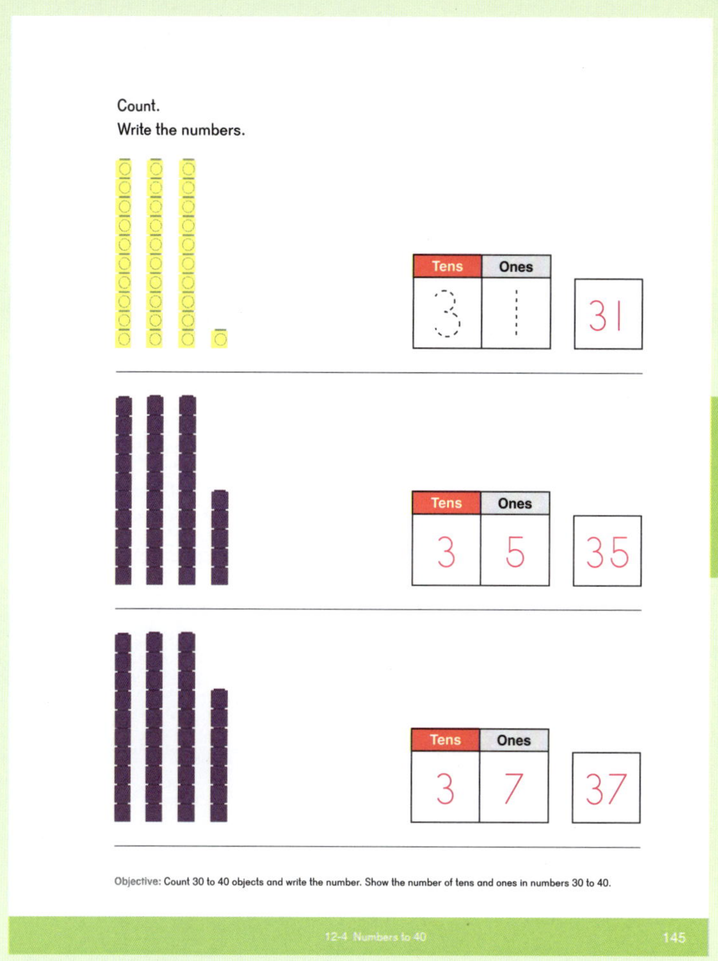

Count.
Write the numbers.

Tens	Ones	
3	1	31

Tens	Ones	
3	5	35

Tens	Ones	
3	7	37

Objective: Count 30 to 40 objects and write the number. Show the number of tens and ones in numbers 30 to 40.

12-4 Numbers to 40 145

145

▲ Stack the Cups

Materials: 40 small drinking cups

Have students stack the cups to make a tower using 40 cups. Challenge them to count and see how many cups they can stack before the tower falls over.

Exercise 4 • page 123

Extend

★ Greatest Wins

Materials: Number Cards (BLM) 1 to 40

Provide students with a deck of Number Cards (BLM) for 1 to 40. Shuffle the deck and deal out half to each player. The players flip their top cards faceup.

The player with the greatest number wins the cards. Play continues until one player is out of cards.

146

Write the numbers.

31

37

40

33

Objective: Count 30 to 40 objects and write the number.

Exercise 4 • page 123

146 12-4 Numbers to 40

Objectives

- Count a set of 40 to 50 objects.
- Understand which digit represents tens and which digit represents ones in two-digit numbers to 50.

Lesson Materials

- Linking cubes in rods of ten and single cubes

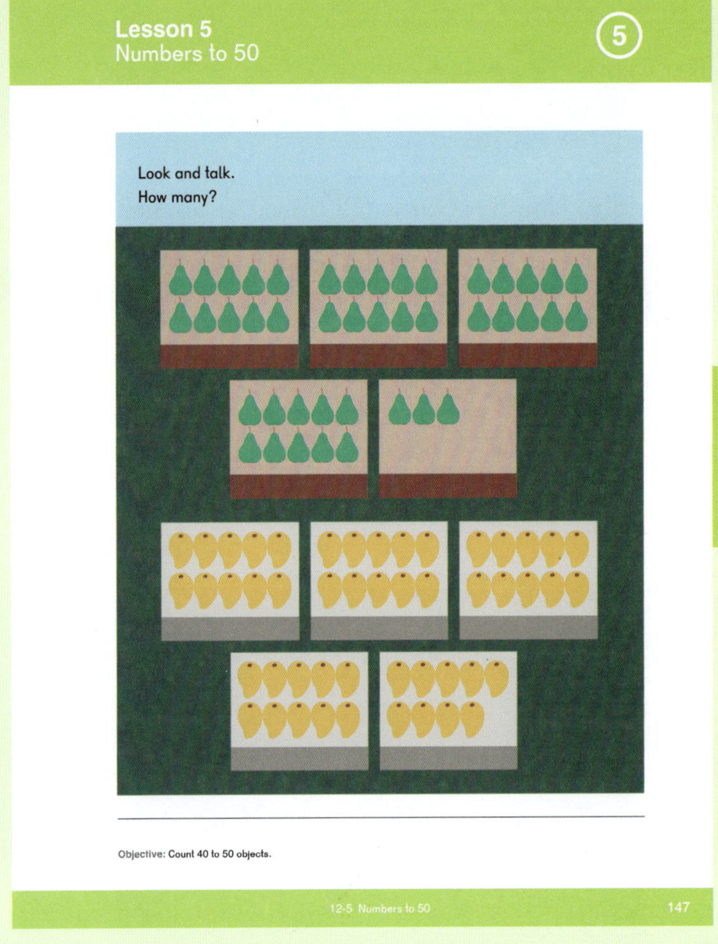

147

Explore

Provide small groups of students with linking cubes in rods of tens and additional single cubes. Write the number 41 on the board and ask students to build the number with the cubes. Repeat with several numbers up to 50.

Learn

Discuss textbook page 147. Have students identify the number of pears and mangoes and write the number on a whiteboard.

Whole Group Activities

▲ Show Me the Number!

Say a number between 20 and 50. Have students write the number on a whiteboard, then hold it up to show. Begin with numbers, "Show me 48," then progress to tens and ones, "Show me 3 tens and 2 ones."

▲ Show and Say

Materials: Number Cards (BLM) 1 to 9, Tens Number Cards (BLM) 10 to 50

Show students a Number Card (BLM) and a Tens Number Card (BLM). Have students say or write the number. For example, you show an 8 and a 30. Students say or write 38.

▲ Claps and Taps

Write a number on the whiteboard, for example, 46. Students clap and say the tens to 40 and tap the ones from 41 to 46.

Take it Outside

▲ Number Walk

Have students walk in a gymnasium or outside and count off 50 steps from a starting line. Students can make estimates of how far they will go in 50 steps. Repeat with other numbers.

Small Group Activities

Textbook Page 148

▲ Draw and Make

Materials: Number Cards (BLM) 20 to 50, linking cubes, Bean or Bead Counters from Lessons 1 and 2

Shuffle the cards and place them facedown. Students draw a card and make the number on the card with linking cubes by showing rods of ten and some more.

Alternatively, students can build the number drawn with Bean or Bead Counters.

▲ How Many Will it Hold?

Materials: 3 slightly different sized containers, small objects to count (cubes, cotton balls, marbles)

Student make estimates on how many objects each container will hold. Students fill the containers with one type of object and then dump them out to count, organizing the objects into groups of tens and ones. They compare their estimates to the actual count and repeat with other objects.

▲ Guess and Count

Materials: Around 50 objects in a clear container (clear jars or plastic bags), Optional: Blank Ten-frames (BLM), plates, cups

Have students guess how many objects are in the container. Students can share the strategies they used to guess number of objects. Have them dump out objects and count them to see how close their guesses were to the correct total. Students can organize their count on Blank Ten-frames (BLM), plates, or in cups labeled with 10.

Exercise 5 • page 125

Extend

★ Number Riddles

Materials: Index cards

Students create a set of cards with number riddles. For example, the front of the card might read, "I am made up of 3 tens and 6 ones. What am I?" and the back of the card reads, "36."

For a greater challenge, students can describe the number out of order. For example, "I am made up of 5 ones and 4 tens."

For an even greater challenge, students can write, "I am made up of 3 tens and 12 ones."

The cards can be used for play in upcoming lessons.

Objectives

- Count a set of 50 to 80 objects.
- Understand which digit represents tens and which digit represents ones in two-digit numbers to 80.

Lesson Materials

- Ten-frame Cards (BLM) 50 to 80
- Number Cards (BLM) 50 to 80

Explore

By this lesson, students should be familiar with the patterns that the tens and ones make when counting.

Shuffle and pass out Ten-frame Cards (BLM) and Number Cards (BLM) so that each student has one card. Tell students to find other students who have cards that match their own. For example, a student with a ten-frame card showing 62 will match up with students holding the 62 number card.

Learn

Discuss textbook page 149. Ask students what they notice about the way the baked goods are arranged. Count the items together. Write the numbers on the board.

Whole Group Activities

▲ Circle Count

Play as in previous lessons, but begin at a random number and count on to 80, then back to the start number.

▲ Show Me the Number!

Say a number between 20 and 80. Have students write the number on a whiteboard, then hold it up to show. Begin with numbers ("Show me 48") then progress to tens and ones ("Show me 3 tens and 2 ones").

▲ Show and Say

Materials: Number Cards (BLM) 1 to 9, Tens Number Cards (BLM) 10 to 80

Show students a Number Card (BLM) and a Tens Number Card (BLM). Have students say or write the number. For example, you show a 7 and a 60. Students say or write 67.

Take it Outside

▲ Number Walk

Have students walk in a gymnasium or outside and count off 80 steps from a starting line. Students can make estimates of how far they will go in 80 steps. Repeat with other numbers.

Small Group Activities

Textbook Page 150

▲ Draw and Make

Materials: Number Cards (BLM) 30 to 80, linking cubes, Bean or Bead Counters from Lessons 1 and 2

Shuffle the cards and place them facedown. Students draw a card and make the number on the card with linking cubes by showing rods of ten and some more.

Alternatively, students can build the number drawn with Bean or Bead Counters.

▲ How Many Will it Hold?

Materials: 3 slightly different sized containers, small objects to count (cubes, cotton balls, marbles)

Student make estimates on how many objects each container will hold. Students fill the containers with one type of object and then dump them out to count, organizing the objects into groups of tens and ones. They compare their estimates to the actual count and repeat with other objects.

▲ Guess and Count

Materials: Around 80 objects in a clear container (clear jars or plastic bags)

Have students guess how many objects are in the container. Students can share the strategies they used to guess the number of objects. Have them dump out objects and count them to see how close their guesses were to the correct total. Students can organize their count on ten-frames, plates, or in cups labeled with 10.

Exercise 6 • page 127

150

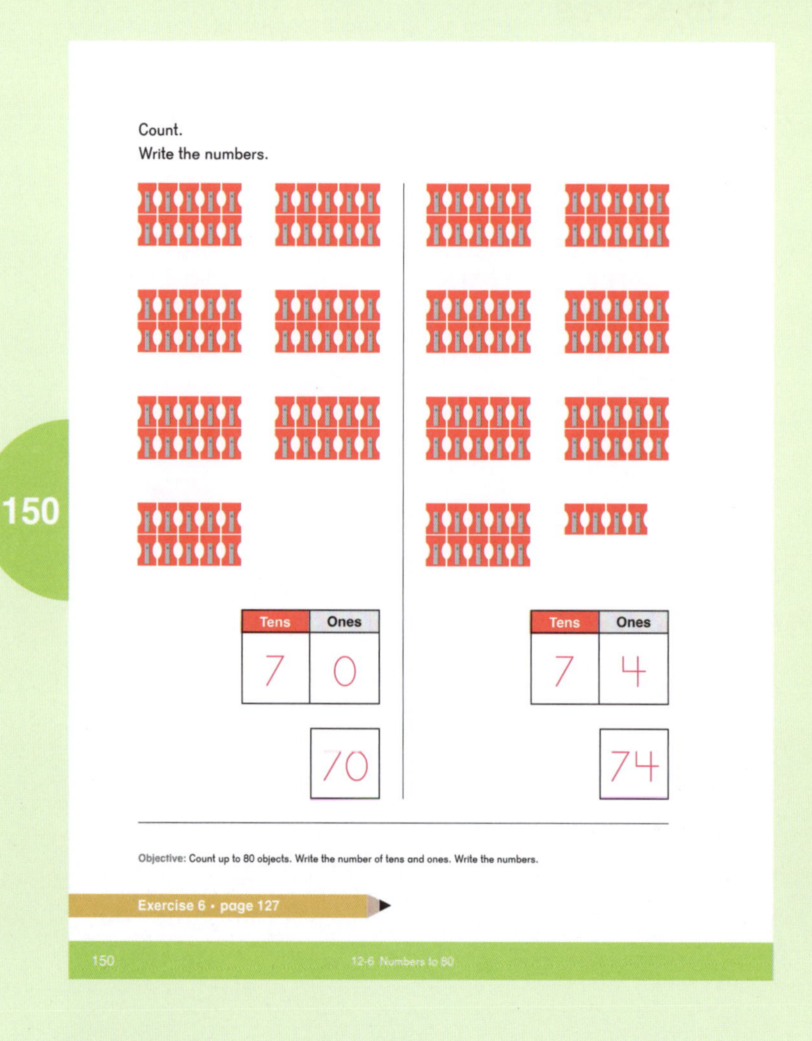
Count.
Write the numbers.

Tens	Ones
7	0

70

Tens	Ones
7	4

74

Objective: Count up to 80 objects. Write the number of tens and ones. Write the numbers.

Exercise 6 • page 127

150 12-6 Numbers to 80

Extend

★ Rock-Paper-Scissors-Math! (Place Value)

Designate one student as the tens and one student as the ones. Students say, "Rock-Paper-Scissors-Math!" and shoot out 0 to 10 fingers. The first students to say the correct number wins the round. For example, "The student in the green shirt will show tens and the student in the blue shirt will show ones." Students show 86 as in this illustration:

Objectives

- Count a set of up to 100 objects.
- Understand which digit represents tens and which digit represents ones in two-digit numbers to 99.

Lesson Materials

- Counters in bags or cups, 30–100 counters in each

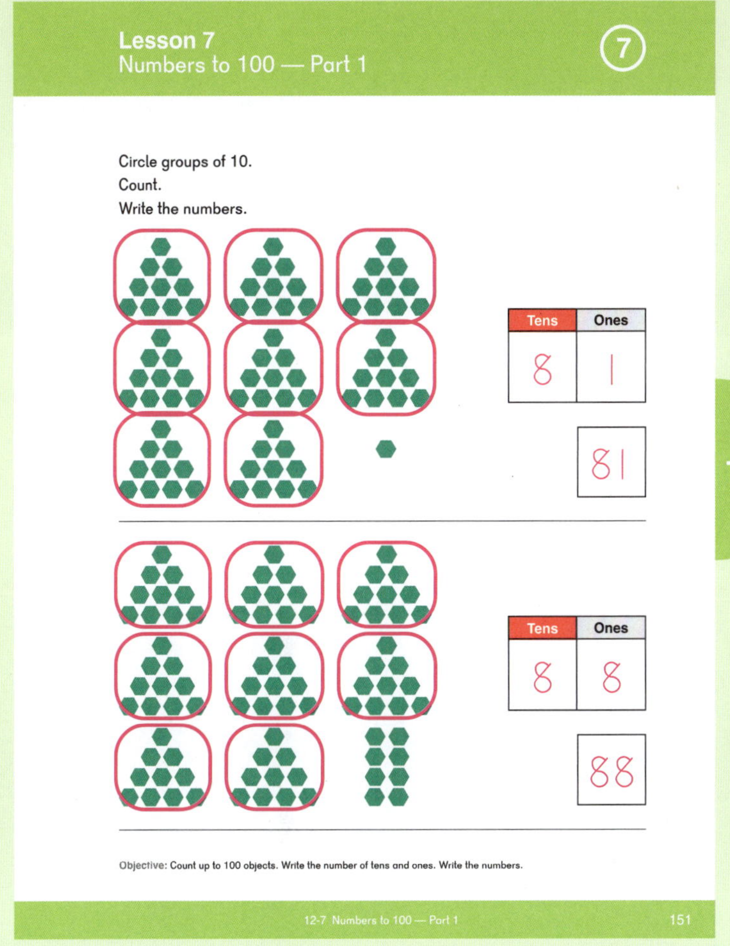

151

Explore

Provide groups of students with bags or cups of small objects or counters. Students work together to group the objects by 10 to determine the total number in their bags.

Learn

Ask students to share their work from **Explore**. Record the number of objects in a place value chart and as digits on the board.

Whole Group Activities

▲ Stand Up for Your Number

Materials: Number Cards (BLM) 40 to 100

Hand out cards so that each student has one. Give clues, such as, "This number has 4 tens and 5 ones" or, "This number is the number I say after 50 when I'm counting by tens." Each student listens to the clues, stands up, and shows her Number Card (BLM) at the appropriate time. The riddle cards from the Lesson 6 **Extend** activity could be used.

▲ Show Me the Number!

Say a number between 20 and 100. Have students write the number on whiteboards, then hold them up to show. Begin with numbers ("Show me 48") then progress to tens and ones ("Show me 3 tens and 2 ones").

▲ Show and Say

Materials: Number Cards (BLM) 1 to 9, Tens Number Cards (BLM) 10 to 100

Play as previously described for numbers to 100.

Take it Outside

▲ Number Walk

Have students walk in a gymnasium or outside and count off 100 steps from a starting line. Students can make estimates of how far they will go in 100 steps. Repeat with other numbers.

Small Group Activities

Textbook Pages 151–152. On page 152, ask students to locate numbers on the chart. Ask them what they notice in each row as numbers go from left to right. Repeat for each column as numbers go from top to bottom.

▲ Paper Chains

Materials: Strips of construction paper in a variety of colors

Have students create paper chains by stapling strips into interconnecting loops. Students create a chain of 10 loops of one color, then continue the chain with 10 loops of a different color until they have a chain of 100 loops.

Alternatively, this can be done by counting the days of school. Have students create a chain for the classroom that matches the number of days in school thus far and add a chain each day from this point forward.

▲ Race to 100

Materials: Hundred Chart (BLM), game marker for each player, die with modified sides: 1, 1, 2, 2, 3, 3

Players start with their markers on "0" of the Hundred Chart (BLM). Players take turns rolling the die and moving their markers forward the corresponding number of spaces. The first player to cross "100" wins.

Variation A: Start at 100 and move the marker back the number of spaces on the die. The first player to "0" is the winner.

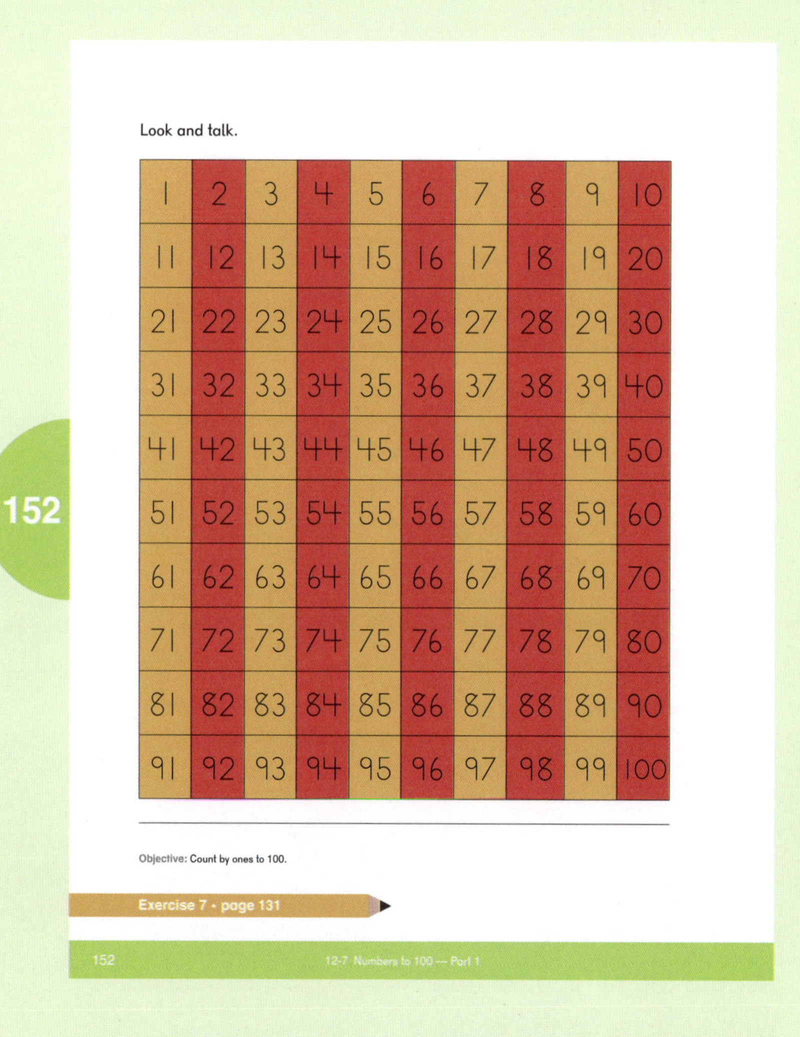

152

Look and talk.

1	2	3	4	5	6	7	8	9	10
11	12	13	14	15	16	17	18	19	20
21	22	23	24	25	26	27	28	29	30
31	32	33	34	35	36	37	38	39	40
41	42	43	44	45	46	47	48	49	50
51	52	53	54	55	56	57	58	59	60
61	62	63	64	65	66	67	68	69	70
71	72	73	74	75	76	77	78	79	80
81	82	83	84	85	86	87	88	89	90
91	92	93	94	95	96	97	98	99	100

Objective: Count by ones to 100.

Exercise 7 • page 131

152 12-7 Numbers to 100 — Part 1

Variation B: Students start at 50 and roll the die and a modified die with +, +, +, −, −, −. The first player to exit the board at either 0 or 100 wins the game.

Exercise 7 • page 131

Extend

★ Hidden Numbers

Materials: Hundred Chart (BLM), counters

Player 1 covers up to 5 numbers on the Hundred Chart (BLM) with the counters. Player 2 figures out the numbers that are covered based on the surrounding numbers.

Objective

- Sequence numbers 1 to 100.

Lesson Materials

- Number Cards to 100 (BLM) sorted into sequences of 10 numbers (1 to 10, 11 to 20, 21 to 30, etc.)

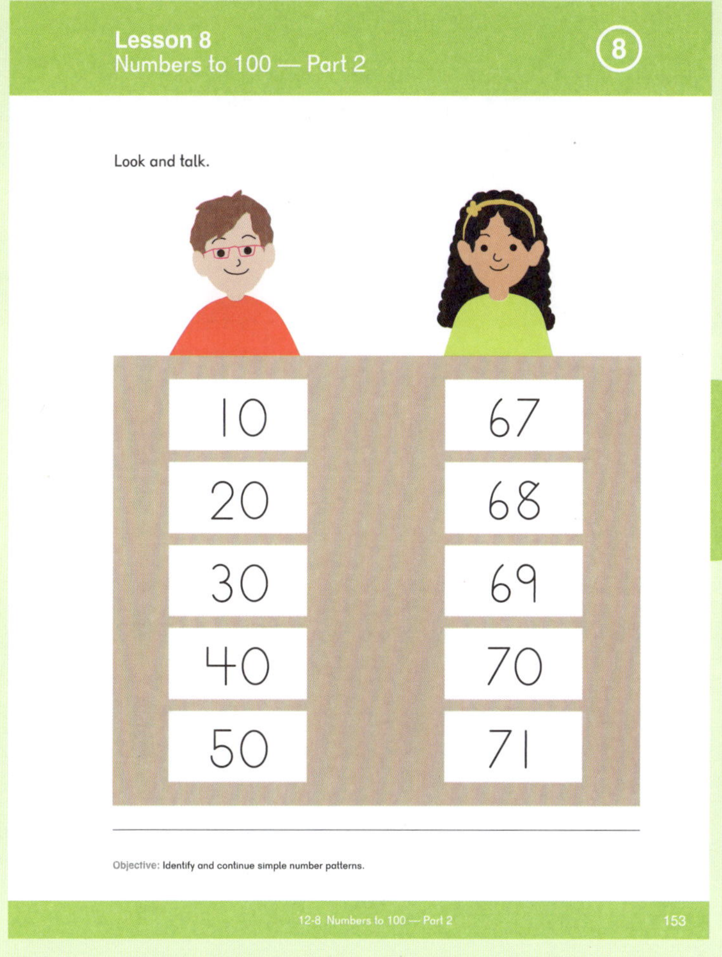

153

Explore

Provide pairs of students with sets of Number Cards (BLM) and have them put the cards in order from least to greatest. The sets of numbers can then be put in order from 1 to 100.

Learn

On textbook page 153, discuss how Emma and Sofia have laid out their cards. Provide additional examples of ordering the numbers by ones or by tens.

Whole Group Activity

▲ Order Up – Number Cards

Materials: Number Cards (BLM) 70 to 100

Give each student a Number Card (BLM). Have them line up in order from least to greatest. As it is unlikely that all cards will be used, have students figure out the missing numbers.

Activities from previous lessons in this chapter can be played again in this lesson for practice.

Take it Outside

▲ Outdoor Hundred Chart

Materials: Chalk, Number Cards (BLM) 1 to 100

Draw a 10 × 10 grid with chalk on the ground, and fill in the tens column and some of the numbers. Give each student a card with a number that is missing on the grid. Their task is to stand in the correct space, then tell how they knew where to stand.

▲ Number Walk

Have students walk in a gymnasium or outside and count off 100 steps from a starting line.

Small Group Activities

Textbook Page 154

154

Fill in the missing numbers.

1	2	3	4	5	6	7	8	9	10
11	12	13	14	15	16	17	18	19	20
21	22	23	24	25	26	27	28	29	30
31	32	33	34	35	36	37	38	39	40
41	42	43	44	45	46	47	48	49	50
51	52	53	54	55	56	57	58	59	60
61	62	63	64	65	66	67	68	69	70
71	72	73	74	75	76	77	78	79	80
81	82	83	84	85	86	87	88	89	90
91	92	93	94	95	96	97	98	99	100

Objective: Sequence and write numbers 1 to 100.

Exercise 8 • page 133

154 12-8 Numbers to 100 — Part 2

▲ Five in a Row

Materials: 4 sets of Number Cards (BLM) 0 to 9, Hundred Chart (BLM), 2 different colored counters or linking cubes

Each player chooses a color for her marker. The stack of Number Cards (BLM) is shuffled and placed facedown.

Students take turns drawing 2 cards, creating a double digit number and covering the corresponding number on the Hundred Chart (BLM) with their markers. Students can switch the digits around to create numbers that are close to each other.

The first player to get 5 of her markers in a row is the winner.

▲ How Many Will it Hold?

Materials: 3 slightly different sized containers, small objects to count (cubes, cotton balls, marbles)

Student make estimates on how many objects each container will hold. Students fill the containers with one type of object and then dump them out to count, organizing the objects into groups of tens and ones. They compare their estimates to the actual count and repeat with other objects.

Exercise 8 • page 133

Extend

★ Hundred Chart Puzzle

Materials: Multiple copies of the Hundred Chart (BLM), printed on different colored paper

Print multiple copies of Hundred Chart (BLM) on different colored paper and cut random shapes along the lines to form puzzle pieces. Have students put the puzzle back together. If needed, provide a complete Hundred Chart (BLM) as a template for students to arrange the puzzle pieces.

Lesson 9 Count by Fives — Part 1

Objective

- Count by fives to 50.

Lesson Materials

- Hundred Chart (BLM)

155

Explore

Have 5 students hold up their hands and have the other students count the fingers by tens. Tell students that they can also count by fives.

Count their fingers again, this time by fives.

Ask students what things they know come in groups of 5. Some ideas:

- Fingers and toes
- Points on a star
- Sides on a pentagon
- Players on a basketball team
- Tally marks

Learn

Show students a Hundred Chart (BLM) and point to the numbers as they skip count aloud by fives to 50. Ask students what they notice about the pattern on the chart as we count by fives to 50.

Whole Group Activities

Modify previous activities for skip counting by 5 using Number Cards (BLM) for multiples of 5 to 50.

▲ Circle Count

Play as in previous lessons, but have students put up one hand as they count around a circle by fives to 50.

▲ Order Up – Number Cards

Materials: Number Cards (BLM) 0 to 100

Give each student a Number Card (BLM) that is a multiple of 5 and play as in the previous lesson.

▲ Stand Up for Your Number

Materials: Number Cards (BLM) 0 to 100

Hand out cards that are multiples of 5 so that each student has one. Play as in previous lessons, but give clues for multiples of 5, such as, "This number has 4 tens and 5 ones."

▲ Magic Thumb

As described in previous lessons, use your thumb to point up or down and have students chorally count on and back to 100 by fives.

Take it Outside

▲ Outdoor Hundred Chart

Materials: Chalk, Number Cards (BLM) 1 to 100

Play as directed in the previous lesson.

Small Group Activities

Textbook Pages 155–156

▲ Give Me 5!

Materials: Paint, art paper

Make handprints on a large sheet of paper. Write the numbers of fingers underneath each hand. Alternatively, students could trace pentagon shapes and count the number of sides.

▲ Guess and Count

Materials: Put up to 50 objects (a multiple of 5) in a clear container (clear jars or plastic bags)

Have students guess how many objects are in the container. Tell them their guess should be a number we say when we count by fives. Extend by using up to 100 objects (use multiples of 5 only). Have students make guesses, record them, and then count the objects by organizing them into groups of five.

156

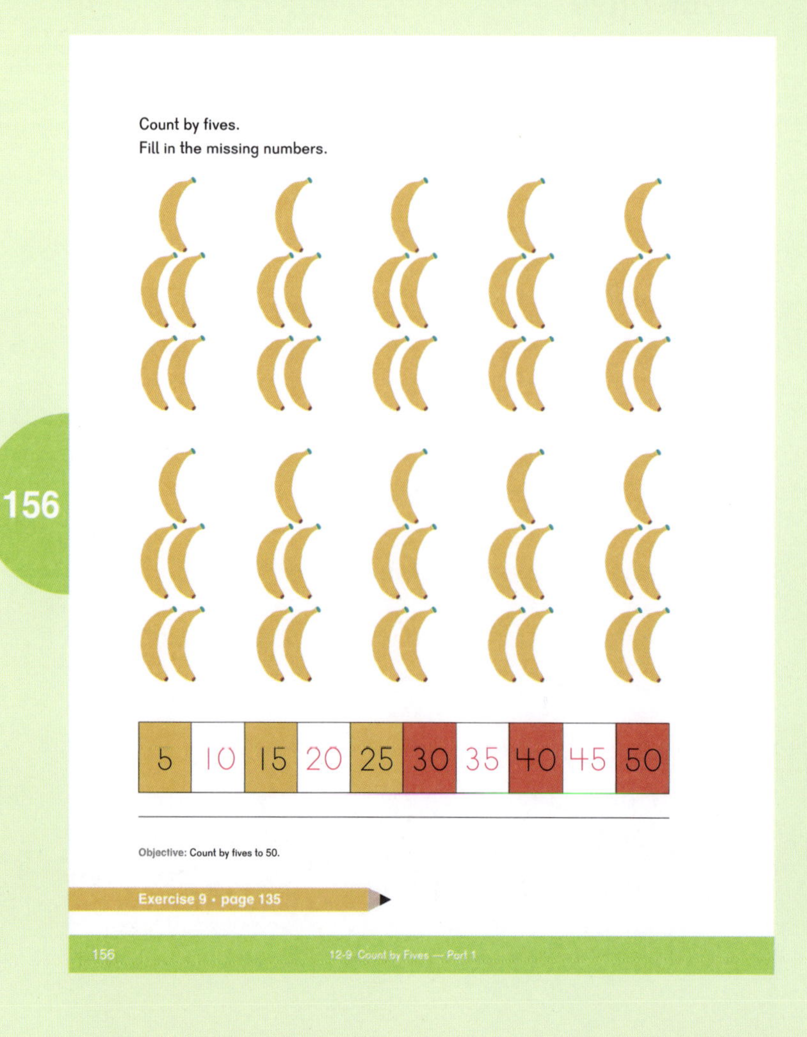
Count by fives.
Fill in the missing numbers.

5	10	15	20	25	30	35	40	45	50

Objective: Count by fives to 50.

Exercise 9 • page 135

156 12-9 Count by Fives — Part 1

Exercise 9 • page 135

Extend

★ Catch and Count

Materials: Soft ball or bean bag

Have students toss the ball to each other. The player who catches the ball says the next number when counting by fives.

The first player starts with "zero" then tosses the ball. The catcher says, "Five," then tosses the ball, etc.

Count back from 100 by fives for a greater challenge.

Lesson 10 Count by Fives — Part 2

Objective

- Count by fives to 50.

Lesson Materials

- Numbers to 50 Chart (BLM)
- Linking cubes, 50 per student or pair of students

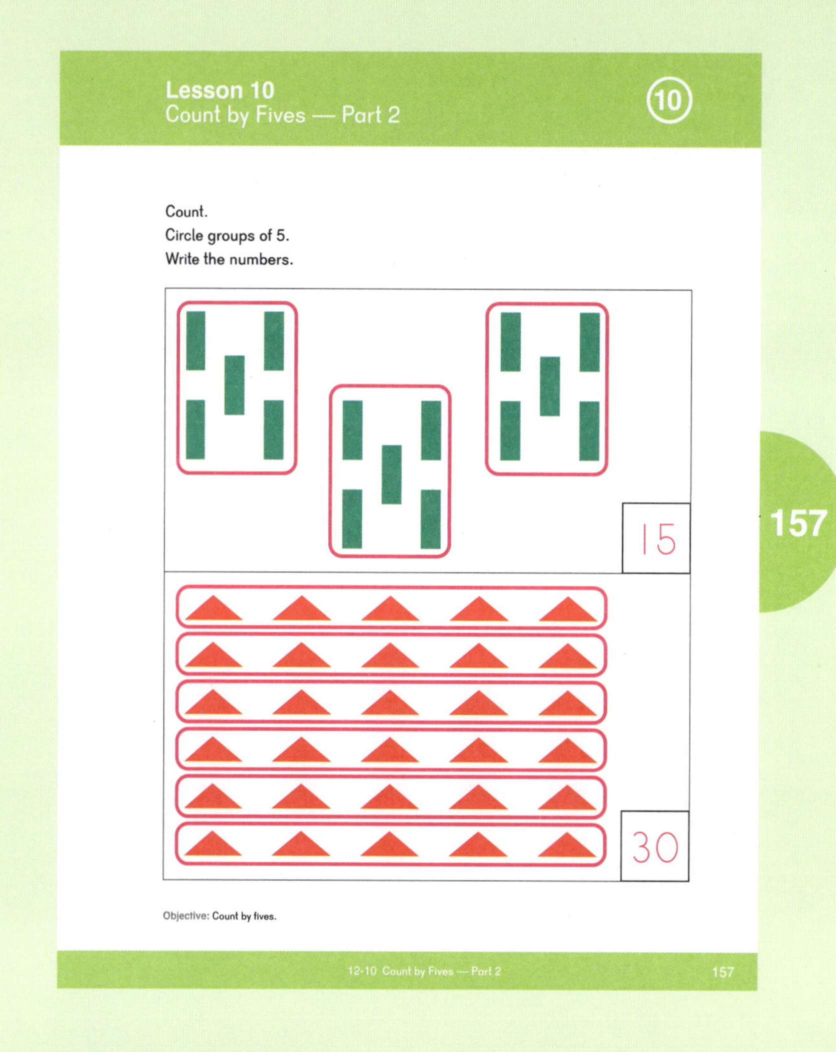

157

Explore

Have partners make towers of 5 linking cubes using 50 cubes in all. Students practice counting by 5 to 50, then counting back from 50 to 5.

Learn

Provide each student with a Numbers to 50 Chart (BLM). Ask them to place a counter or color the square for each number they say when they count by fives to 50.

Discuss the numbers on the chart that have counters or are colored. Students may note that they all have 5 or 0 in the ones place.

Whole Group Activities

▲ Reading Time

Materials: *Lots of Ladybugs* by Michael Dahl

▲ Order Up – Number Cards

Materials: Number Cards (BLM) 0 to 100

Give each student a Number Card (BLM) that is a multiple of 5 and play as in previous lessons.

▲ Stand Up for Your Number

Materials: Number Cards (BLM) 0 to 100

Play as directed in the previous lesson.

▲ Magic Thumb

Play as directed in the previous lesson to 100.

Take it Outside

▲ Hopscotch

Materials: Chalk, game marker

Create a hopscotch court that counts by 5 to 50.

Small Group Activities

Textbook Pages 157–158

▲ What is Missing?

Materials: Number Cards (BLM) 5 to 50 that are multiples of 5, 1 set per player

Players remove 2 numbers from their decks of Number Cards (BLM). They shuffle the cards and trade their deck with another player. Players lay the cards in their new deck in order and say the missing numbers from their decks.

▲ Craft Stick Count by Fives

Materials: Craft sticks with numbers 0 to 50 in multiples of 5 written on one end

Independent activity: Students line the craft sticks in order from 0 to 50.

Partner activity: Player 1 holds the sticks in a bundle. Player 2 pulls a stick and counts on by fives from that number.

158

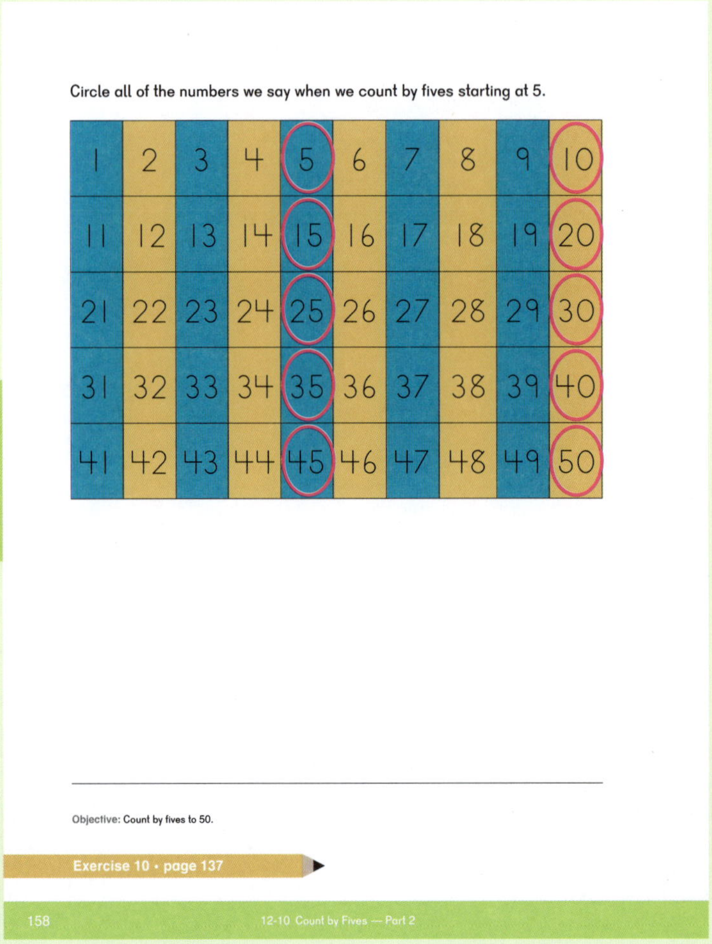

Circle all of the numbers we say when we count by fives starting at 5.

1	2	3	4	5	6	7	8	9	10
11	12	13	14	15	16	17	18	19	20
21	22	23	24	25	26	27	28	29	30
31	32	33	34	35	36	37	38	39	40
41	42	43	44	45	46	47	48	49	50

Objective: Count by fives to 50.

Exercise 10 • page 137

Exercise 10 • page 137

Extend

★ I'm thinking of a number...

Materials: Numbers to 50 Chart (BLM), counters

Provide students with a Numbers to 50 Chart (BLM). Students place two counters on the chart, one at 1 and one at 50. Player 1 chooses a number that is a multiple of 5 between 1 and 50, but does not tell Player 2 what he is thinking.

For example, the student may think of 25. Player 2 guesses the number 45. Player 1 says, "The number I'm thinking of is less than 45." Player 2 moves the counter from 50 to 45. Guesses continue until the unknown number is either guessed or squeezed between the counters.

Objective

- Practice skills from the chapter.

Practice lessons are designed for further practice and assessment as needed.

Students can complete the textbook pages and workbook pages as practice and/or as an assessment.

Use activities and extensions from the chapter for additional review and practice.

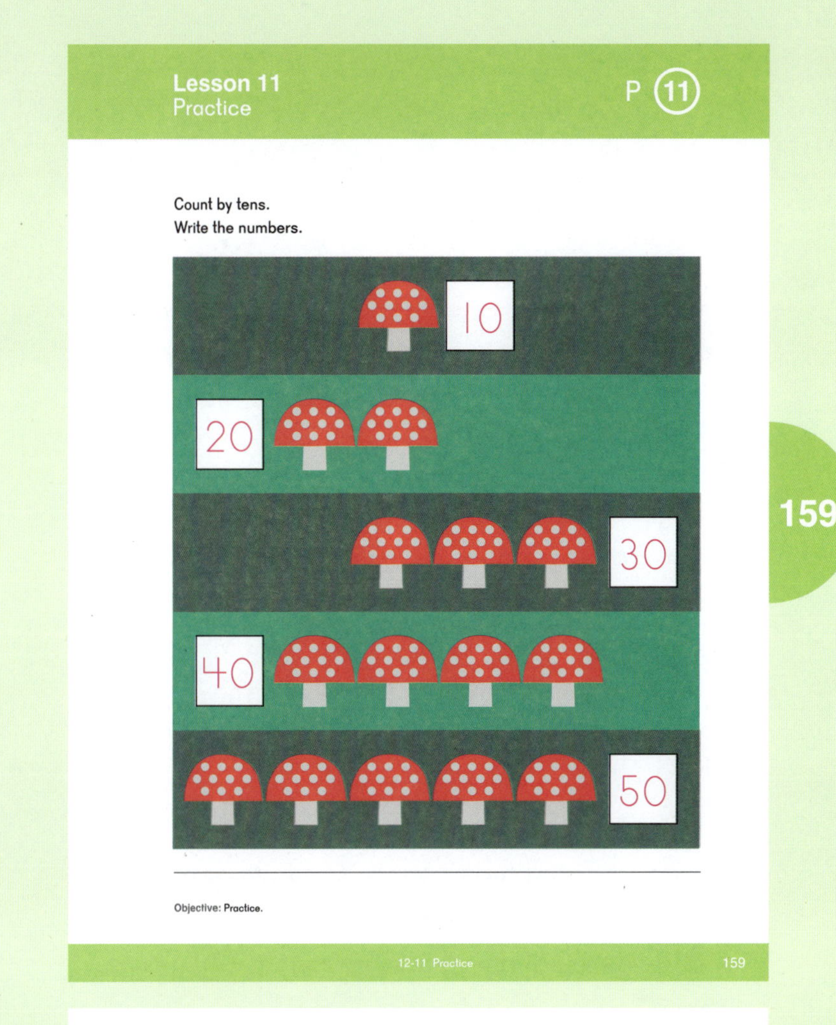

159

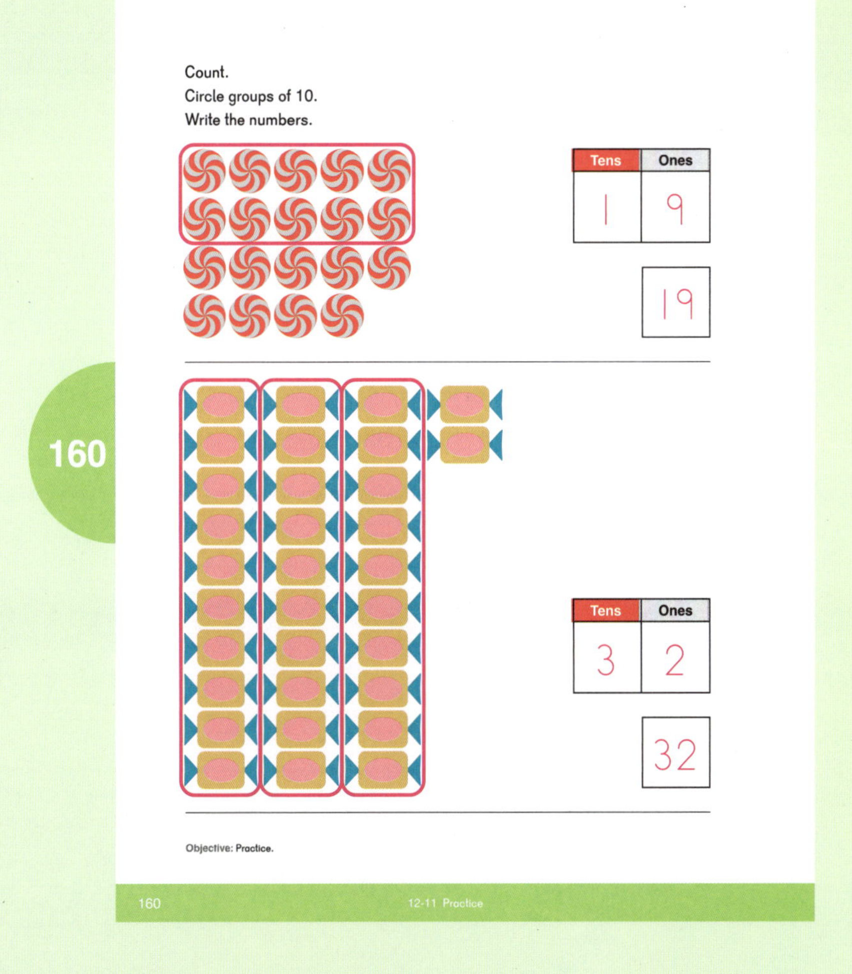

160

Exercise 11 • page 139

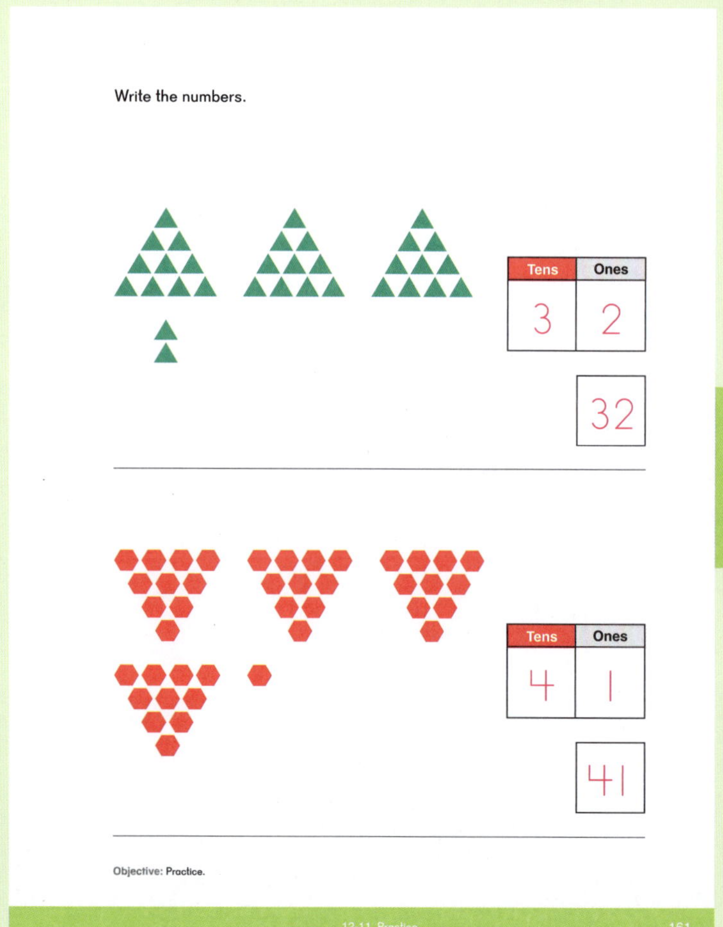

161

Extend

★ Roll to 100

Materials: Hundred Chart (BLM), crayons or markers – a different color for each player, 2 die modified with 0s in place of the 6s

Play in groups up to 4. Players take turns. On each turn, a player rolls the dice and adds the two numbers together. The player then counts on that many squares on the Hundred Chart (BLM) from the previous number and marks the new number.

Sample play:

- Player 1 rolls a 4 and a 3. He adds them to get a total of 7 and then counts 7 places on the hundred board and marks that square.
- Player 2 rolls a 5 and 5. She counts and marks the 10.
- Player 1 rolls a 4 and 0. He counts on 4 from the 7 and marks his new number, 11.
- Player 2 rolls a 4 and 4. She counts on 8 from the 10 and marks her new number, 18.

The first player to pass 100 is the winner.

Note: Using crayon marks of different colors instead of game tokens helps students see where their numbers were, and how many they have added on. A Hundred Chart (BLM) in a dry erase sleeve makes an excellent gameboard.

1	2	3	4	5	6	7	8	9	10
11	12	13	14	15	16	17	18	19	20
21	22	23	24	25	26	27	28	29	30
31	32	33	34	35	36	37	38	39	40
41	42	43	44	45	46	47	48	49	50
51	52	53	54	55	56	57	58	59	60
61	62	63	64	65	66	67	68	69	70
71	72	73	74	75	76	77	78	79	80
81	82	83	84	85	86	87	88	89	90
91	92	93	94	95	96	97	98	99	100

★ Hundred Chart Battleship

Materials: 7 counters, Hundred Chart (BLM), divider (box lid or privacy folder)

Play in pairs. Use the divider to keep players from seeing each other's game boards. Each player claims a group of 2, 3, and 4 numbers either across or vertically on his own Hundred Chart (BLM) with the counters. Those are their "ships."

Players then take turns calling numbers to find opponents' ships. Opponents call out, "Hit," (marked with a circle) or, "Miss."

On the sample board shown below, the player has laid out his three ships. He has 4 misses and has 2 hits on his opponent.

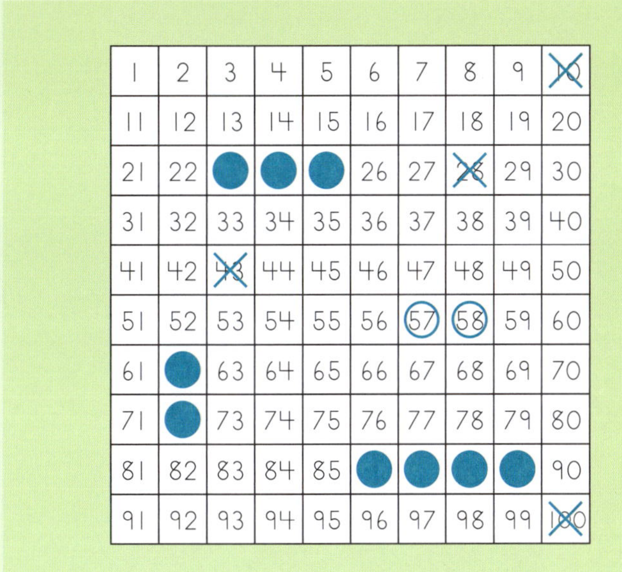

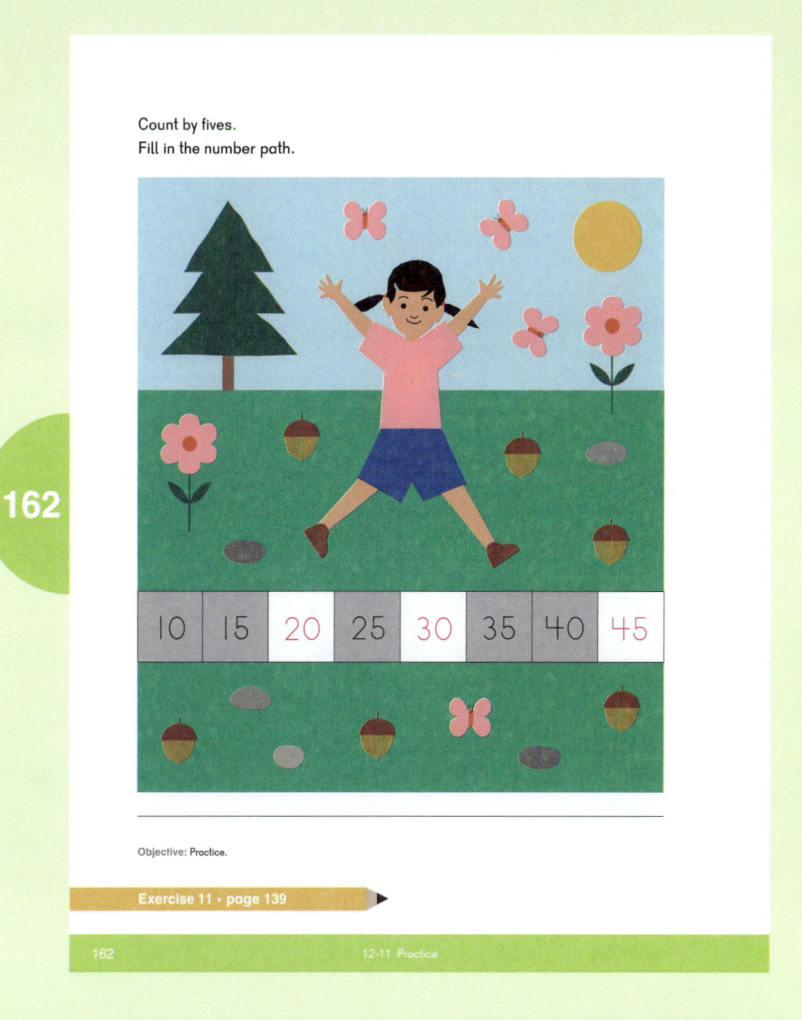

Exercise 1 • pages 117–118

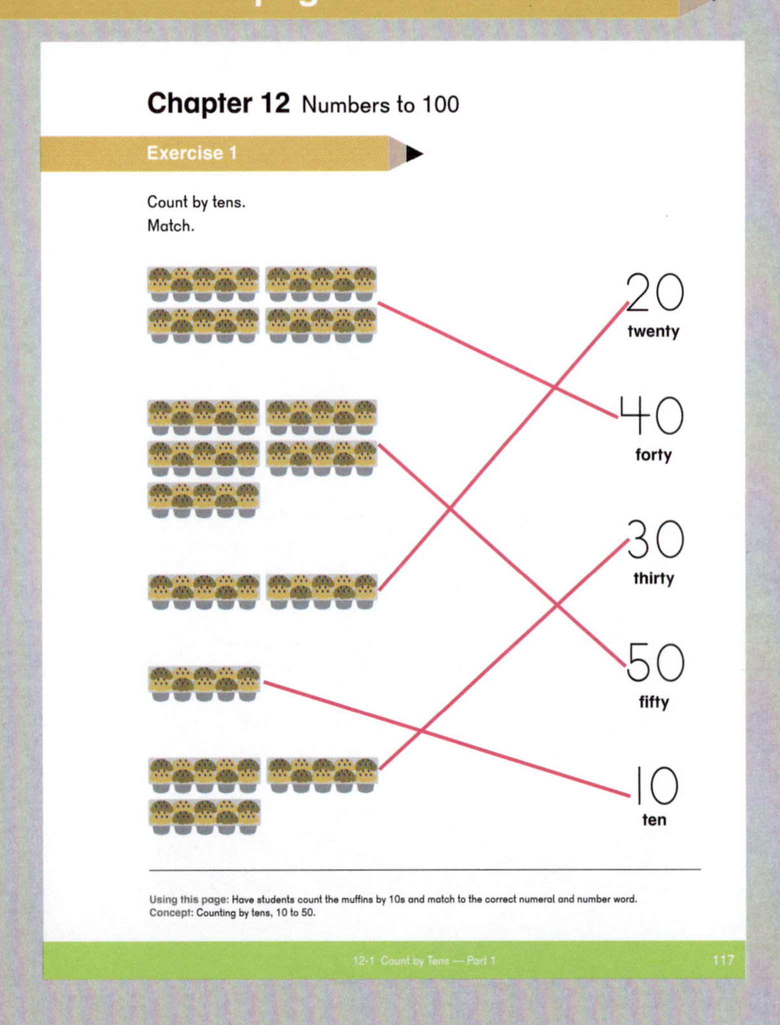

Chapter 12 Numbers to 100

Exercise 1

Count by tens.
Match.

20 twenty

40 forty

30 thirty

50 fifty

10 ten

Using this page: Have students count the muffins by 10s and match to the correct numeral and number word.
Concept: Counting by tens, 10 to 50.

12-1 Count by Tens — Part 1 117

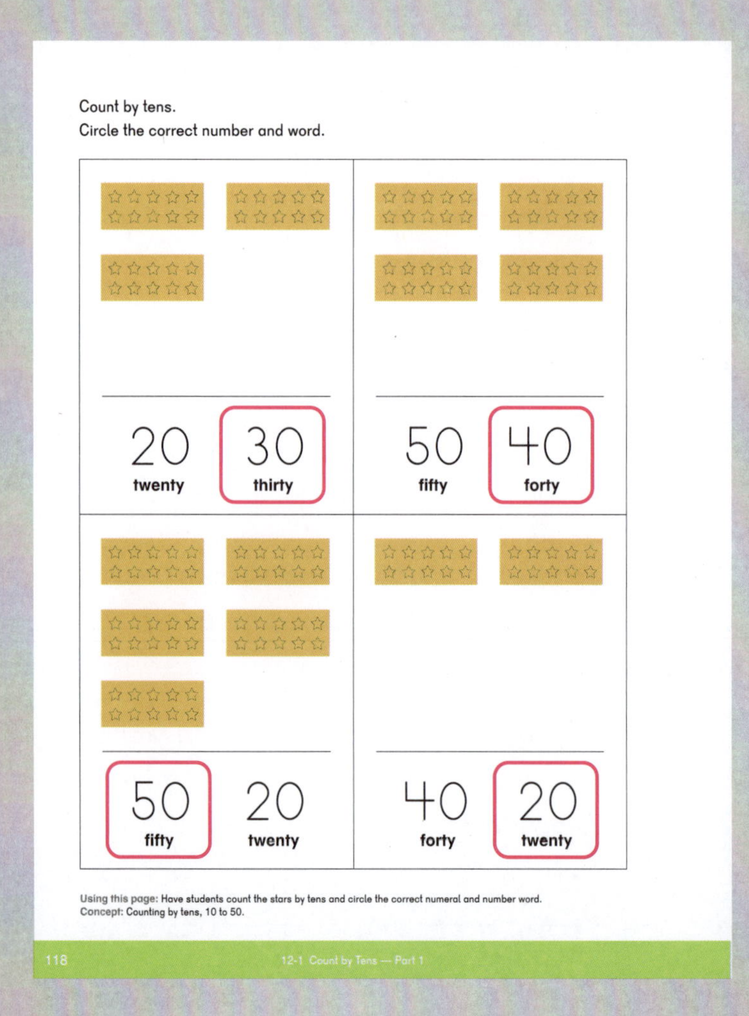

Count by tens.
Circle the correct number and word.

20 twenty | 30 thirty

50 fifty | 40 forty

50 fifty | 20 twenty

40 forty | 20 twenty

Using this page: Have students count the stars by tens and circle the correct numeral and number word.
Concept: Counting by tens, 10 to 50.

118 12-1 Count by Tens — Part 1

Exercise 2 • pages 119–120

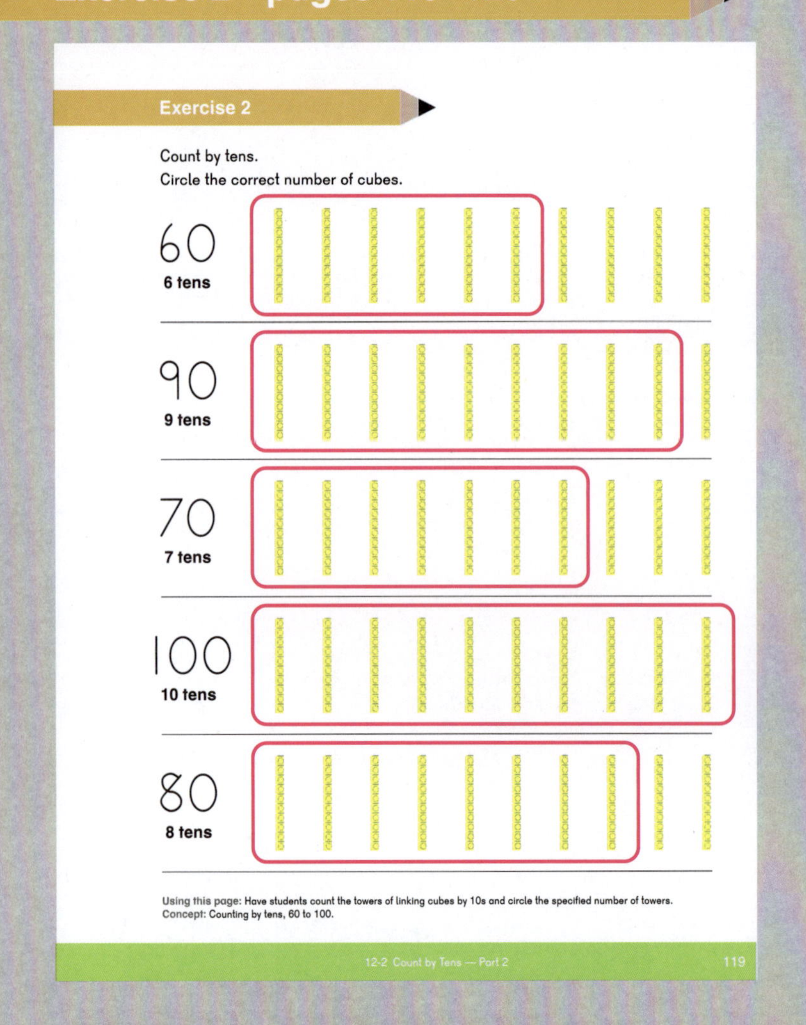

Exercise 2

Count by tens.
Circle the correct number of cubes.

60 6 tens

90 9 tens

70 7 tens

100 10 tens

80 8 tens

Using this page: Have students count the towers of linking cubes by 10s and circle the specified number of towers.
Concept: Counting by tens, 60 to 100.

12-2 Count by Tens — Part 2 119

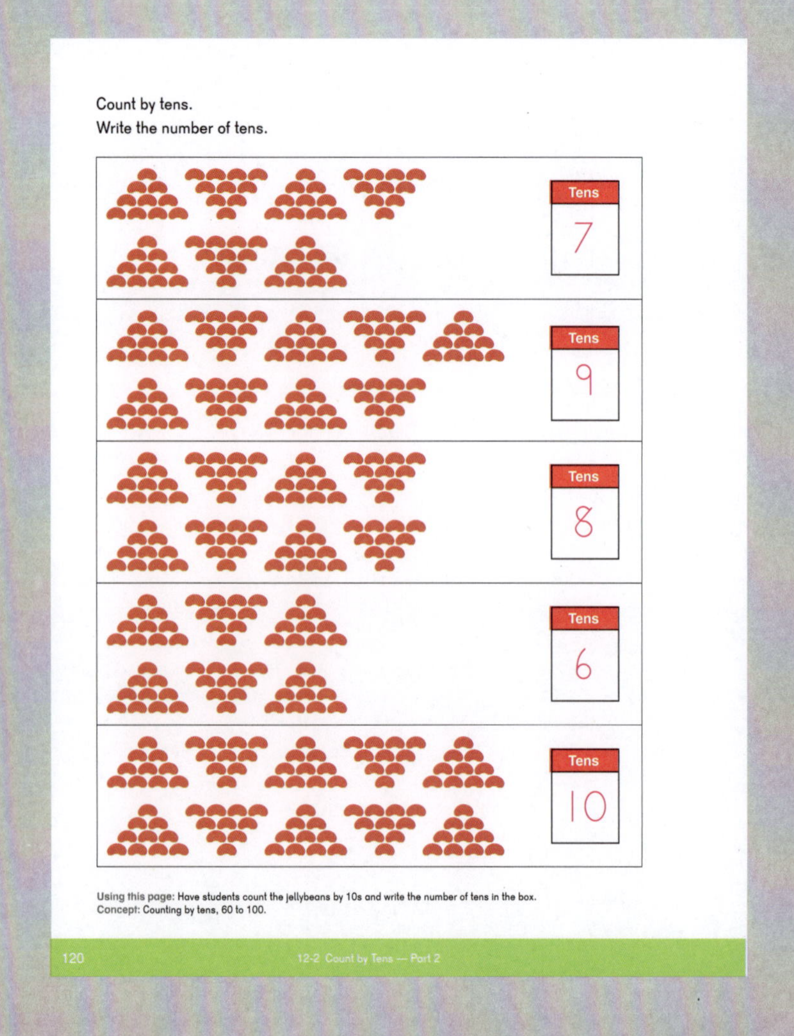

Count by tens.
Write the number of tens.

Tens 7

Tens 9

Tens 8

Tens 6

Tens 10

Using this page: Have students count the jellybeans by 10s and write the number of tens in the box.
Concept: Counting by tens, 60 to 100.

120 12-2 Count by Tens — Part 2

Exercise 3 • pages 121–122

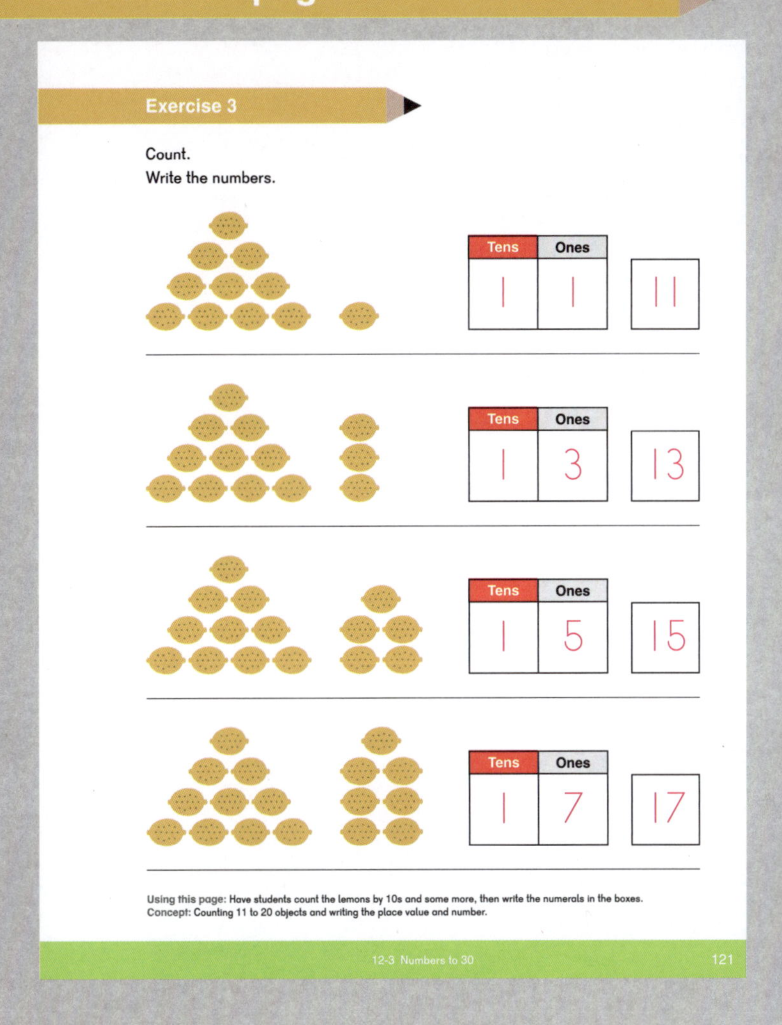
Exercise 3

Count.
Write the numbers.

Tens	Ones	
1	1	11
1	3	13
1	5	15
1	7	17

Using this page: Have students count the lemons by 10s and some more, then write the numerals in the boxes.
Concept: Counting 11 to 20 objects and writing the place value and number.

12-3 Numbers to 30 121

Count.
Write the numbers.

Tens	Ones	
2	3	23
2	2	22
2	6	26
3	0	30

Using this page: Have students count the cubes by 10s and some more, then write the numerals in the boxes.
Concept: Counting 20 to 30 objects and writing the place value and number.

122 12-3 Numbers to 30

Exercise 4 • pages 123–124

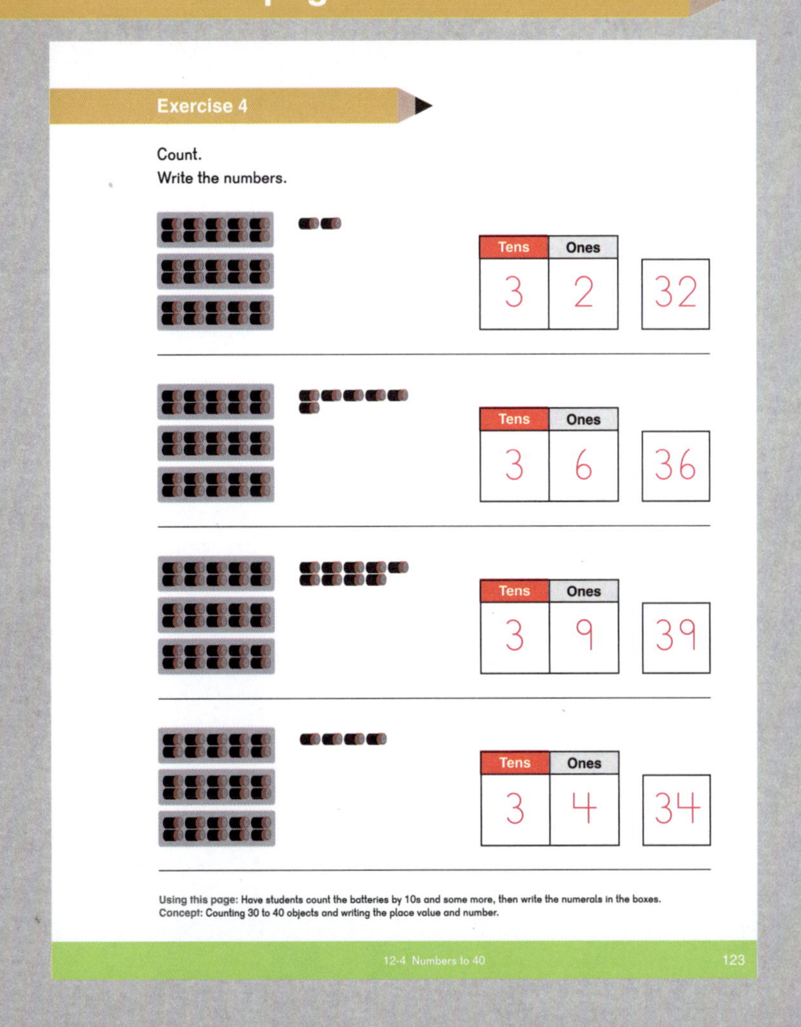
Exercise 4

Count.
Write the numbers.

Tens	Ones	
3	2	32
3	6	36
3	9	39
3	4	34

Using this page: Have students count the batteries by 10s and some more, then write the numerals in the boxes.
Concept: Counting 30 to 40 objects and writing the place value and number.

12-4 Numbers to 40 123

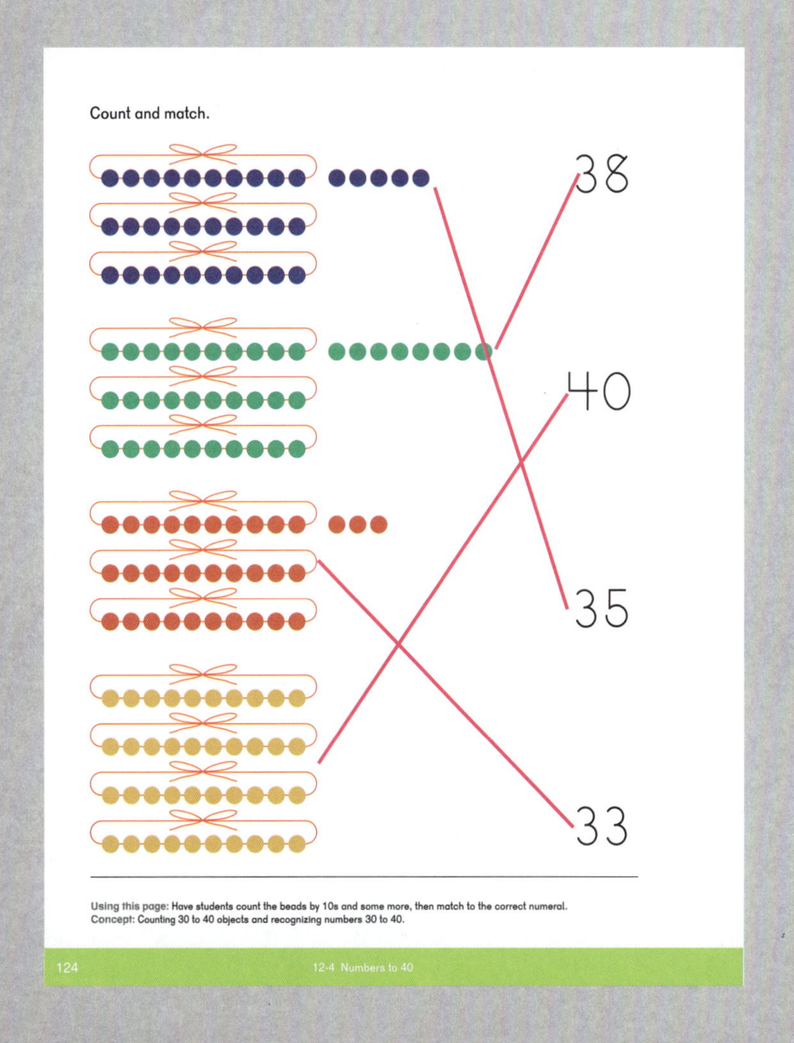
Count and match.

Using this page: Have students count the beads by 10s and some more, then match to the correct numeral.
Concept: Counting 30 to 40 objects and recognizing numbers 30 to 40.

124 12-4 Numbers to 40

Exercise 5 • pages 125–126

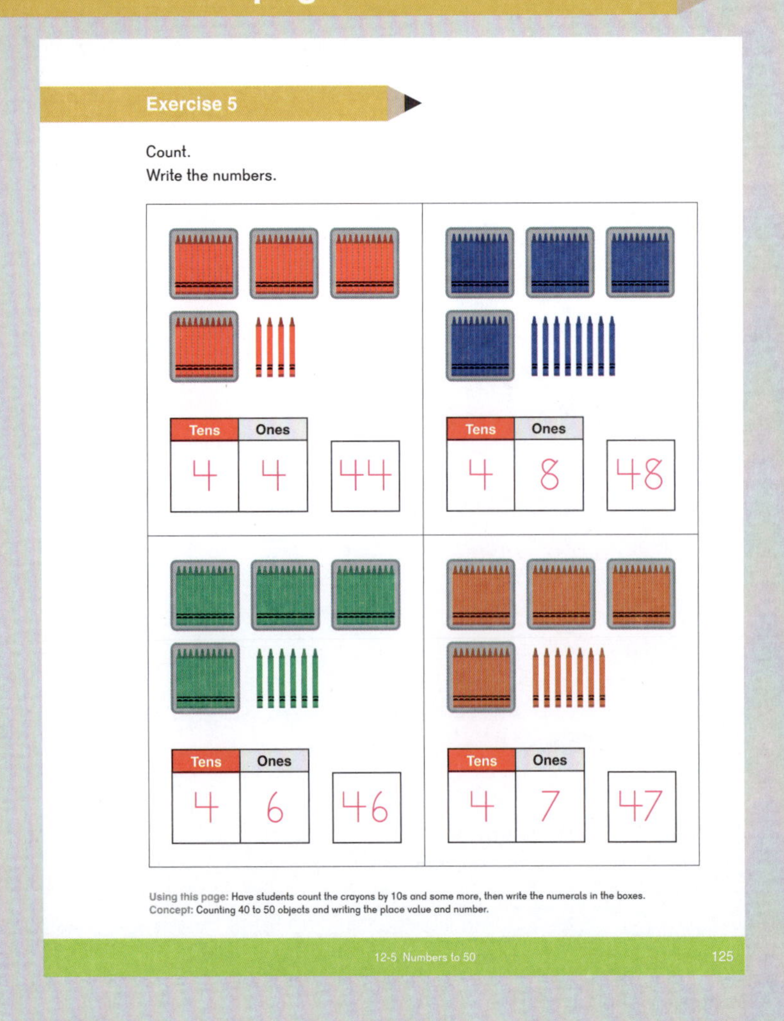

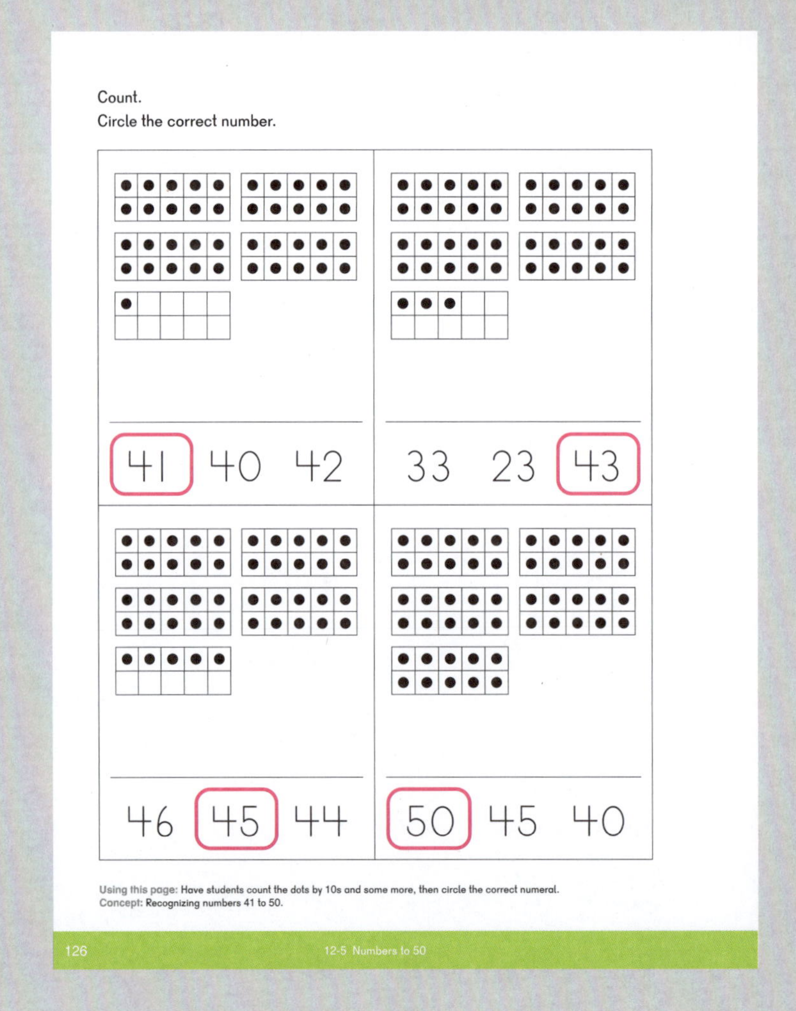

Exercise 6 • pages 127–130

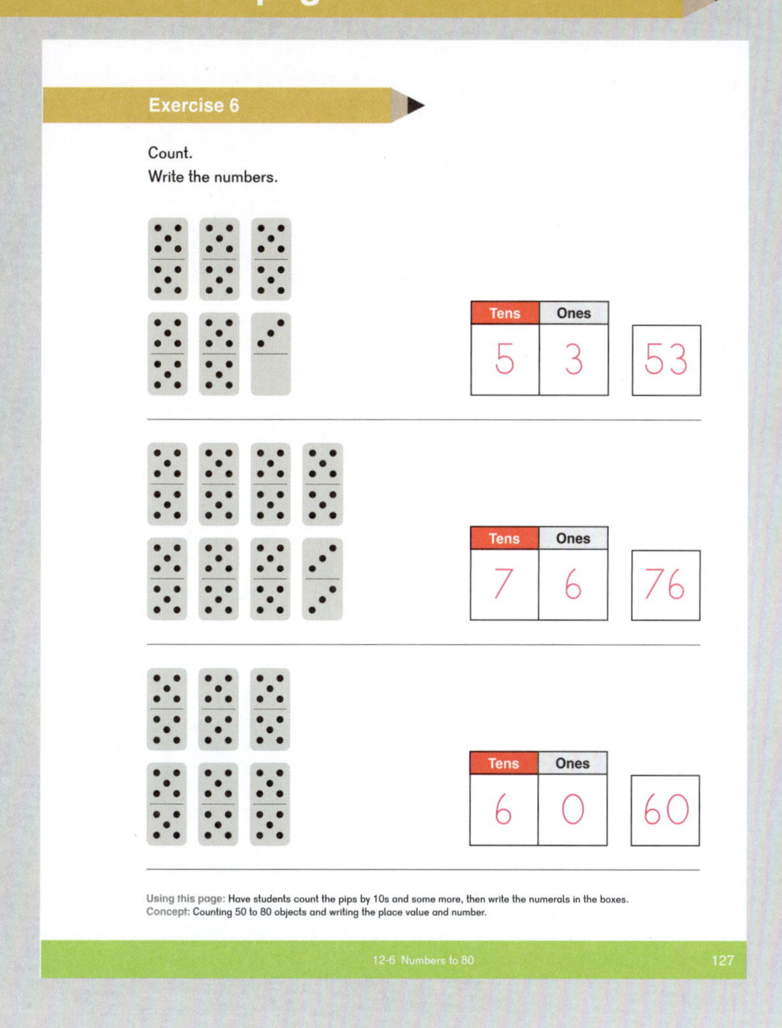

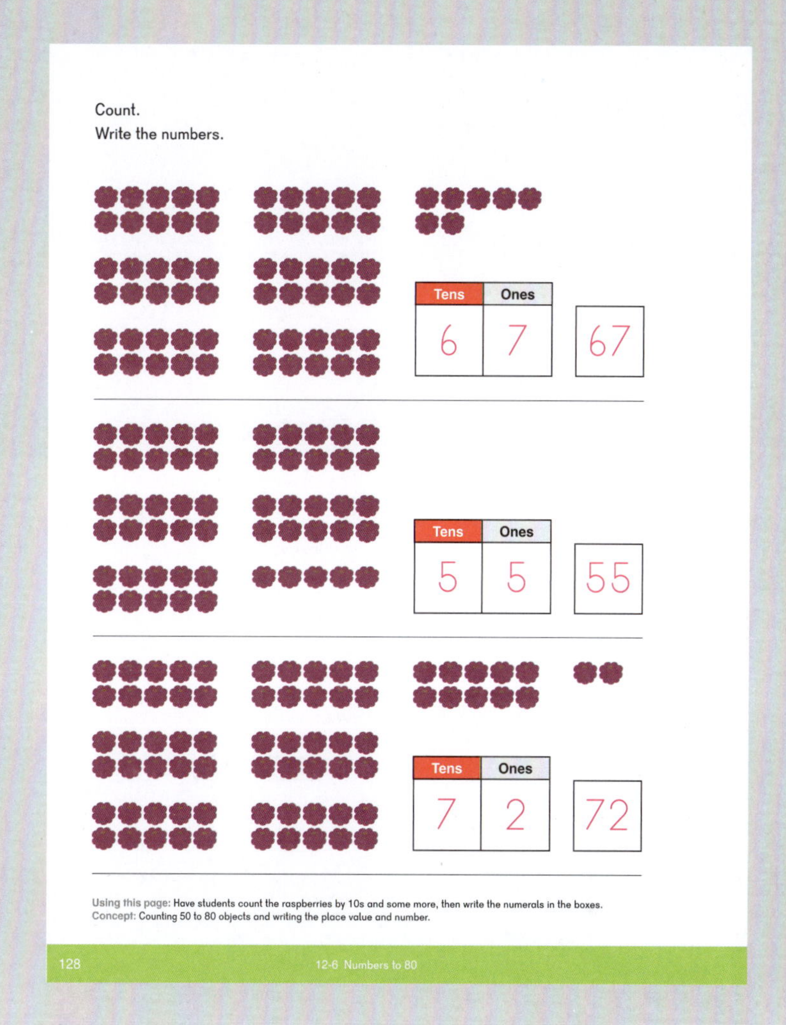

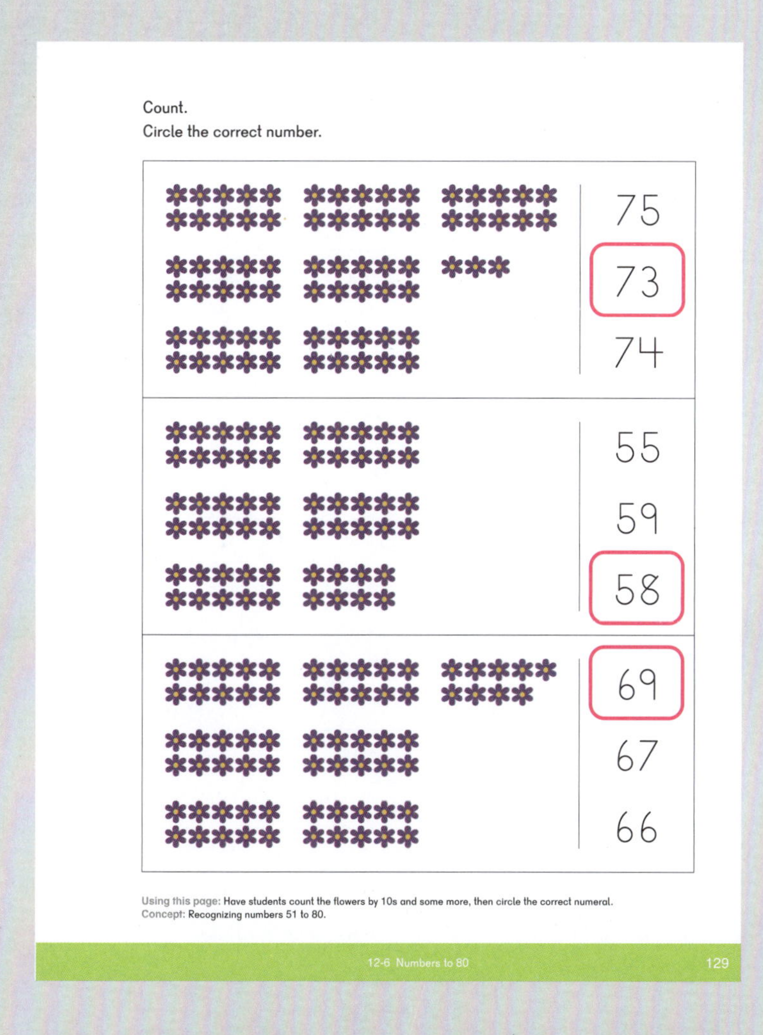

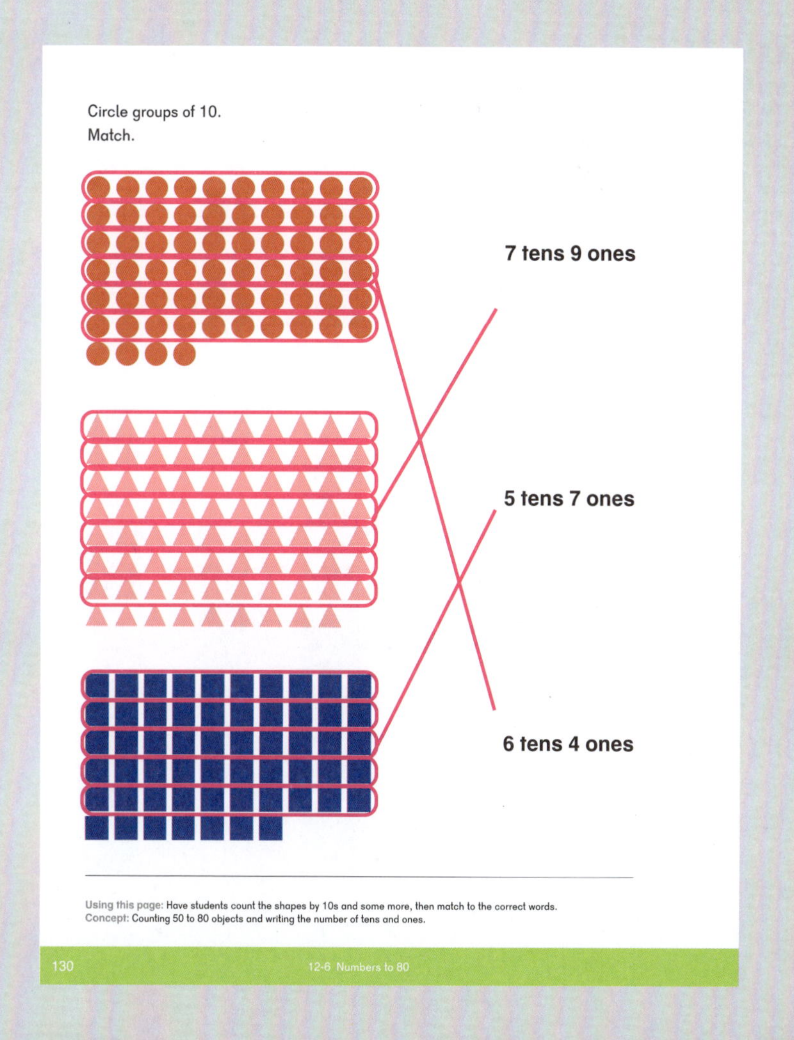

Exercise 7 • pages 131–132

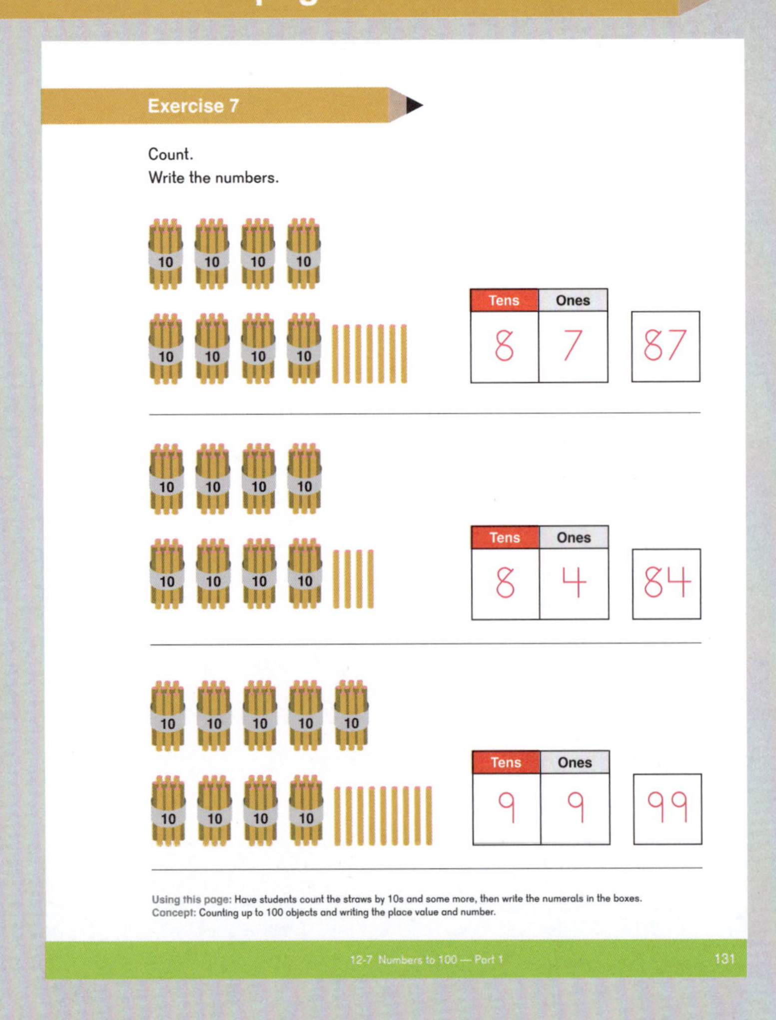

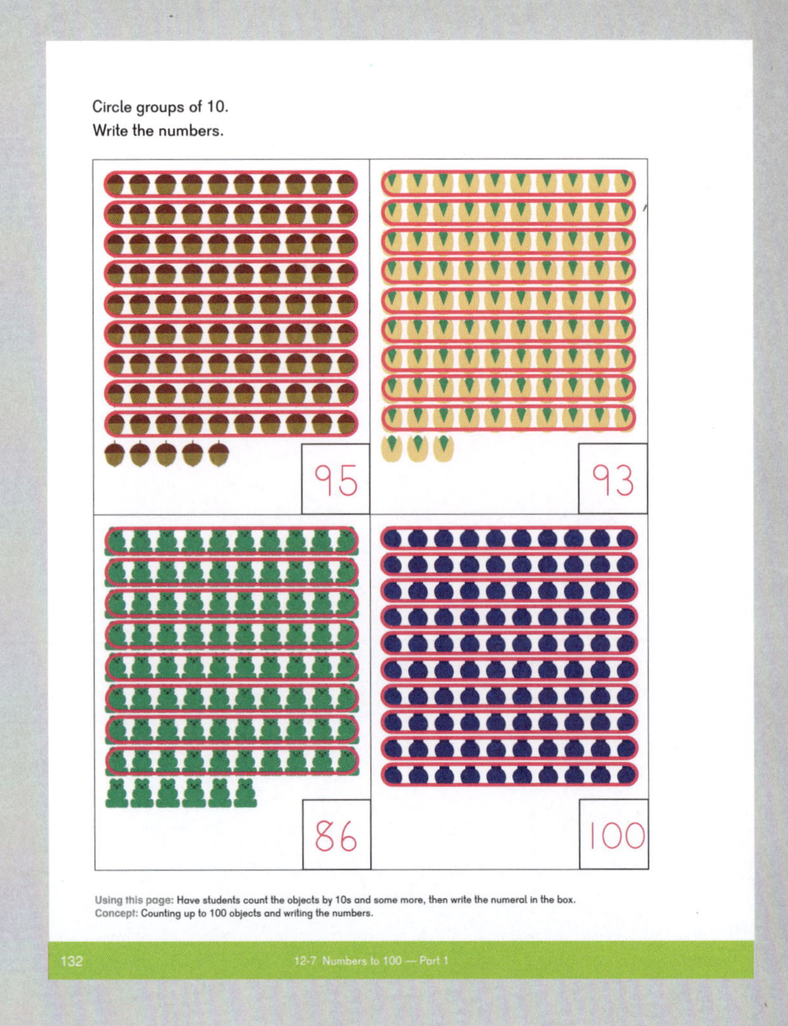

Exercise 8 • pages 133–134

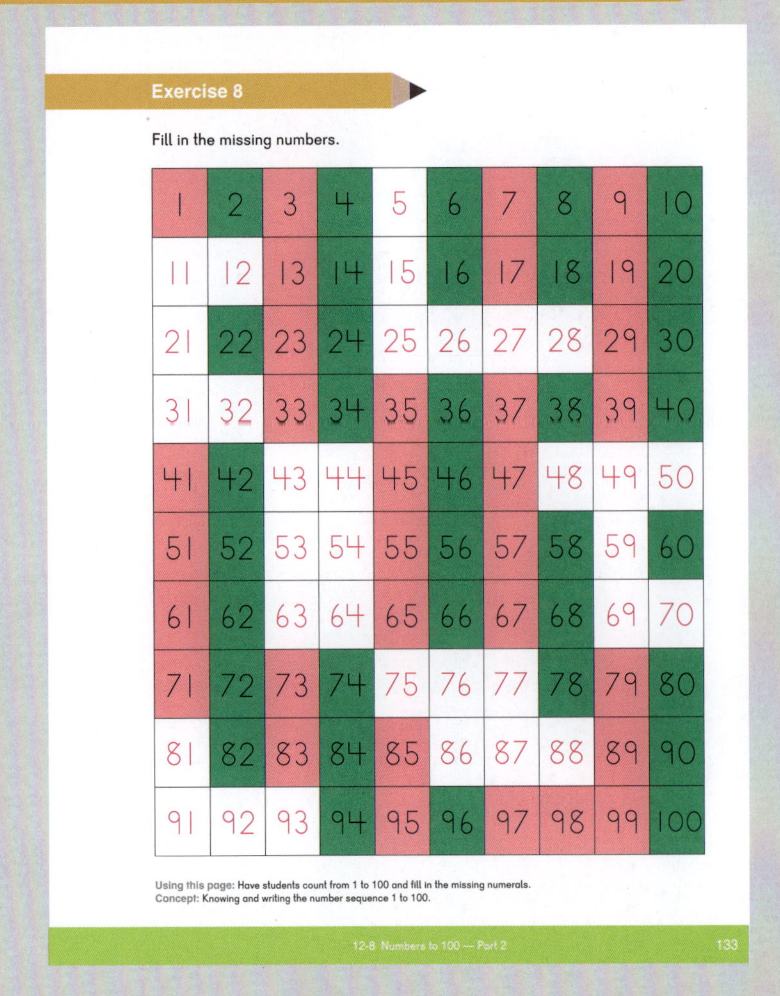

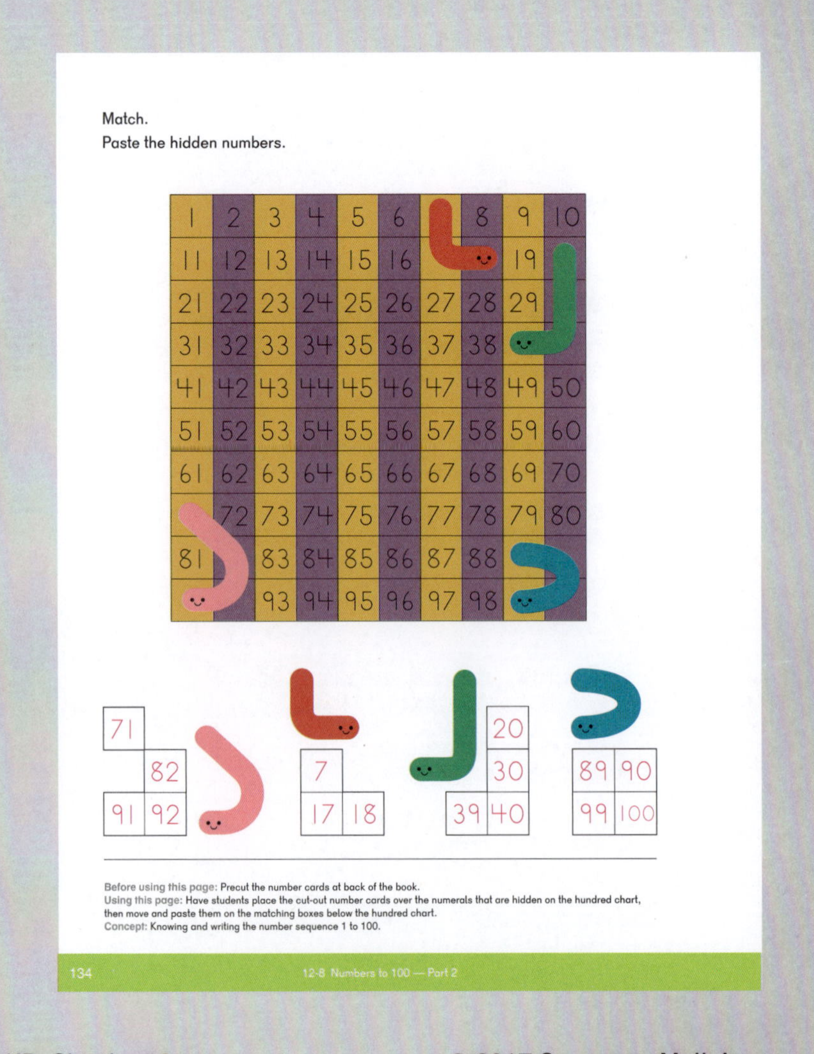

Exercise 9 • pages 135–136

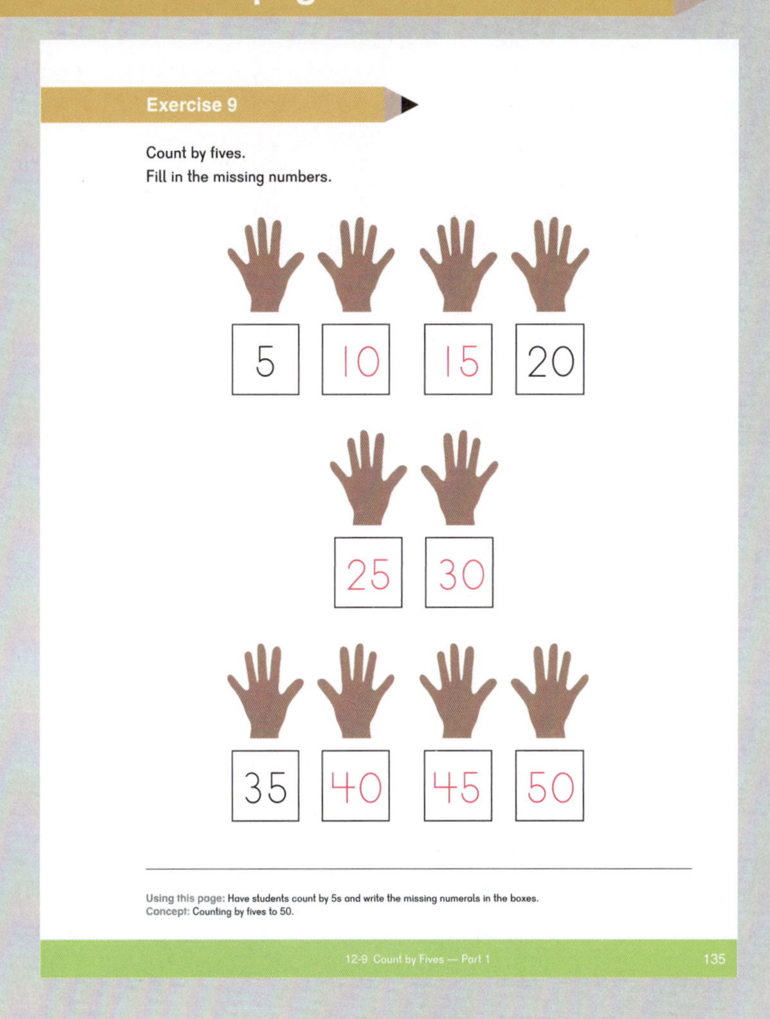

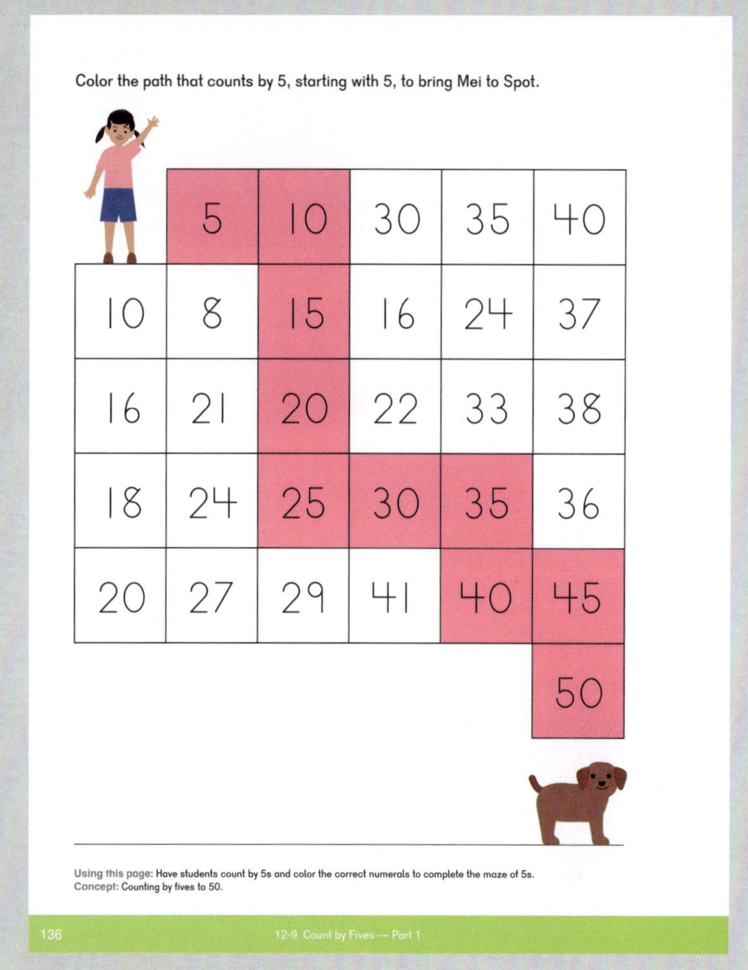

Exercise 10 • pages 137–138

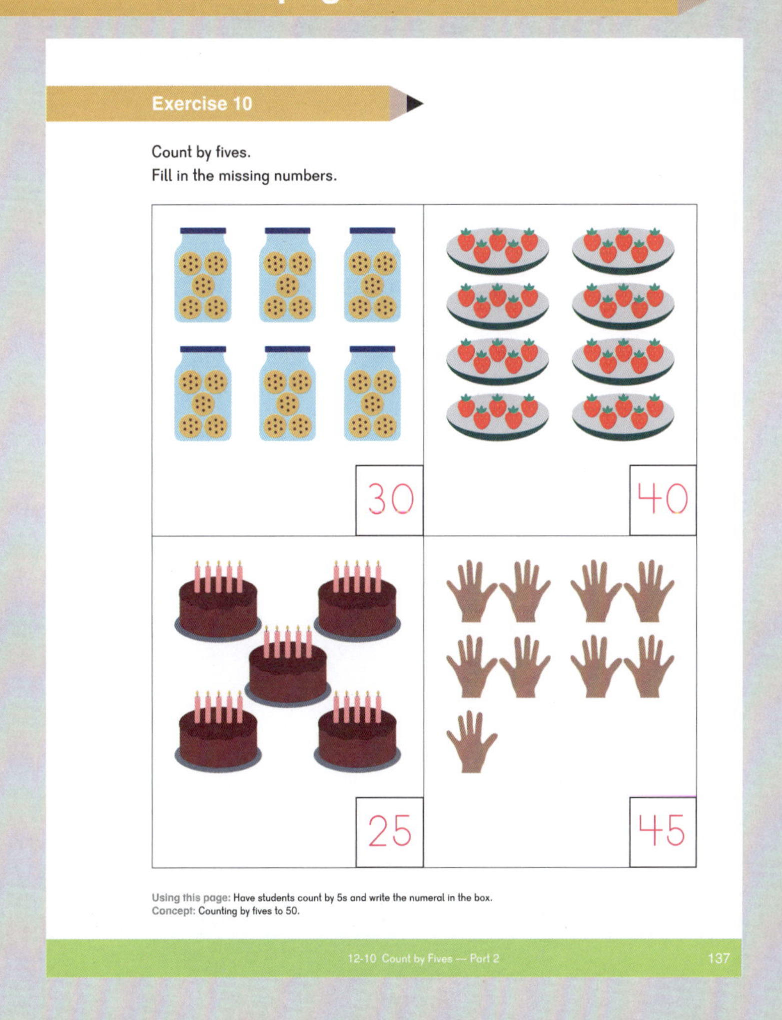

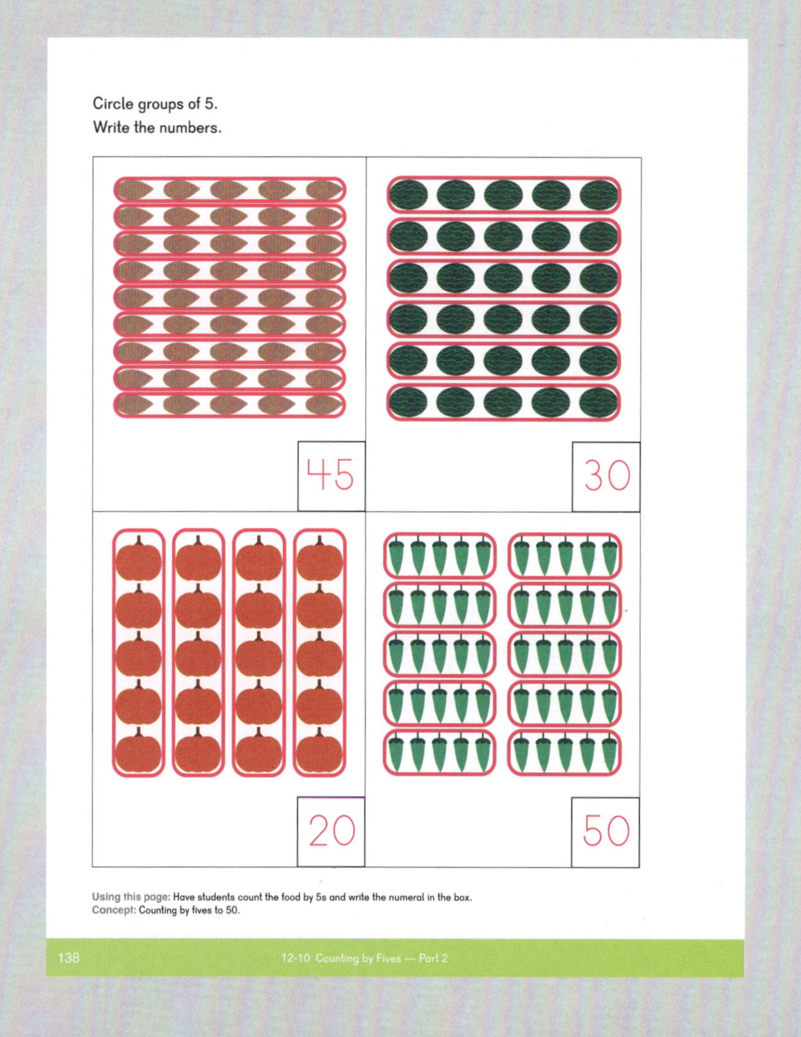

Exercise 11 • pages 139–140

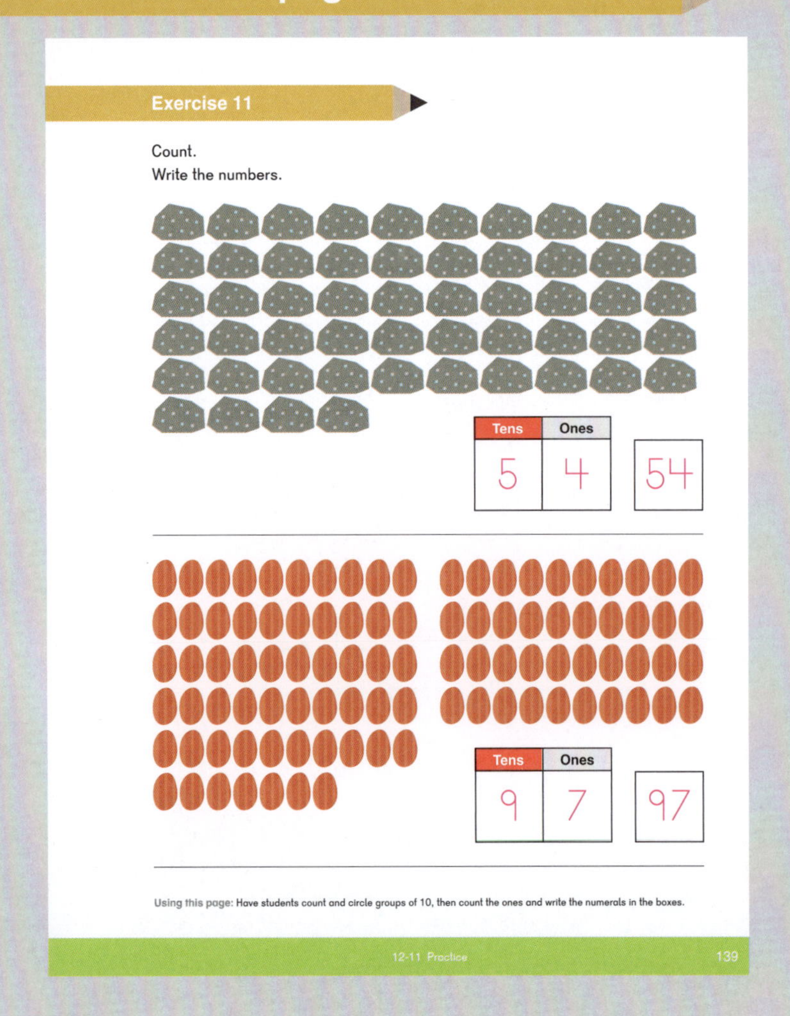

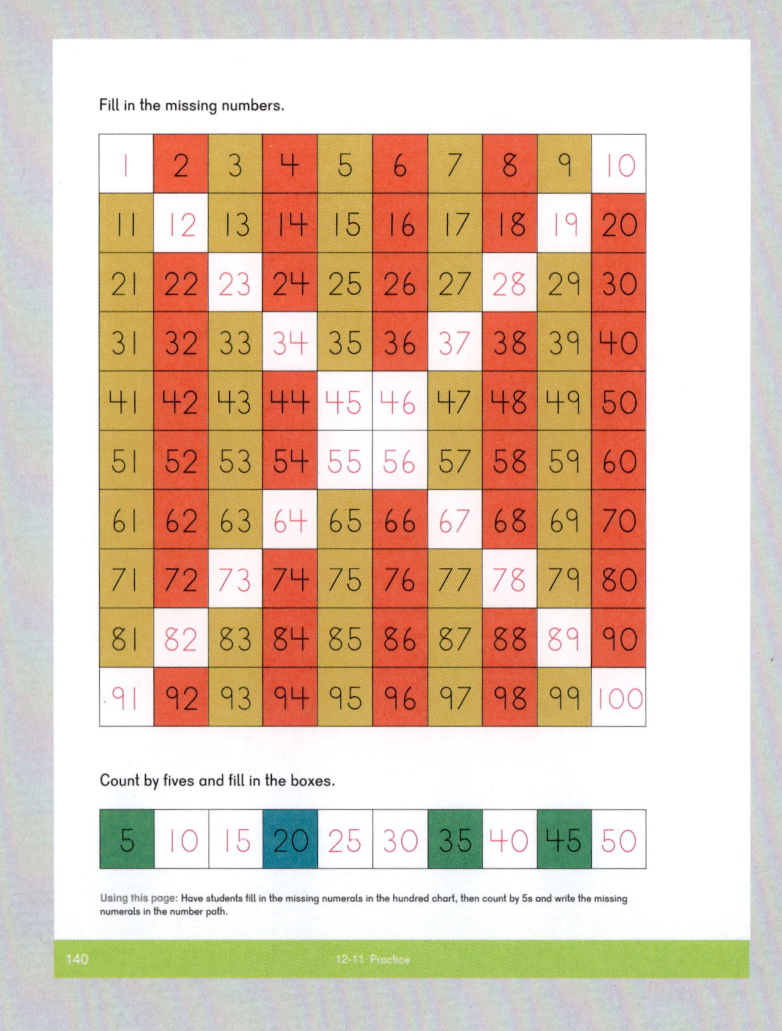

Suggested number of class periods: 5–6

	Lesson	Page	Resources	Objectives
	Chapter Opener	p. 213	TB: p. 163	
1	Day and Night	p. 214	TB: p. 164 WB: p. 141	Understand the passage of time including night, day, morning, afternoon, and evening.
2	Learning About the Clock	p. 216	TB: p. 166 WB: p. 143	Learn which numbers are on an analog clock face and their order.
3	Telling Time to the Hour — Part 1	p. 218	TB: p. 168 WB: p. 145	Tell time to the hour.
4	Telling Time to the Hour — Part 2	p. 220	TB: p. 170 WB: p. 147	Show time to the hour on a clock.
5	Practice	p. 222	TB: p. 172 WB: p. 149	Practice reading clocks and telling time to the hour.
	Workbook Solutions	p. 225		

Notes

In this chapter, students will learn to:

- Tell time to the hour on both analog and digital clocks.
- Understand the concept of time passing and recognize different times of the day, such as morning, afternoon, evening, day, and night.

Measuring time was originally based on an analog clock, which reflects the idea that the hours cycle through each day and repeat the next day. This concept is not as evident on digital clocks. Students therefore first learn how to tell time with an analog clock where they can observe the movement of the hands as time passes.

There is only 1 minute out of every 60 that students will actually see the time as exactly on the hour, or that a digital clock will show time to the hour. Since they are learning to tell time to the hour, they will learn which hand is the hour hand on an analog clock. Also, since time is rarely on the hour, students will also need to understand how the minute hand moves in relation to the hour hand, and recognize its position before the hour or after the hour.

Time can be difficult for students to grasp since it is something they cannot see. Although students will be using analog clocks with geared hands in order to observe the movement of the hands, just moving the hands around does not give them a feel for how long an hour is or what an hour means. Provide opportunities for students to look at a classroom analog clock throughout the day and look at the position of the hands at different times, particularly when the time is "about" on the hour. Help them become aware of activities that take about an hour.

It is assumed that all students will have access to recording tools. When a lesson refers to a whiteboard, any writing materials can be used.

Materials

- Art paper
- Brass fasteners (brads)
- Cardstock
- Classroom game board
- Crayons or markers
- Demonstration clock
- Dice
- Geared clocks
- Paper plates
- Ruler
- Yardstick

Blackline Masters

- Alligator Cards
- Analog Clock Face
- Analog Time Cards
- Digital Time Cards
- Number Cards
- Story Template
- Time of Day Cards
- Word Time Cards

Storybooks

- *Telling Time With Big Mama Cat* by Dan Harper, Barry Moser, and Cara Moser
- *The Clock Struck One: A Time-Telling Tale* by Trudy Harris and Carrie Hartman
- *It's About Time* by Stuart Murphy
- *The Completed Hickory Dickory Dock* by Jim Aylesworth
- *Bats Around the Clock* by Kathi Appelt
- *What Time Is It, Mr. Crocodile?* by Judy Sierra

Letters Home

- Chapter 13 Letter

Notes

Lesson Materials

- Picture books (see suggestions on page 211 of this Teacher's Guide)

Look at page 163. Ask students what they notice about the pictures:

- What is happening?
- When during the day could these activities be happening?
- Invite students to tell stories about the pictures.

To extend, ask students to tell what could have happened between when Dion wakes up and when he has a snack, between snack and basketball practice, etc.

Ask students to share if they know what time they wake up and when they go to bed.

Use this discussion to assess students' understanding of time.

Read a picture book in which different things happen throughout a day, pausing to ask questions about general times during the day.

Chapter 13

Time

163

163

Extend

★ Order the Day

Materials: Time of Day Cards (BLM)

Pass out 3 or 4 Time of Day Cards (BLM) to pairs of students. Ask students to order the cards based on what they think happens first, second, and last. Ask students to tell stories about their pictures.

Objective

- Understand the passage of time including night, day, morning, afternoon, and evening.

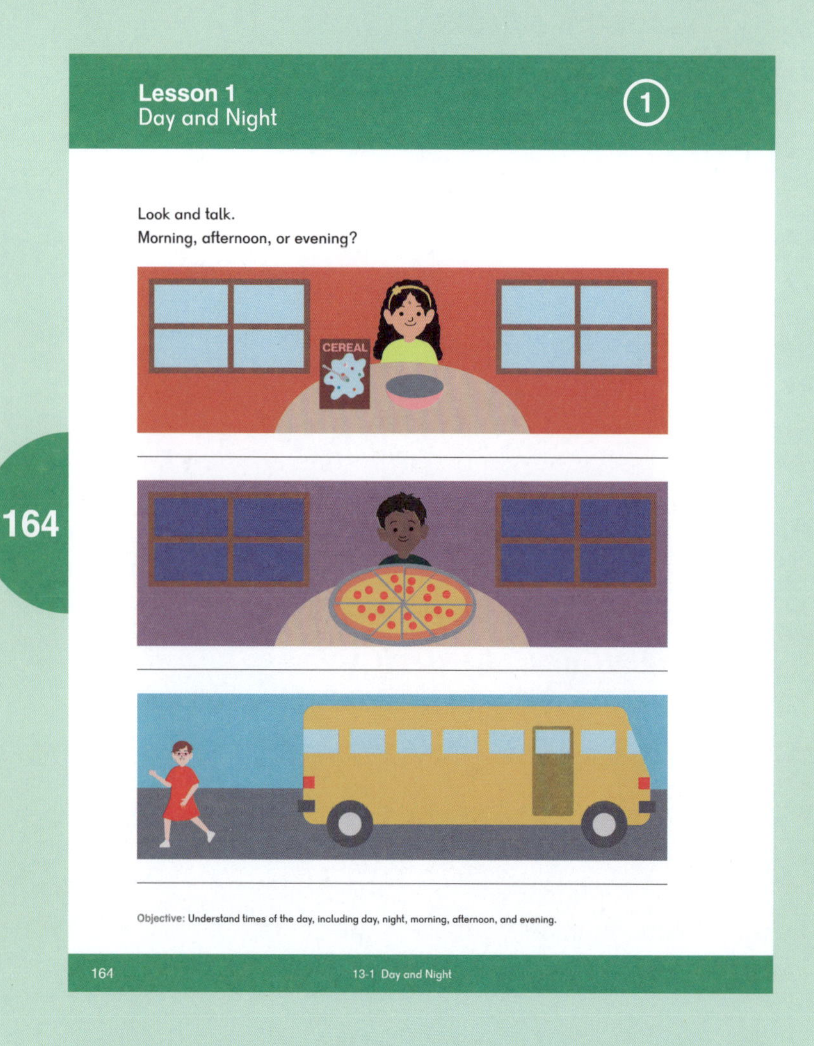

Explore

Discuss activities that students do during the morning, afternoon, and evening. These ideas could be organized on a chart divided into three sections: Morning, Afternoon, and Evening.

Learn

Have students discuss the events on textbook page 164. Ask them how they can tell from the picture what time of day the event is happening. Use the terms:

- Day and night
- Morning
- Afternoon
- Evening

Whole Group Activity

Act it Out

Have students act out activities they do during a day. Have them start asleep, then announce, "It's morning." Students can get up and act out activities they do in the morning. Continue with afternoon and evening activities, ending when students fall asleep at night.

Small Group Activities

Textbook Page 165

▲ Time of Day Sort

Materials: Time of Day Cards (BLM)

Provide students with Time of Day Cards (BLM). Students should sort the cards under the headers Morning, Afternoon, and Night. Using the cards to guide them, students can then tell stories about what happens in a day.

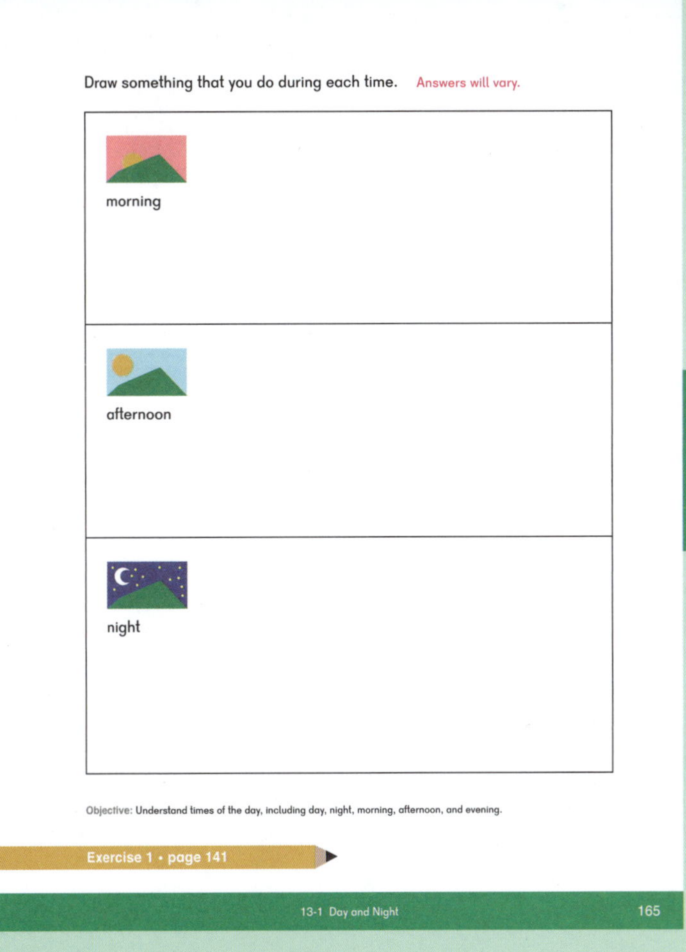
Draw something that you do during each time. Answers will vary.

morning

afternoon

night

Objective: Understand times of the day, including day, night, morning, afternoon, and evening.

Exercise 1 • page 141

13-1 Day and Night 165

165

Exercise 1 • page 141

Extend

★ Book of My Day — Part 1

Materials: Story Template (BLM), art paper, crayons or markers

Students write or illustrate stories about their days at school or a favorite day (birthday, holiday, etc.) including the words "morning," "afternoon," and "night." This can be organized into a book to use in a later lesson.

Objective

- Learn which numbers are on an analog clock face and their order.

Lesson Materials

- Demonstration clock
- Picture books (see suggestions on page 211 of this Teacher's Guide)

Explore

Take students on a clock hunt around the school. Have students notice the different types of clocks they see.

Alternatively, read a picture book about time, pausing on pages with clocks to ask students what they notice about the clocks.

Discuss students' prior knowledge of clocks and how they work. Students may say:

- I've seen clocks with just numbers on my dad's phone.
- The short hand is the hour hand, the long hand is the minute hand.
- I have a watch.
- We use clocks to tell time so we know when we need to do things.

Learn

Show students a large demonstration clock or a classroom clock. Ask students what they notice about the clock shown on page 166 and the demonstration clock. Ask how the clocks are the same or different. Have students say the numbers as you point to them in clockwise order, starting at 1. Discuss how the numbers are spaced around the clock.

Although telling time to the hour is in the next lesson, students may be curious about the hands. The textbook clock face has 2 hands. As is typical of many analog clocks, one hand is shorter and one is longer. The hands point to numbers to indicate the hours and the minutes. Minutes will formally be addressed in **Dimensions Math® 1B**.

Whole Group Activity

▲ Clock Face Simon Says

Play Simon Says using commands like:

- Simon says, "Point both arms to the 12 on the clock."
- Simon says, "Point your head at the 6 on the clock."
- Simon says, "Bend at the waist to show 9 on the clock."

Draw or display a clock face on the board for students to reference or refer to.

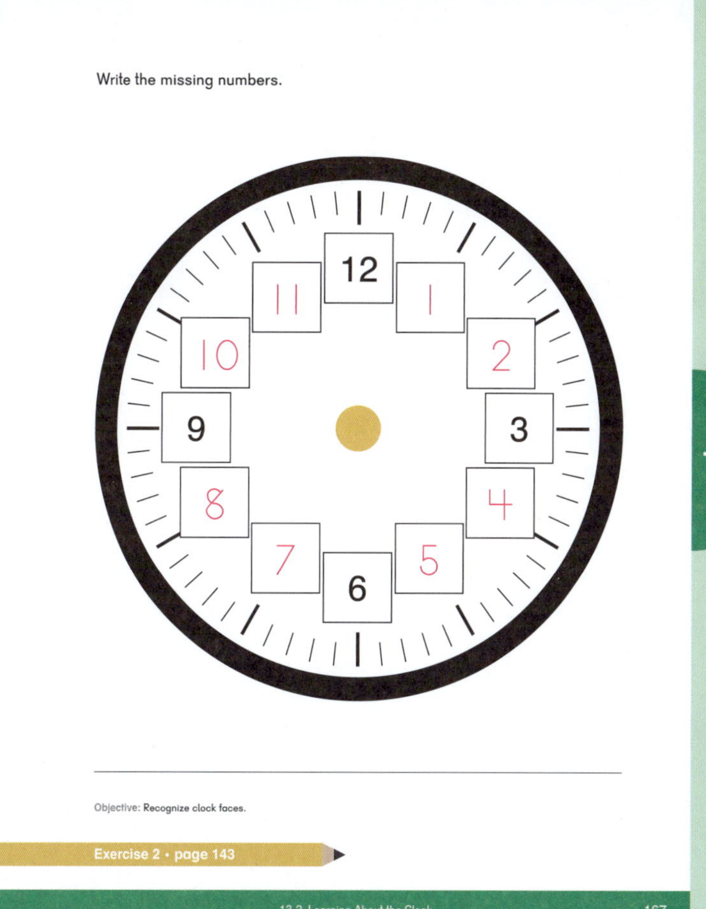

167

Small Group Activities

Textbook Page 167

▲ Paper Plate Clocks

Materials: Paper plates, markers, brass fasteners, hour hands and minute hands cut from cardstock

Give each student a paper plate. Have them make clocks by adding numbers around the clock face. Use brass fasteners to attach hour hands and minute hands.

As an extension, as students complete their clocks, have them partner up and tell stories about their days using the clocks as a guide. A student may say, "I wake up at 7 o'clock," and point the hour hand to the number 7.

Exercise 2 • page 143

Extend

★ Elapsed Time

Materials: Time of Day Cards (BLM)

Have students create elapsed time problems for each other to solve. For example, a student may write or say:

- It is now 9 in the morning. What time is 1 hour later?
- Soccer practice starts at 4 o'clock and lasts 2 hours. What time does soccer practice end?

Objective

- Tell time to the hour.

Lesson Materials

- Geared clocks or student clocks from the previous lesson
- Demonstration clock

Explore

Have students set their clocks so that the minute hand points to the 12 and the hour hand points to the 1. Explain to students that we read this time as "1 o'clock."

Learn

Write "1 o'clock" on the board.

If using geared clocks, have students move the minute hand clockwise around the clock until it is pointing to the 12 again and the hour hand points to the 2. If using student clocks, ask students to show 2 o'clock with the minute hand pointing to the 12 and the hour hand pointing to the 2. Ask students how we would say this time.

Continue through several different times. At each hour, ask where the two hands are pointing and which one tells us the hour.

Discuss what happens when the minute hand goes around the clock again and the time changes from 12 o'clock to 1 o'clock. Note that a.m. and p.m. will be introduced in **Dimensions Math® 2B**.

Show 1 o'clock again on the demonstration clock. Write "1:00" on the board and explain that the number on the left of the colon tells the hour and the numbers on the right of the colon tell the minutes. Tell students that when there are 2 zeros written this way, we read it as "o'clock," for example, "1 o'clock."

Discuss the different clocks on page 169.

168

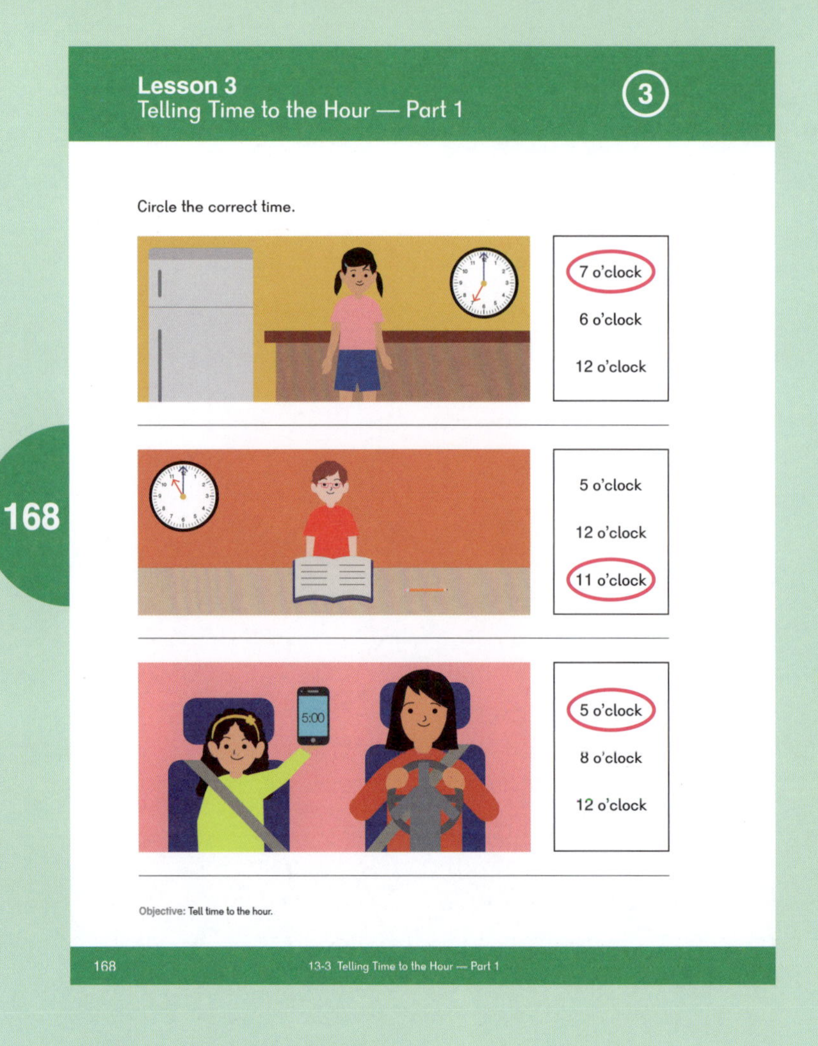
Lesson 3
Telling Time to the Hour — Part 1
3

Circle the correct time.

7 o'clock
6 o'clock
12 o'clock

5 o'clock
12 o'clock
11 o'clock

5 o'clock
8 o'clock
12 o'clock

Objective: Tell time to the hour.

168 13-3 Telling Time to the Hour — Part 1

Whole Group Activity

▲ Clock Exercises

Materials: Geared clock

Display a time that is on the hour with no minutes on a geared clock. For example, 5 o'clock. Ask students to tell the time and then do an activity, i.e., hop, clap, stomp, stretch, etc., as they count, 1 o'clock, 2 o'clock, 3 o'clock, and so on. Display another time, choose another exercise, and repeat.

Small Group Activities

Textbook Page 167

▲ What Time Could It Be?

Materials: Time of Day Cards (BLM), Analog Time Cards (BLM)

Using the Time of Day Cards (BLM), and a set of Analog Time Cards (BLM), students match a picture card with its possible time. For example, a picture card that shows a child taking a bath could match the 7 o'clock card.

As an extension, ask students to find a match and write or illustrate an additional activity that could occur at that same time.

▲ Match

Materials: Analog Time Cards (BLM), Digital Time Cards (BLM), and Word Time Cards (BLM)

Students arrange the cards faceup in a grid. Students take turns finding two cards that go together. Start out with Analog Time Cards (BLM) and Digital Time Cards (BLM). As the students improve at the game, add in Word Time Cards (BLM) and tell them that they need all 3 cards to have a match.

★ Memory

Materials: Analog Time Cards (BLM), Digital Time Cards (BLM), and Word Time Cards (BLM)

Play as described in **Match** in this lesson, but start with the cards arranged facedown, in a grid.

Exercise 3 • page 145

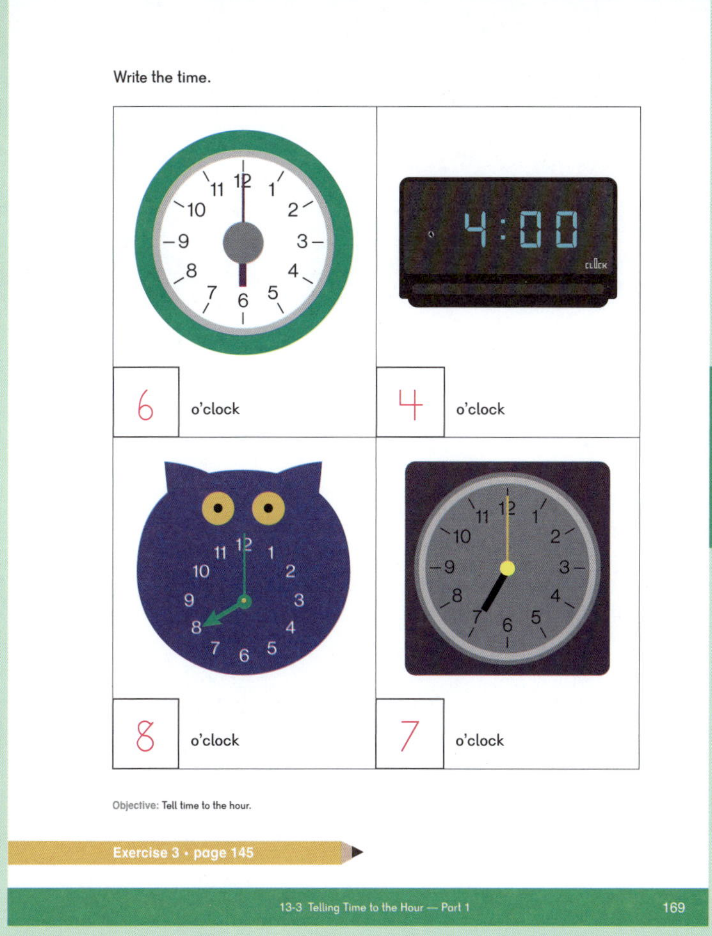

Extend

★ Longest Time

Materials: 4 sets of Analog Time Cards (BLM)

Deal all the cards to each player equally. Each player flips 2 cards and finds how many hours there are between the earlier time and the later time drawn. (The higher number of hours is always the later time.)

The player with the greater time difference keeps the cards. Play continues until a player is out of cards.

If a student knows the difference between a.m. and p.m., she could say that between 11:00 a.m. and 12:00 a.m. is 13 hours.

Lesson 4 Telling Time to the Hour — Part 2

Objective

- Show time to the hour on a clock.

Lesson Materials

- Geared clocks
- Number Cards (BLM) 1 to 12
- Ruler
- Yardstick

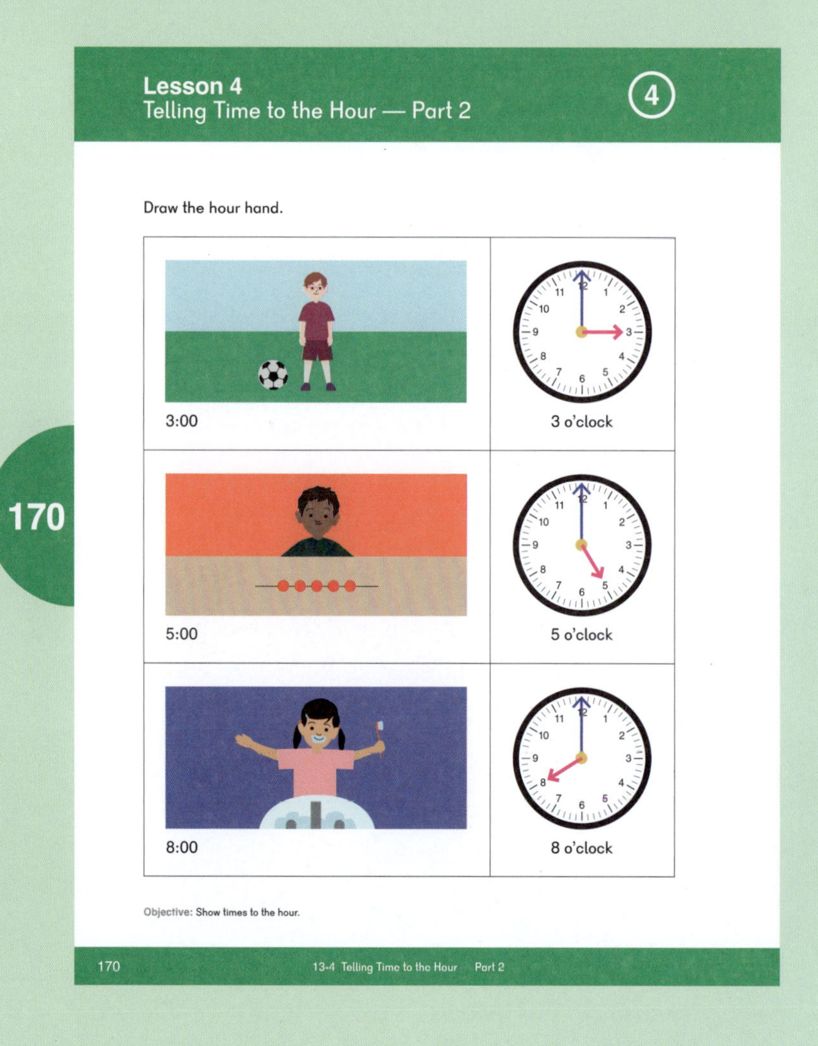

Explore

Provide each student with a geared clock. Have students share stories about things that they do at certain times on the hour. As each story is told, write the time on the board and have children show the time on their geared clocks.

Learn

Create a human clock. Give 12 students a Number Card (BLM) and have them sit in a circle like a clock. Have students take turns showing the time by laying the ruler and yardstick on the ground pointing to the correct numbers. Use the yardstick for the minute hand and the ruler for the hour hand.

Whole Group Activity

▲ Alligator, Alligator, Alligator!

Materials: Geared clock, Alligator Cards (BLM), Word Time Cards (BLM)

Have students sit in a circle. Arrange Word Time Cards (BLM) facedown in the center of the circle, plus at least 3 Alligator Cards (BLM). Students take turns choosing a card, reading the time, and displaying the time on a geared clock. If an alligator is chosen, all students jump up and shout, "Alligator, alligator, alligator!" then sit again. Have another student choose a card and continue.

Small Group Activities

Textbook Pages 170–171

▲ Takeover!

Materials: Analog Time Cards (BLM), Digital Time Cards (BLM), Word Time Cards (BLM), game board from the classroom

Give each group of 2–4 students a game board from the classroom, and a deck of cards made of Analog Time Cards (BLM), Digital Time Cards (BLM), and Word Time Cards (BLM). Students choose a card and move the number of spaces on the board to match the time on the card. For example, if the card shows 5 o'clock, the student moves ahead 5 spaces.

To extend, students move ahead the time shown on the analog and word cards, but move back on the time shown on the digital cards.

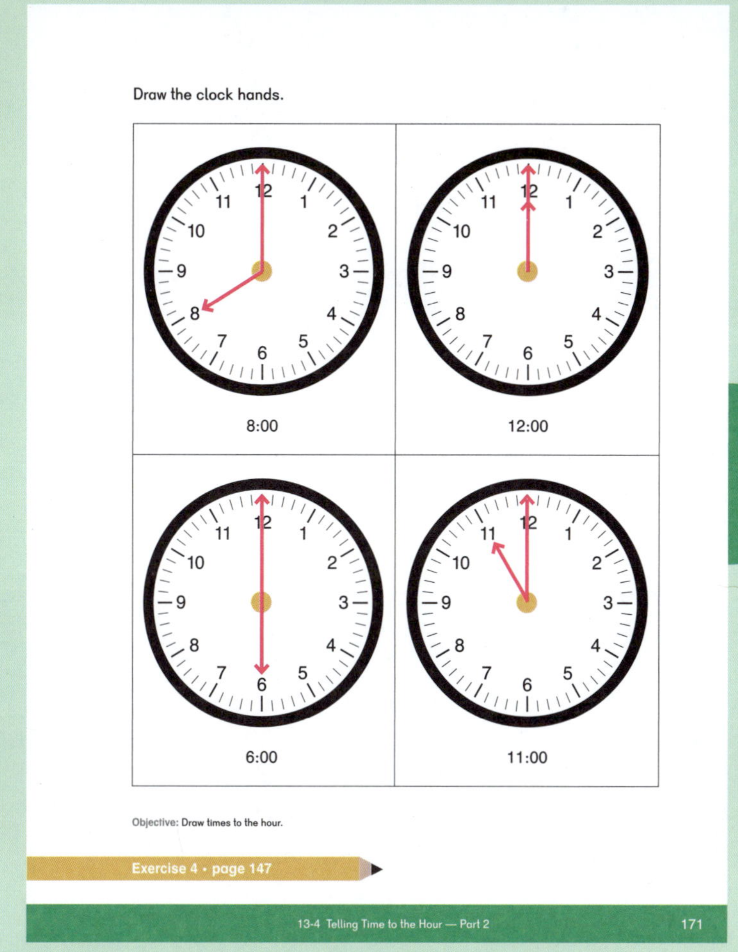
Draw the clock hands.

8:00

12:00

6:00

11:00

Objective: Draw times to the hour.

Exercise 4 • page 147

13-4 Telling Time to the Hour — Part 2 171

171

Exercise 4 • page 147

Extend

★ Book of My Day — Part 2

Materials: Analog Clock Face (BLM)

Provide students with Analog Clock Faces (BLM). Students add clock times to their stories about their day from Lesson 1. The clock faces can be glued into the book.

Objective

- Practice reading clocks and telling time to the hour.

Lesson Materials

- Picture books (see suggestions on page 211 of this Teacher's Guide)

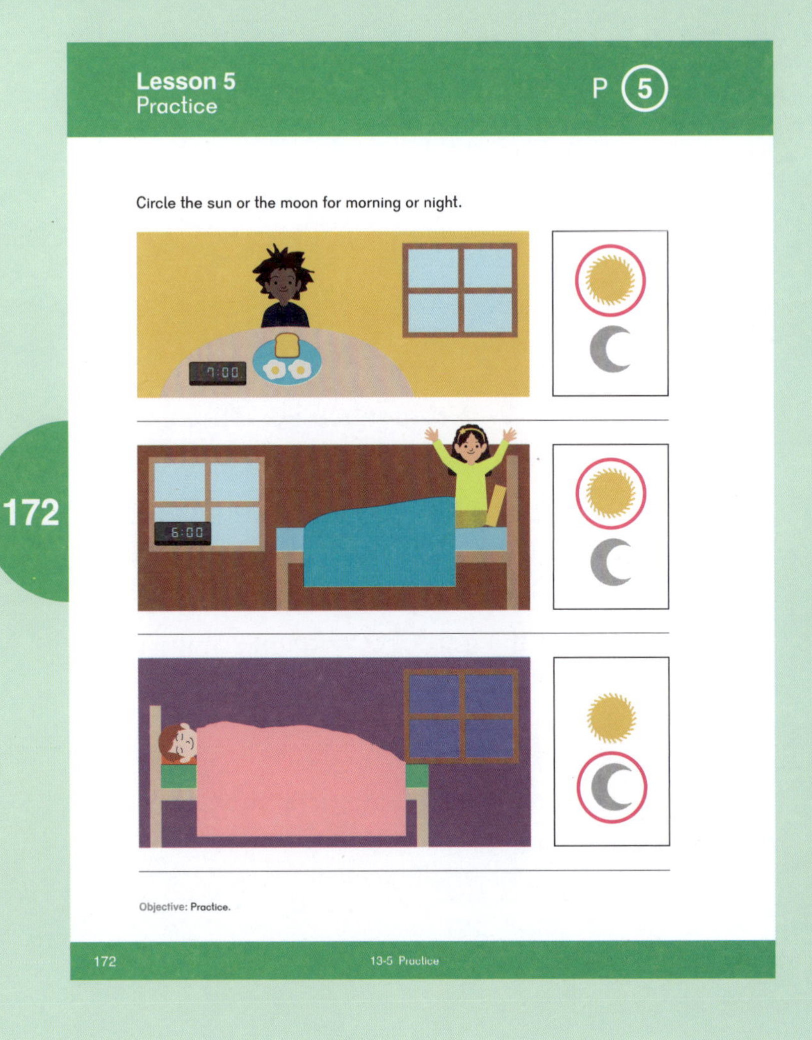

172

Explore

Read a picture book, pausing to ask questions about time as you read. For a challenge, ask elapsed time questions.

Learn

This lesson is designed to provide practice with telling time to the hour. Choose activities from the previous lessons to consolidate learning.

Whole Group Activity

▲ Clock Exercises

Materials: Geared clock

Display a time that is on the hour with no minutes on a geared clock. For example, 5 o'clock. Ask students to tell the time and then do an activity, i.e., hop, clap, stomp, stretch, etc., as they count, 1 o'clock, 2 o'clock, 3 o'clock, and so on. Display another time, choose another exercise, and repeat.

Small Group Activities

Textbook Pages 172–174

▲ Match

Materials: Analog Time Cards (BLM), Digital Time Cards (BLM), and Word Time Cards (BLM)

Students arrange the cards faceup in a grid. Students take turns finding two cards that go together. Start out with Analog Time Cards (BLM) and Digital Time Cards (BLM). As the students improve at the game, add in Word Time Cards (BLM) and tell them that they need all 3 cards to have a match.

★ Memory

Materials: Analog Time Cards (BLM), Digital Time Cards (BLM), and Word Time Cards (BLM)

Play as described in **Match** in this lesson, but start with the cards arranged facedown, in a grid.

Take it Outside

▲ What Time is It, Mr. Fox?

Students line up in a row facing the leader, Mr. Fox, who is standing at least 20 steps away with his back to the students.

Students say altogether, “What time is it, Mr. Fox?” Mr. Fox replies with an hour. Students take the same number of steps as the hour. For example, if Mr. Fox says, “3 o’clock,” the students will all take 3 steps closer to Mr. Fox.

Continue up to 5 times. At any point, Mr. Fox can say, “Lunch time!” and turn and try to tag students before the students can return to the starting line. If a student is tagged, he sits. On the next round, the tagged students join Mr. Fox and play continues until there are no students left to tag. Choose a new Mr. Fox and continue to play.

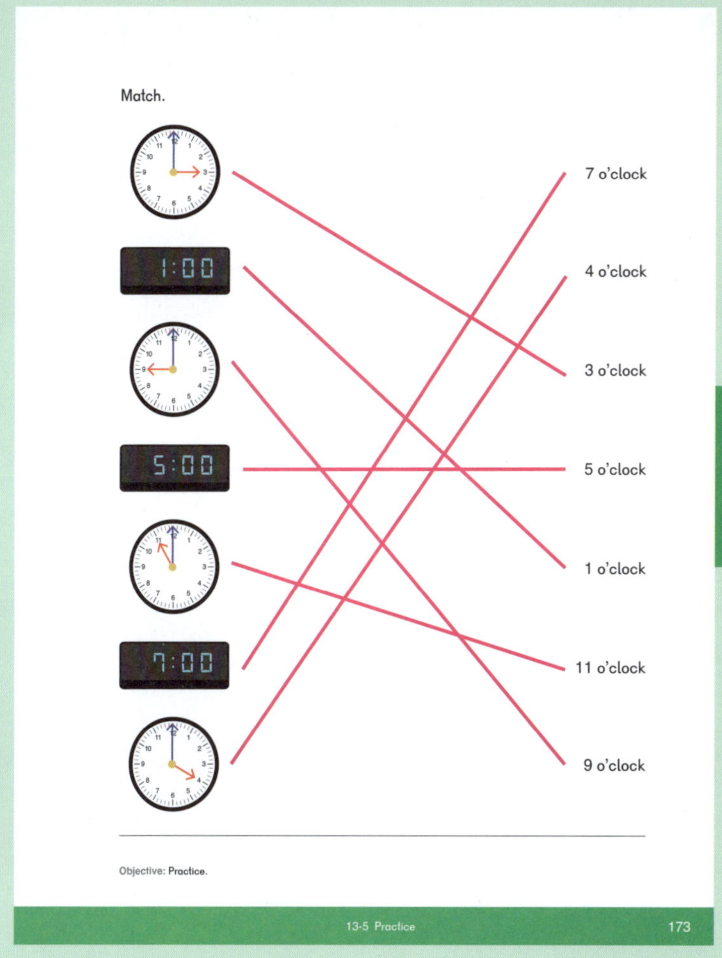

173

Exercise 5 • page 149

Extend

★ What Time is it Now?

Materials: Analog Clock Face (BLM) in dry erase sleeves, Digital Time Cards (BLM), Word Time Cards (BLM), die, geared clocks

Students work with a partner. Each student draws a Digital Time Card (BLM) or Word Time Card (BLM) and rolls the die. The number on the die represents hours past. Each student determines the new time and shares with his partner to check.

For example, the 6 o'clock card is chosen and the number 5 is rolled. The student starts at 6 o'clock and counts on 5 hours to determine the new time. Students can use the Analog Clock Face (BLM) in dry erase sleeves or a geared clock if they need support.

174

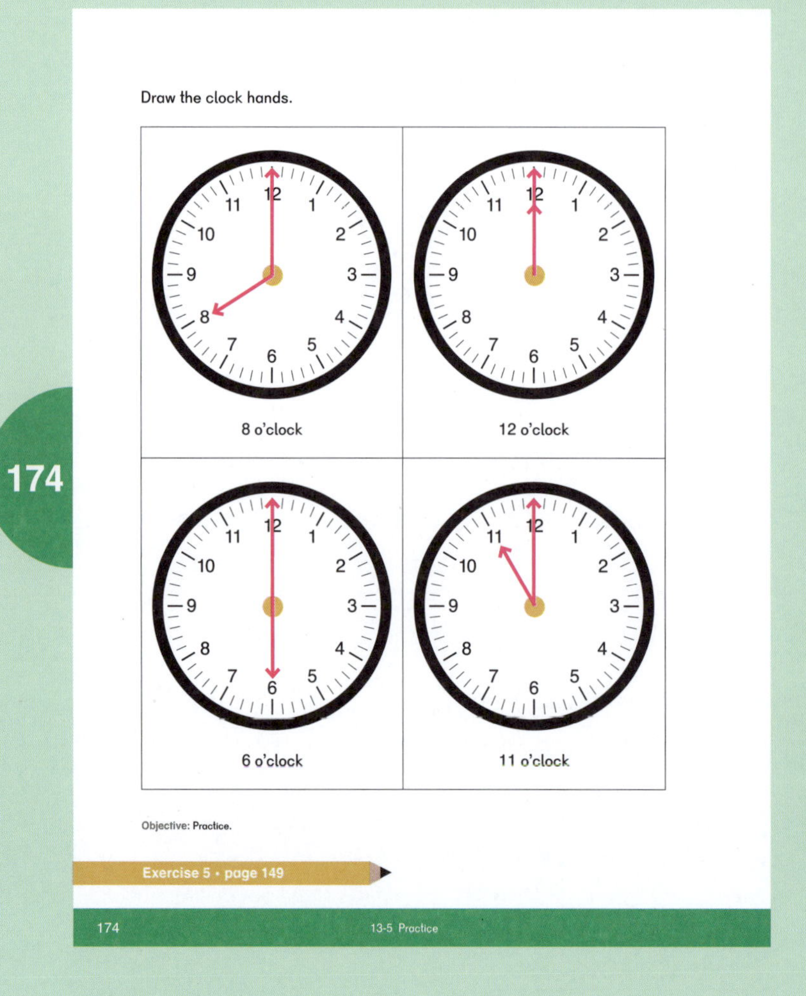

Exercise 1 • pages 141–142

Chapter 13 Time

Exercise 1

Is it day or night?
Circle ☀ or ☾ in each box.

Using this page: Have students look at each picture and determine if it is in the day or night, then circle the sun for day and the moon for night.
Concept: Identify day and night.

13-1 Day and Night 141

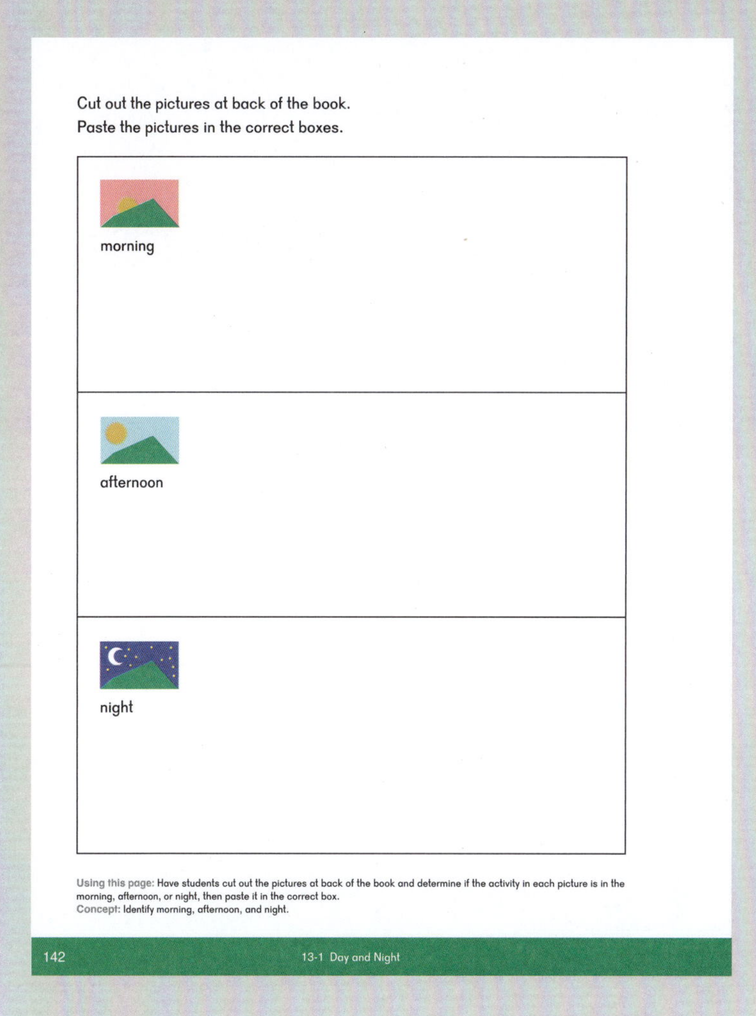

Cut out the pictures at back of the book.
Paste the pictures in the correct boxes.

morning

afternoon

night

Using this page: Have students cut out the pictures at back of the book and determine if the activity in each picture is in the morning, afternoon, or night, then paste it in the correct box.
Concept: Identify morning, afternoon, and night.

142 13-1 Day and Night

Exercise 2 • pages 143–144

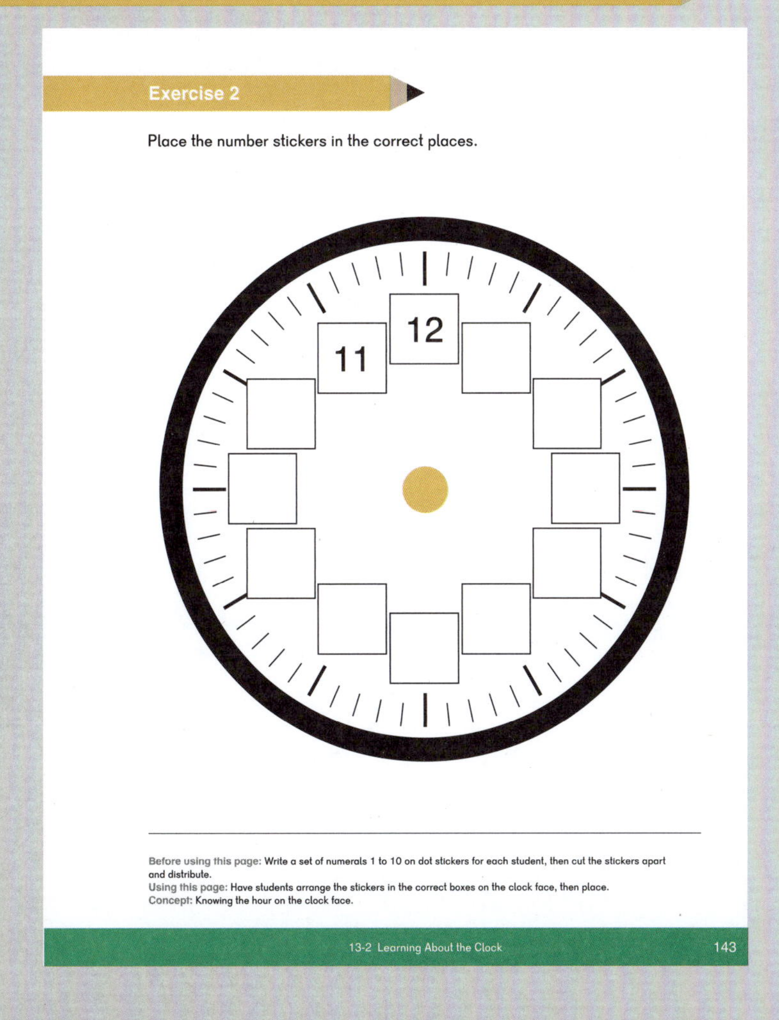

Exercise 2

Place the number stickers in the correct places.

Before using this page: Write a set of numerals 1 to 10 on dot stickers for each student, then cut the stickers apart and distribute.
Using this page: Have students arrange the stickers in the correct boxes on the clock face, then place.
Concept: Knowing the hour on the clock face.

13-2 Learning About the Clock 143

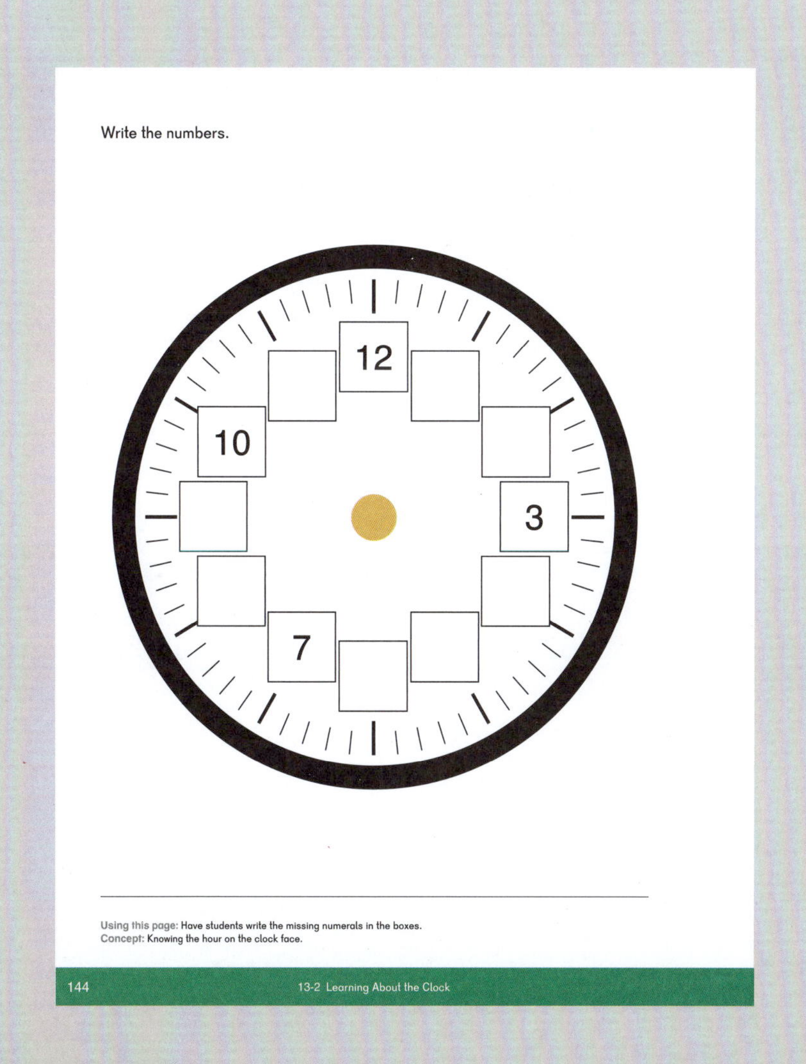

Write the numbers.

Using this page: Have students write the missing numerals in the boxes.
Concept: Knowing the hour on the clock face.

144 13-2 Learning About the Clock

Exercise 3 • pages 145–146

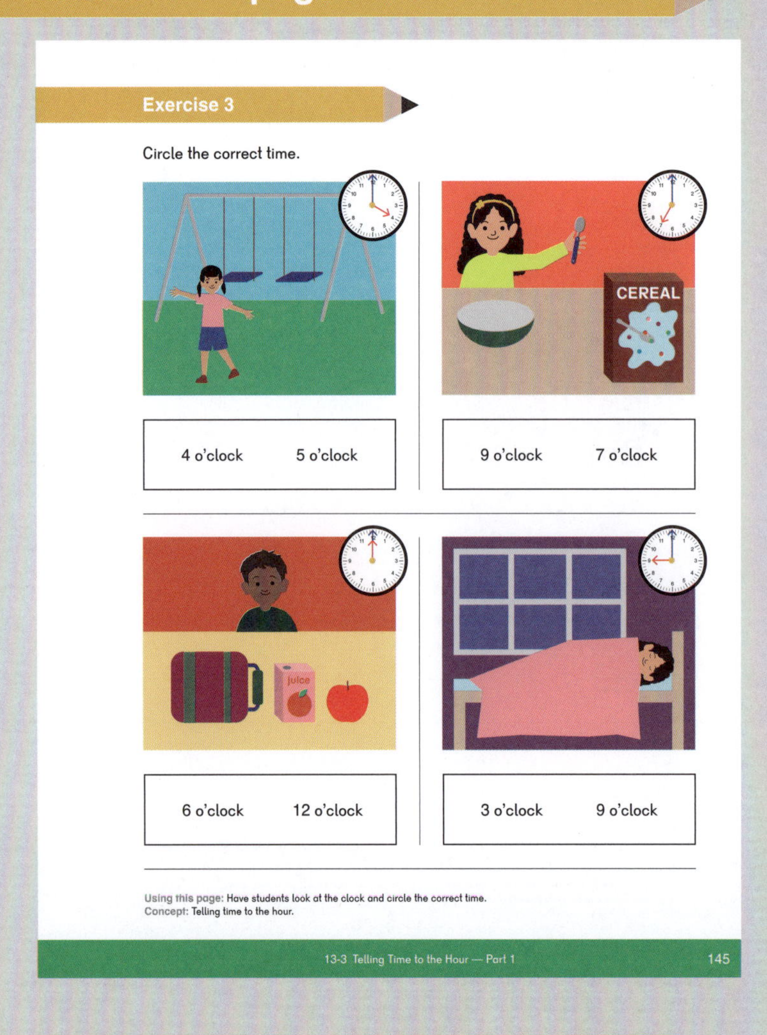

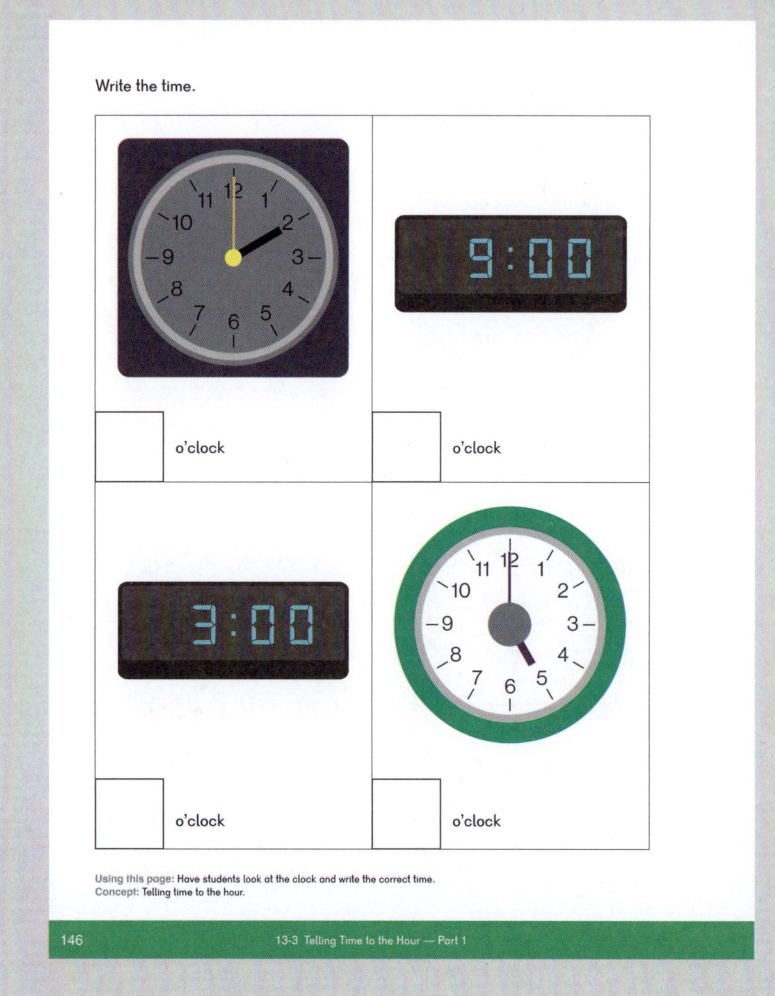

Exercise 4 • pages 147–148

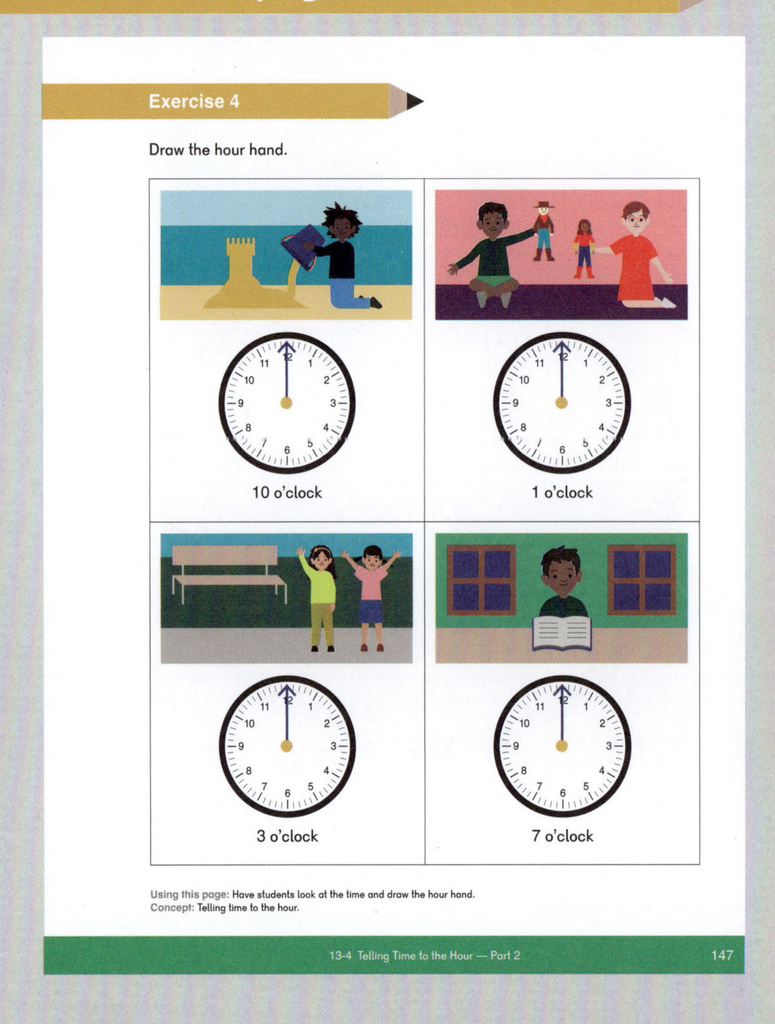

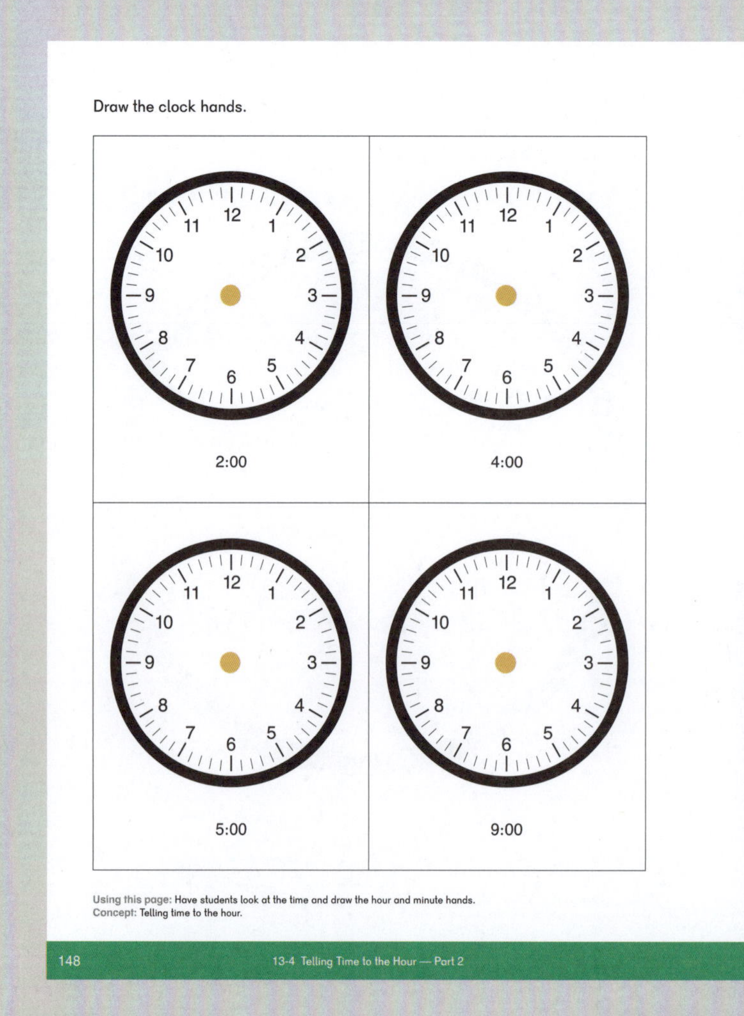

Exercise 5 • pages 149–150

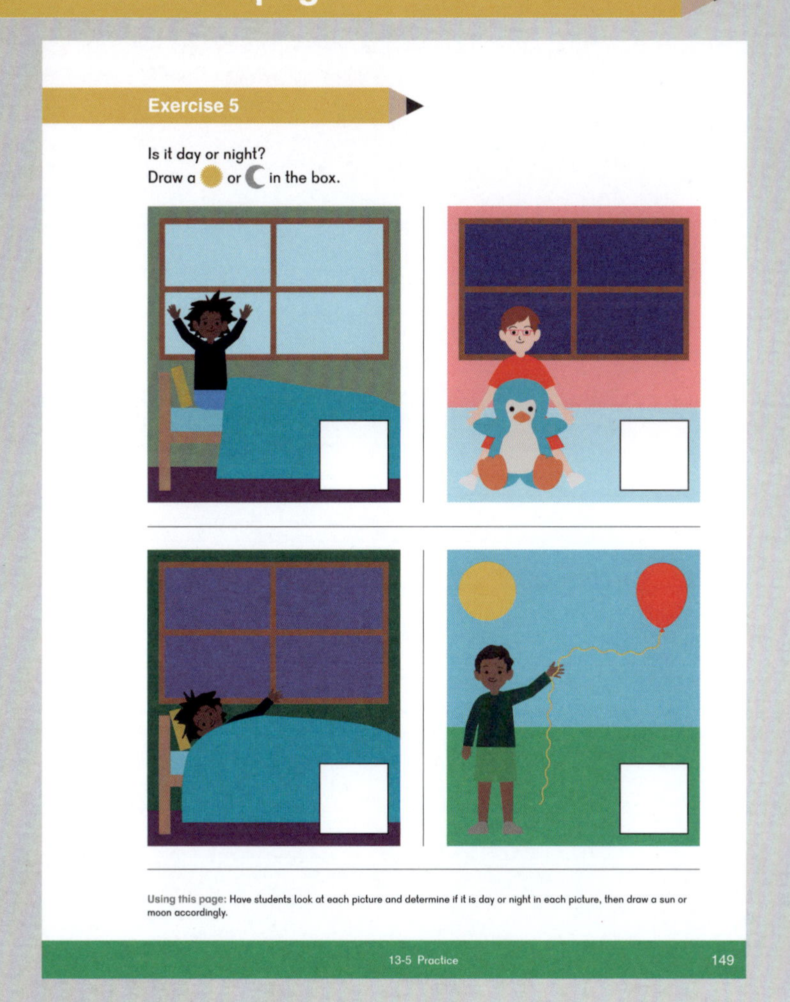

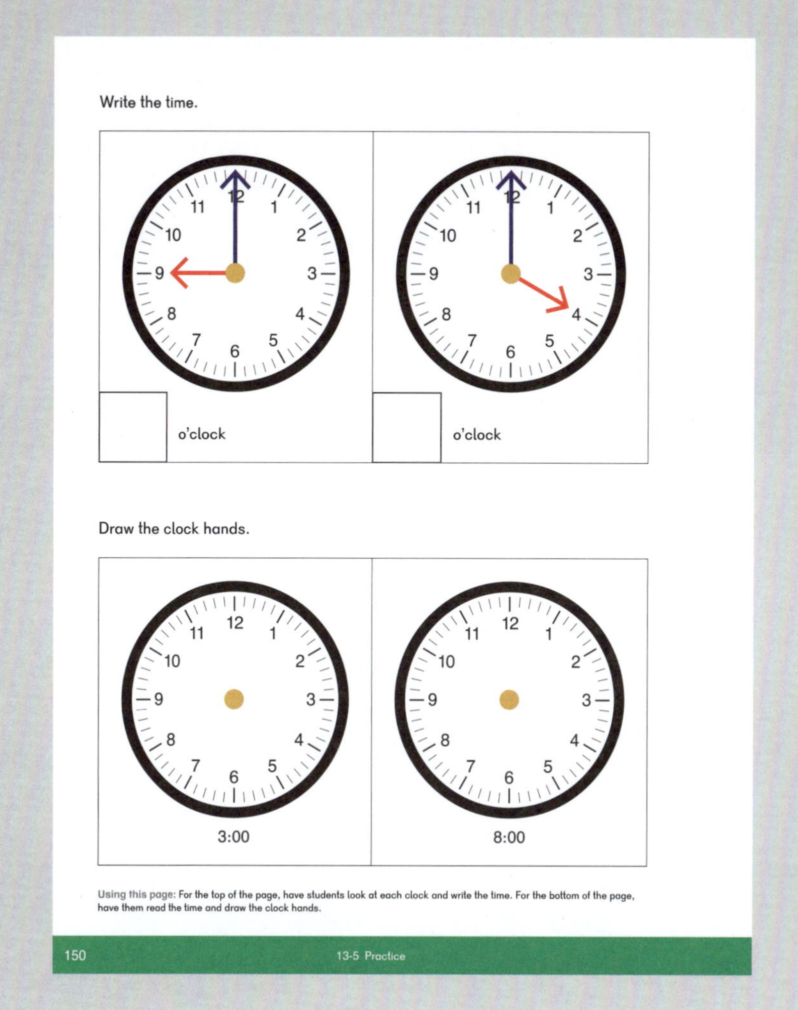

Notes

Suggested number of class periods: 6–7

	Lesson	Page	Resources	Objectives
	Chapter Opener	p. 233	TB: p. 175	
1	Coins	p. 234	TB: p. 176 WB: p. 151	Identify pennies, nickels, dimes, and quarters. Read the names for penny, nickel, dime, and quarter.
2	Pennies	p. 236	TB: p. 178 WB: p. 153	Learn the value of a penny as 1 cent. Count pennies to find their total value in cents.
3	Nickels	p. 238	TB: p. 180 WB: p. 155	Learn the value of a nickel as 5 cents. Determine the value in cents of a set of up to 10 nickels.
4	Dimes	p. 241	TB: p. 184 WB: p. 157	Learn the value of a dime as 10 cents. Determine the value in cents of a set of up to 10 dimes.
5	Quarters	p. 244	TB: p. 188 WB: p. 159	Learn the value of a quarter as 25 cents. Determine the value in cents of a set of up to 4 quarters.
6	Practice	p. 246	TB: p. 190 WB: p. 161	Practice skills from the chapter.
	Workbook Solutions	p. 248		

Notes

Automatic recognition of money, particularly coins, is a basic skill in society. Money provides students with an immediate reason to master addition and subtraction. Working with money provides ample opportunities to reinforce number sense and mental math. Understanding money helps students recognize that quantities other than 1 can be considered a unit and prepares students for decimals and fractions.

As physical money is becoming less common in transactions, children are having fewer experiences with money outside of the classroom. Students may enter Kindergarten without basic coin recognition. Providing students with as many opportunities to reinforce their experiences with coins should be a priority for the classroom.

In this chapter, students will learn the name and value of a penny, nickel, dime and quarter. They will practice skip counting by fives and tens using coins.

Minimal supplemental activities are provided as students should spend most of the lessons counting money and making change. Students will revisit counting money in **Dimensions Math® 1B**.

While students may work at school with play money, it is important that students handle real coins. Typically, less than $2 in real change for each student is sufficient.

Students should recognize that regardless of the image on coins, a coin shares common attributes: color, size, ridged edges or not, and thickness. While the image on the back of various quarters can be different, a quarter is always silver colored, larger than pennies, nickels, and dimes, and has ridged edges.

Note: This chapter does not require students to know 100 cents is a dollar but most students will be familiar with the term “dollar.”

In preparation for **Coin Rubbings** in Lesson 1, use rubber cement and adhere two coins (one heads up and one tails up) to a cardboard card.

It is assumed that all students will have access to recording tools. When a lesson refers to a whiteboard, any writing materials can be used.

Materials

- Cardboard
- Coin rubbing cards as described on page 230 of this Teacher's Guide
- Crayons
- Cup or can
- Dice, regular and modified with sides labeled: 1¢, 1¢, 5¢, 5¢, 10¢, 25¢
- Dimes
- Magnifying glasses
- Nickels
- Objects to label with price tags
- Pennies
- Price tags/labels
- Quarters
- Rubber cement

Blackline Masters

- Blank 5 × 5 Grid
- Blank Double Ten-frames
- Blank Graph
- Blank Number Bond Template
- Blank Ten-frame
- Hundred Chart

Storybooks

- *Jenny Found a Penny* by Trudy Harris
- *The Great Pet Sale* by Mick Inkpen
- *Once Upon a Dime: A Math Adventure* by Nancy Kelly Allen and Adam Doyle
- *The Coin Counting Book* by Rozanne Lanczak Williams
- *The Penny Pot* by Stuart J. Murphy
- *A Chair for My Mother* by Vera B. Williams

Letters Home

- Chapter 14 Letter

Notes

Lesson Materials

- Coins, including quarters, dimes, nickels, and pennies

Provide each student with some quarters, dimes, nickels, and pennies.

Have students look at the different coins and discuss their attributes. How are the coins similar and different in size, color, and images on the front and back? Do students see any numbers on the coins? What else is on the coins?

Have students sort the coins and then discuss how they sorted the coins.

Ask students if they know the names of the coins. Some students may also know the value of the coins. Use this introduction to assess students' prior knowledge of coins.

If students will be using play coins during the unit, allow them to compare real coins to play coins.

Discuss the coins in Sofia's piggy bank on page 175. Ask students, "How many coins does Sofia have? Do you know the value of her coins?"

If students are familiar with coins, continue to Lesson 1, where students will identify the coins by name.

Ask students:

- Which coins do you have the most of?
- Which coins do you have the least of?
- Are there any coins of which you have the same amount?
- How many pennies and dimes, quarters and nickels, etc., do you have?

Extend

★ Coin Sort

Materials: Blank Graph (BLM), coins

Give students ten coins that are a mixture of quarters, dimes, nickels, and pennies.

Ask students to make groups with the coins. They may sort by color, size, value, etc.

Have students put the coins on the Blank Graph (BLM) to create picture graphs of their groups.

Objectives

- Identify pennies, nickels, dimes, and quarters.
- Read the names for penny, nickel, dime, and quarter.

Lesson Materials

- Pennies, nickels, dimes, and quarters

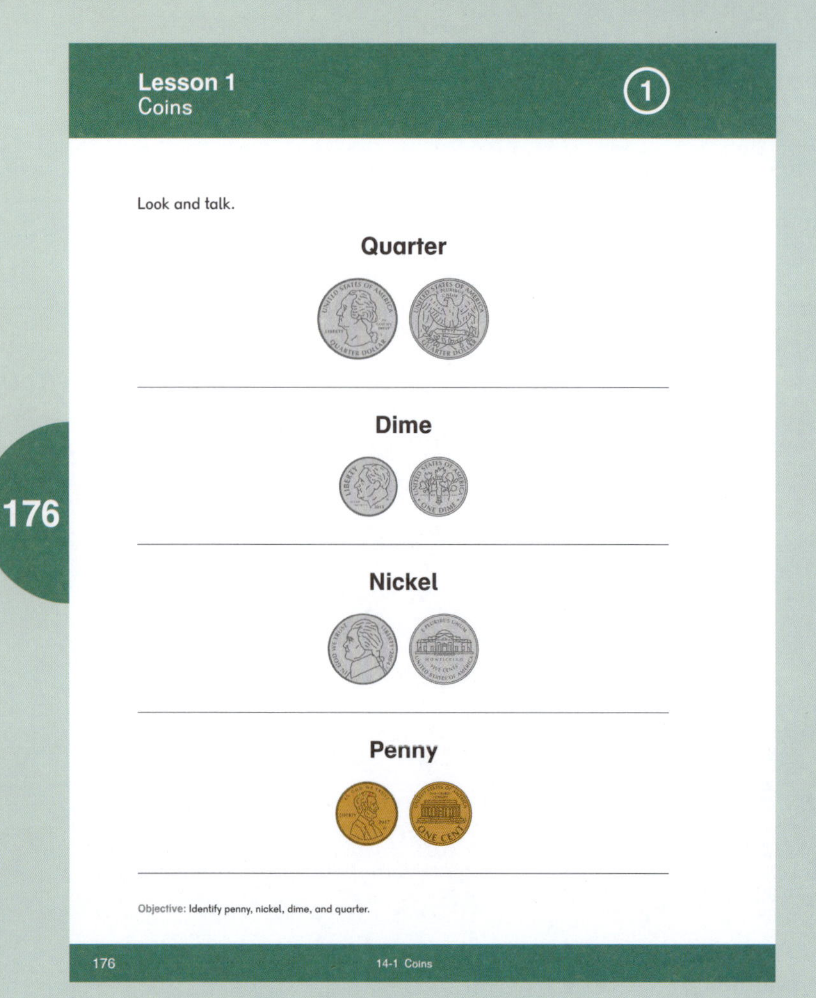

176

Lesson 1
Coins

1

Look and talk.

Quarter

Dime

Nickel

Penny

Objective: Identify penny, nickel, dime, and quarter.

176 14-1 Coins

Explore

Give each student one of each coin: penny, nickel, dime, and quarter.

Discuss the shapes, sizes, colors, pictures and any other noticeable attributes on the different coins.

Learn

Look and discuss page 176. Tell students the names of the coins.

Whole Group Activity

▲ Guess My Coin

Materials: 1 quarter, dime, nickel, and penny for each student

Provide students with a penny, nickel, dime, and quarter. Give students clues and ask them to determine which coin you are describing. For example, “I'm thinking of a coin that is silver colored and smooth along the edges.” Students should hold up the nickel.

To extend, students can take turns giving clues to the group or in pairs.

Small Group Activities

Textbook Page 177

▲ Coin Rubbings

Materials: Coin rubbing cards, brown/copper crayons, gray/silver crayons, magnifying glasses

Students place paper over the coins and use the side of a crayon to make coin rubbings. Have students use a copper or brown crayon for the penny and silver or gray for the nickel, dime, and quarter. Provide magnifying glasses and have students investigate the rubbings up close.

▲ Coin Bonds

Materials: Quarters, dimes, nickels, pennies, Blank Number Bond Template (BLM)

Students grab a small handful of coins and sort the coins into two groups. For example, a student grabs 6 coins. 6 becomes the whole. 4 coins are silver and 2 coins are copper. The student returns the coins to the bag, grabs another handful and repeats the activity. To extend, have students record 2 addition and 2 subtraction sentences for each grab.

▲ Coin Patterns

Materials: Quarters, dimes, nickels, and pennies

Players take turns building and completing a pattern. Player 1 starts a pattern using pennies, nickels, dimes, or quarters. Player 1 describes his pattern and Player 2 builds and extends the pattern. Player 1 checks Player 2's work. Players switch roles and play continues.

Exercise 1 • page 151

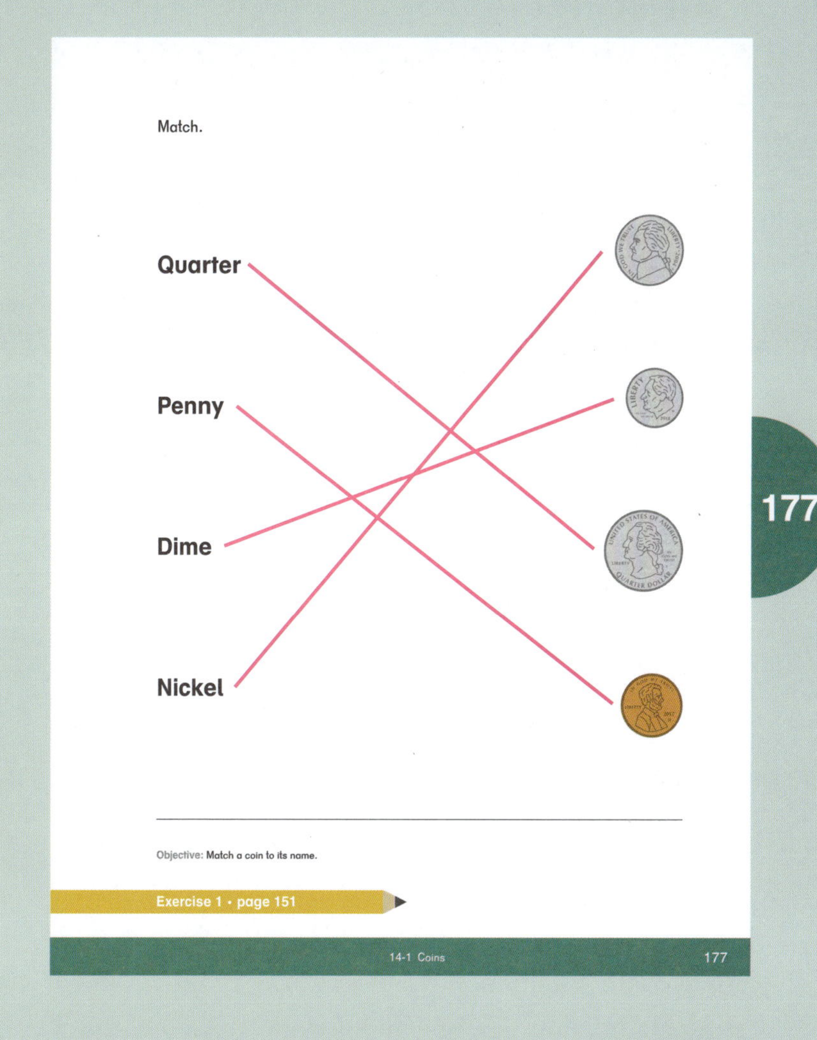

177

Extend

★ Copy Me

Materials: Quarters, dimes, nickels, and pennies

Players sit opposite each other with a barrier in between them. Players take turns choosing up to 5 coins to create an image with the edges of the coins touching. Player 1 describes her image to Player 2 who tries to duplicate it. Players switch roles and play continues.

Objectives

- Learn the value of a penny as 1 cent.
- Count pennies to find their total value in cents.

Lesson Materials

- Pennies
- Blank Ten-frame (BLM)

Explore

Provide each student with 10 pennies. Have them count the pennies and put them on a Blank Ten-frame (BLM) as needed.

Learn

Explain to students that a penny is a coin that has a value of, or is worth, one cent. Have them find the words "one cent" on the coin. Ask students to find the value of the pennies they have in front of them. Ask students to show you how many pennies would equal 3 cents, 5 cents, etc.

Show students the symbol for cents. Discuss that it looks like the letter "c" with a line through it. This will help them remember that the word cents begins with a "c."

Help students understand the ¢ symbol. Write an amount on the board. Have students show you how many pennies make that amount. For example, write 6¢ and students show 6 pennies.

Look at page 178. Count the coins and find the value.

Whole Group Activities

▲ Penny Flash

Materials: 5 pennies

Quickly reveal up to 5 pennies to students. Students write the value of the pennies on their whiteboards.

For example, show 3 pennies. Students write 3¢ on their boards.

▲ Listen and Count

Materials: Cup or can, 10 pennies

Drop up to 10 pennies, one at a time, into a cup or can. Students count the coins mentally as they hear them and write the value of the coins on whiteboards.

Small Group Activities

Textbook Page 179

▲ Ten-frame Fill Up

Materials: Die, 20 pennies per player, 1 Blank Double Ten-frame (BLM) per player

Players take turns rolling the die and adding the corresponding number of pennies to one of the ten-frames on their Blank Double Ten-frame (BLM).

Players must fill a ten-frame with an exact roll. If adding pennies to a ten-frame makes more than 10, the player passes. As a ten-frame is filled, students remove the 10 pennies and score a point. Play continues until time is up. The player with the most points wins.

▲ Shop

Materials: 20 pennies, priced objects similar to those pictured on textbook page 179 or 181

One student begins with 20 pennies. His partner is the shopkeeper. The student spends his coins on items with a price tag until he is out of money. Then they switch roles. Challenge students to find the most or the fewest items they can buy with 20 pennies.

▲ Counting Pennies

Materials: At least 100 pennies per student, Hundred Chart (BLM)

Instruct students to count out 100 pennies. Watch how they count their pennies. Do they put them in groups of 5 pennies or 10 pennies? A pile of 100 pennies? Provide a Hundred Chart (BLM) for students who are counting by ones and have them put one penny on top of each number as they count.

Exercise 2 • page 153

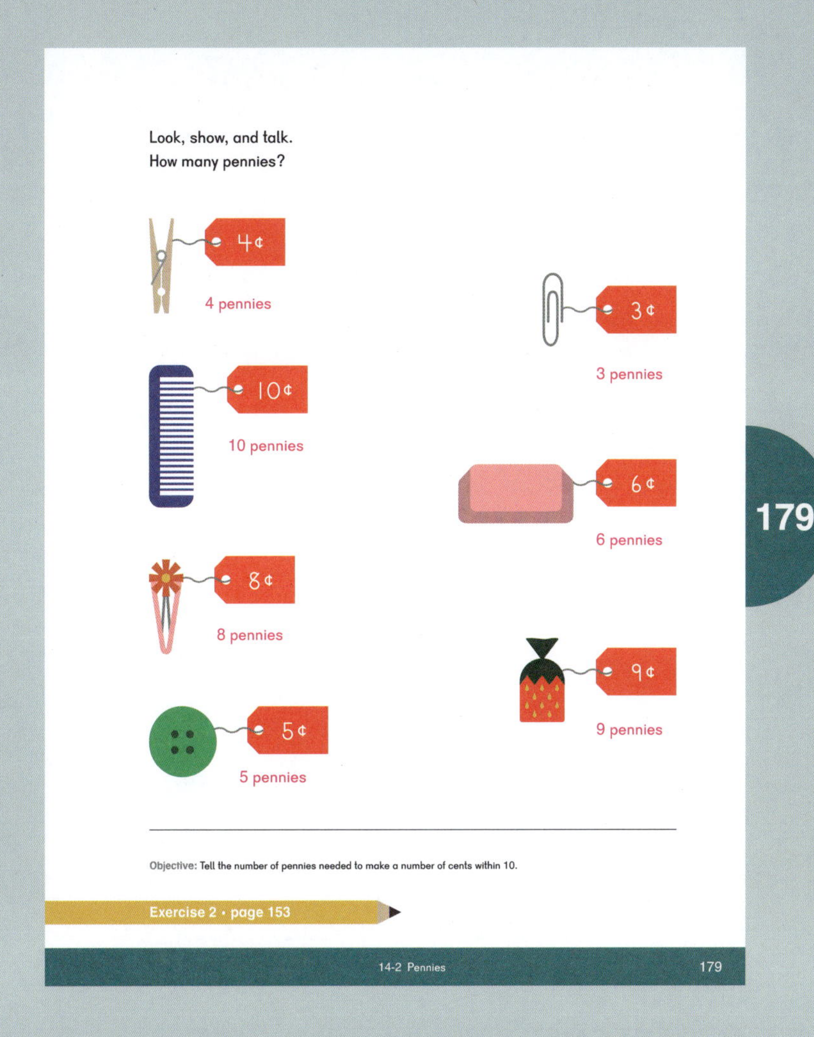

179

Extend

★ Penny Nim

Materials: 1 Blank Ten-frame (BLM) per pair of students, 10 pennies per pair of students

This is an extension of the game introduced in Chapter 7, Lesson 1 on page 7 of this Teacher's Guide.

Players take turns placing one, two, or three pennies on the open squares of a Blank Ten-frame (BLM). Whoever fills up the ten-frame wins.

Questions to ask students:

- How did you/they win that game?
- What do you think your opponent will do if you place three pennies?
- Is there a way to win every time?
- Would you use a different strategy if we used a double ten-frame board?

Objectives

- Learn the value of a nickel as 5 cents.
- Determine the value in cents of a set of up to 10 nickels.

Lesson Materials

- Nickels
- Pennies
- Hundred Chart (BLM) with fives and tens shaded

Explore

Provide students with pennies and nickels. Have them compare the two coins and find the words “five cents” on the nickel. Ask how many pennies have the same value as a nickel.

Ask:

- If you have 5 pennies and 1 nickel, what is the value of the coins?
- How many ways can you make 10 cents?

Learn

Discuss Mei’s comments about the value of a nickel. Show students 5 pennies and one nickel and tell students you could use either to buy something that costs 5 cents.

Have students skip count by fives along with Emma while counting the nickels.

Write an amount (in multiples of 5) up to 50¢ on the board. Have students show you how many nickels are worth the same amount. Students may need a Hundred Chart (BLM) with columns for the fives and tens shaded to help them with the skip counting. Look at page 181 and have students show how many nickels it will take to buy each item.

Extend the activity by asking students to show different ways to pay for items that cost amounts that are not multiples of 5. Ask, how could we buy something that costs 7 cents?

Whole Group Activities

▲ Show Me

Materials: Pennies, nickels

With nickels and pennies in front of them, give clues and have students show you the correct coin. For example:

- Show me a penny.
- Show me the largest coin. What is it called?
- Show me the coin with the greatest value.

Extend: Show two ways to make 6¢.

▲ Listen and Count

Materials: Cup or can, nickels

Drop nickels, one at a time, into a cup or can. Students count the value of the nickels by fives mentally as they hear them and write the value on their whiteboards.

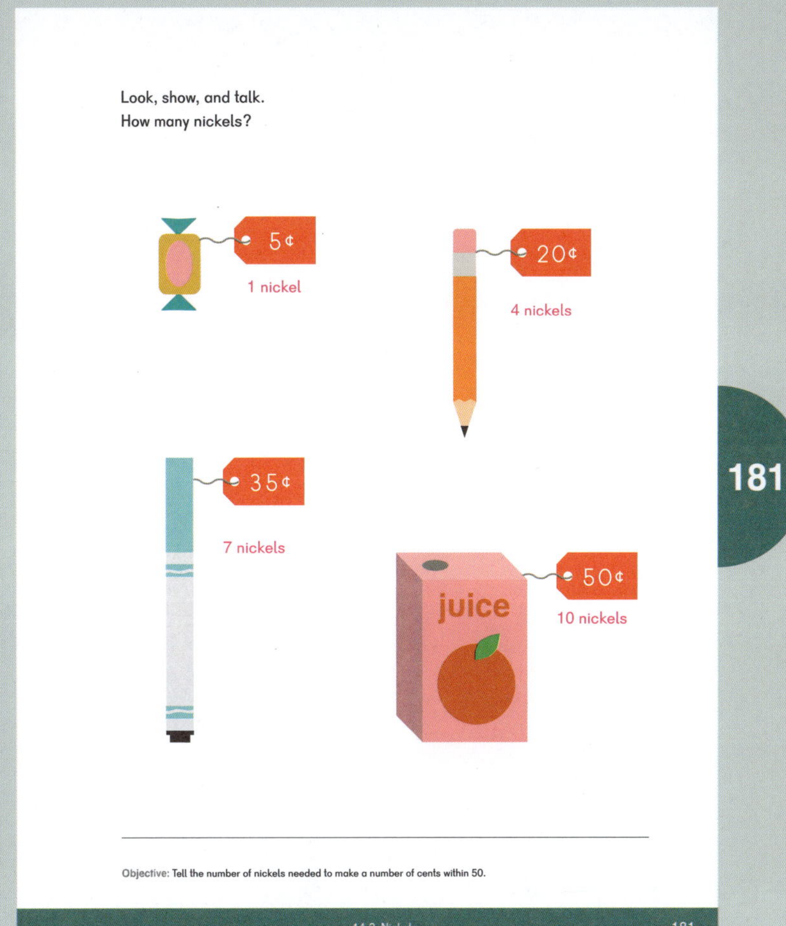

Small Group Activities

Textbook Pages 182–183

▲ Counting Nickels

Materials: At least 20 nickels per student, Hundred Chart (BLM)

Have students count 100 cents in nickels. Provide a Hundred Chart (BLM) and have them put one nickel on all the numbers they say when counting by fives.

★ Extend by having students each grab a handful of nickels and determine their values. The player with the greatest value scores a point. Players can use a Hundred Chart (BLM) to help count the value of the nickels.

▲ Shop

Materials: 10 nickels, 10 pennies, priced objects similar to those pictured on textbook page 181 that are less than 50¢ and in multiples of 5

One student begins with 10 nickels. His partner is the shopkeeper. The student spends his coins on items with a price tag until he is out of money. Then they switch roles.

To extend, provide items with prices up to 20 cents (not only multiples of 5). Students start with 10 nickels and 10 pennies to spend.

Exercise 3 • page 155

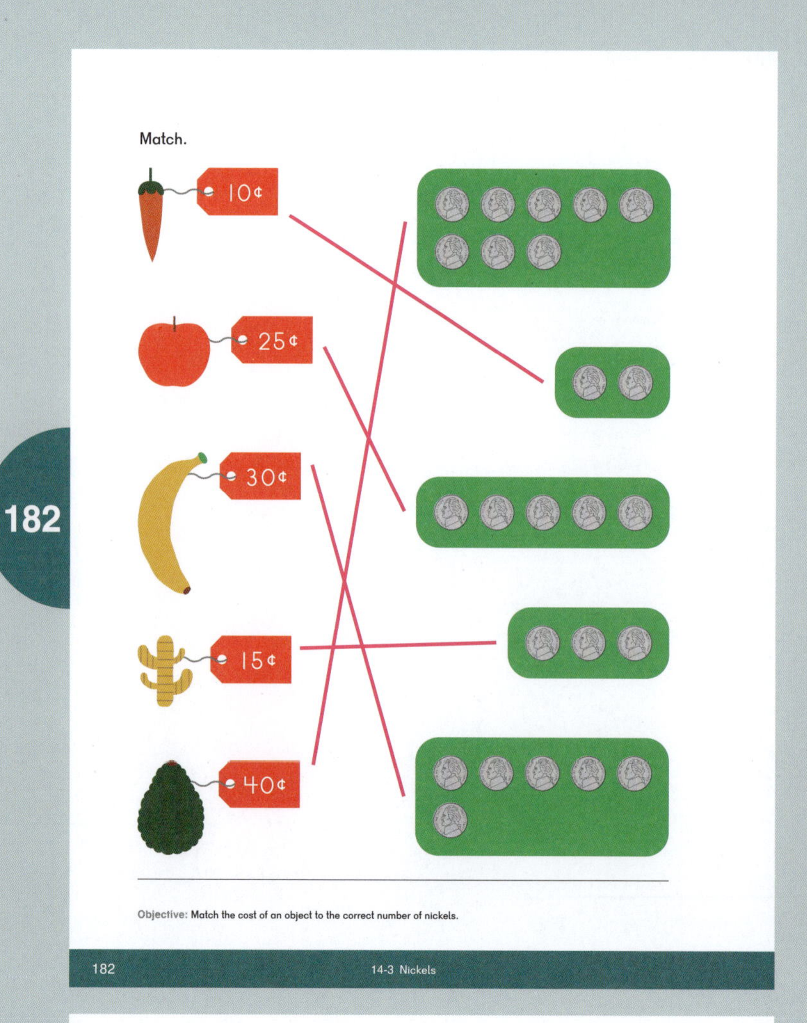

Extend

★ Greatest Value

Materials: Pennies, nickels

Students each grab a small handful of nickels and pennies from a bag or tub. Each student determines the value of the coins grabbed.

The student with the greatest value wins a point for the round. Students return the coins to the tub and play continues.

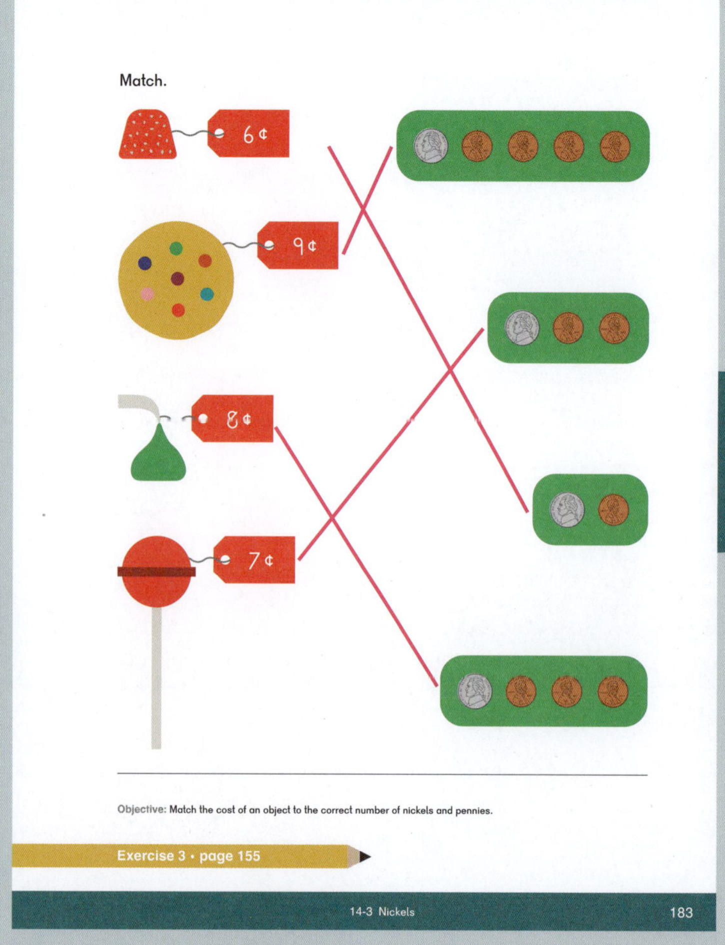

Objectives

- Learn the value of a dime as 10 cents.
- Determine the value in cents of a set of up to 10 dimes.

Lesson Materials

- Pennies
- Nickels
- Dimes
- Hundred Chart (BLM)

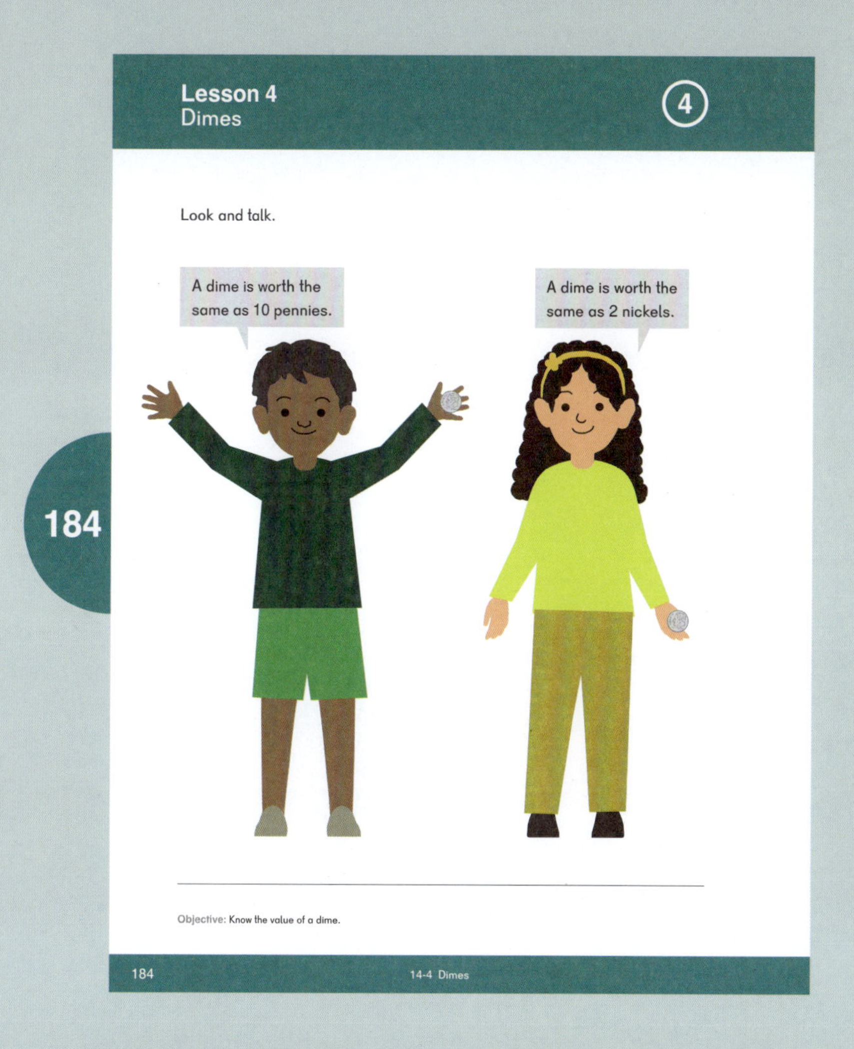

Explore

Provide students with dimes, nickels, and pennies. Have them compare the coins. They may note that although a penny has the words "one cent" and a nickel has "five cents," the dime just has "one dime" written on the back. Explain that a dime is a coin with a value of 10 cents.

Ask:

- How many pennies have the same value as 1 dime?
- How many nickels have the same value as 1 dime?
- Show how many ways you can make 10 cents.

Learn

Discuss Alex and Sofia's comments about the value of a dime on page 184. Show students 10 pennies, 2 nickels, and 1 dime and tell them you could use any of them to buy something that costs 10 cents. It is often difficult for students to understand that the dime, being smaller than the nickel, actually has a greater value.

Have students touch dimes as they skip count by tens to find the value of 10 dimes.

Write an amount up to 90¢ in multiples of 10 on the board. Have students show you how many dimes are worth the same amount. Students may need a Hundred Chart (BLM) to help them with the skip counting.

Have students look at page 185 and have them show how many dimes it will take to buy each item.

Extend the activity by asking students how they can buy the items with other combinations of coins. They may find they can buy the cherry with:

- 2 nickels
- 1 nickel and 5 pennies

Repeat with other items on the page.

Whole Group Activity

▲ Listen and Count

Materials: Cup or can, dimes

Drop up to 10 dimes, one at a time, into a cup or can. Students count the value of the dimes by tens mentally as they hear them and write the value on their whiteboards.

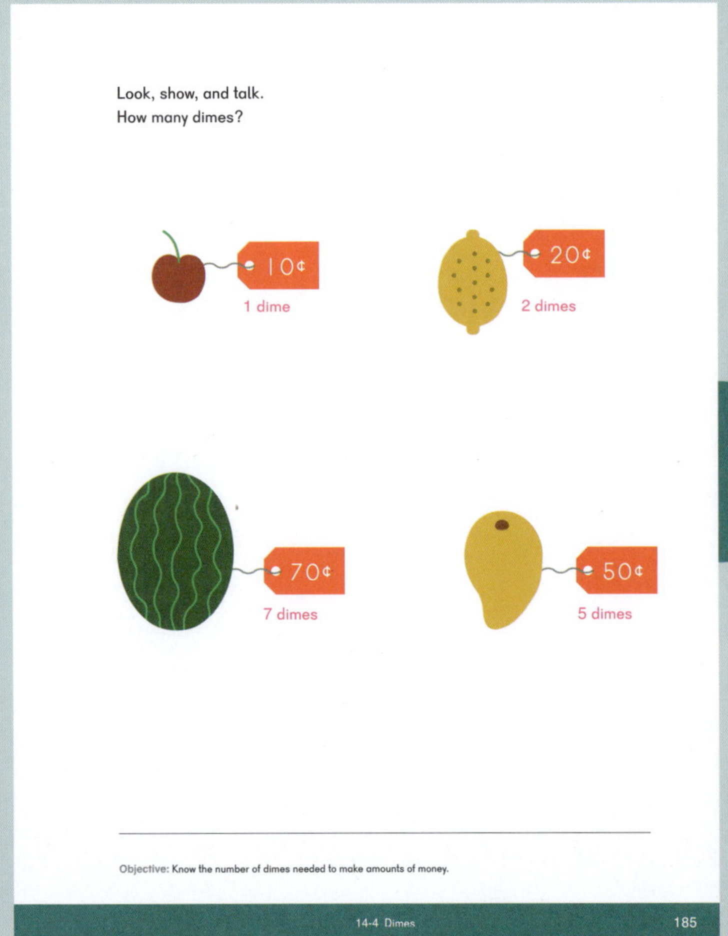

185

Small Group Activities

Textbook Pages 186–187

▲ Counting Dimes

Materials: At least 10 dimes per student, Hundred Chart (BLM)

Have students count 100 cents in dimes. Provide a Hundred Chart (BLM) and have them put one dime on all the numbers they say when counting by 10.

Extend by having students each grab a handful of dimes and determine the values. The player with the greatest value scores a point. Players can use the Hundred Chart (BLM) to help count the value of the dimes.

▲ Shop

Materials: 10 dimes, 10 nickels, 10 pennies, priced objects similar to those pictured on textbook page 181 that are less than 90¢ and in multiples of 10

One student begins with 10 dimes. Her partner is the shopkeeper. The student spends her dimes on items with a price tag until she is out of coins. Then they switch roles.

To extend, provide items with prices up to 95 cents in multiples of 5 and 10. Students start with 10 dimes and 10 nickels to spend.

To extend further, add pennies to the **Shop** activity and price items to 99 cents.

Exercise 4 • page 157

Extend

★ How Many Ways?

Materials: Dimes, nickels, pennies

Have students show different ways of making 50¢. For example, students could show:

- 4 dimes and 10 pennies
- 4 dimes, 1 nickel, and 5 pennies
- 4 dimes and 2 nickels

Ask, "Which way of making 50 cents uses the fewest coins? Which way uses the most coins?"

★ Greatest Value

Materials: Pennies, nickels, dimes

Students each grab a small handful of dimes, nickels, and pennies from a bag or tub. Each student determines the value of the coins grabbed.

The student with the greatest value wins a point for the round. Students return the coins to the tub and play continues.

186

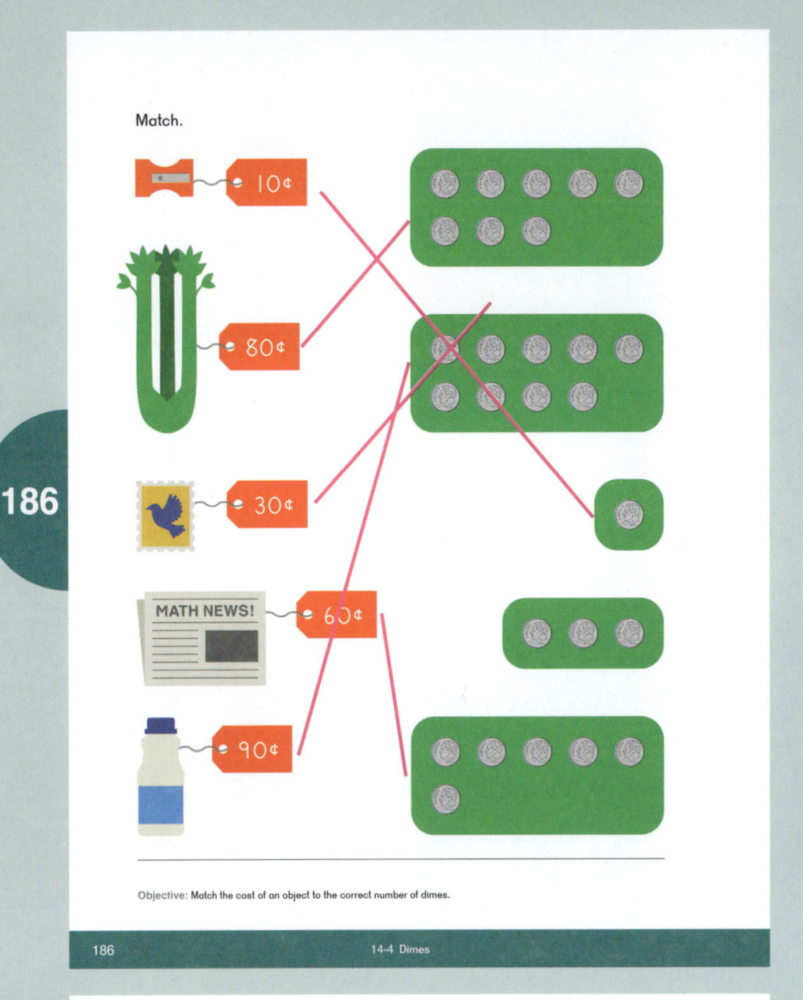

187

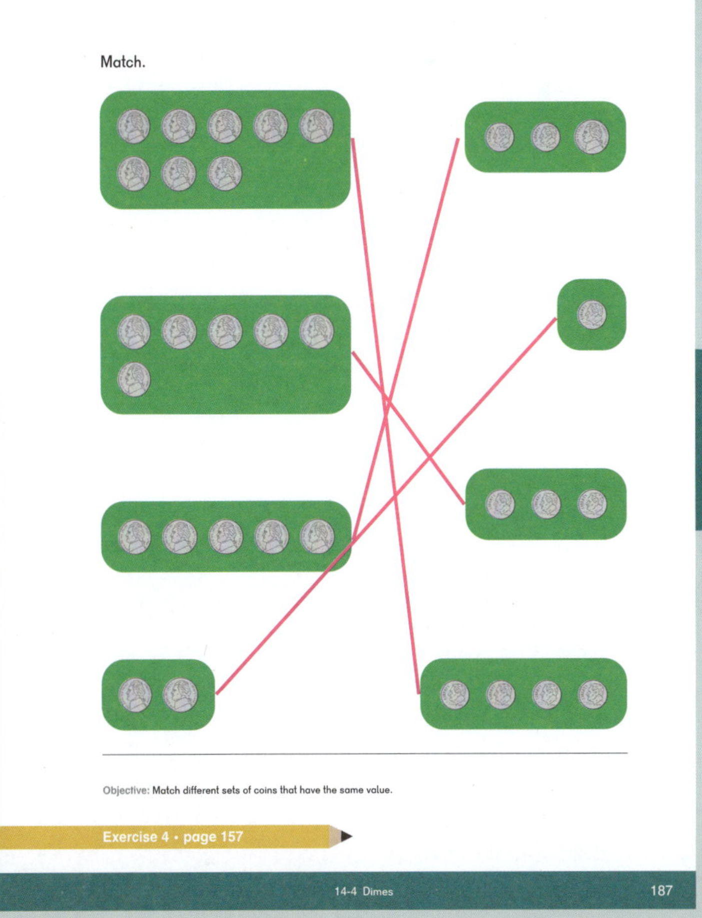

Objectives

- Learn the value of a quarter as 25 cents.
- Determine the value in cents of a set of up to 4 quarters.

Lesson Materials

- Nickels
- Dimes
- Quarters
- Hundred Chart (BLM), 1 per student

188

Lesson 5
Quarters
5

Look and talk.

Established 1925
PARTYVILLE
Doodads & Knick-knacks

I have 5 nickels.
I have some dimes and nickels.
I have a quarter.

Objective: Know the value of the quarter.

188 14-5 Quarters

Explore

Provide students with quarters, dimes, and nickels. Have them compare the coins. They may note that the dime has "one dime" written on the back and the quarter has the words "one quarter," or that both the quarter and dime have ridged edges and the nickel is smooth. Explain to students that one quarter has a value of 25 cents.

Ask them to explore other ways to show a value of 25¢ using pennies, nickels, and dimes.

Learn

Discuss the three friends' comments about the coins they have. Tell students that all three friends have the same total amount. Ask, "How is that possible?"

Have students show Alex and Mei's amount of money with the appropriate coins. Note that students may confuse the number of coins with their value. Mei has more coins, but the value of her coins is the same as the value of Alex's.

The third friend has dimes and nickels. How many of each could he have?

When counting a mixed set of coins, we often start with the coin with the greatest value and count on the value of each additional coin. Students can use a Hundred Chart (BLM) as a guide as you show 2 coins and ask them to find the value. For example, if a quarter and a dime are shown, have students start on 25 and then move their fingers down a row on the chart to add 10. They land on 35. A quarter and a dime is worth 35¢.

Whole Group Activity

▲ Quarter Chant

Skip counting by quarters is challenging. Practice the following chant to help students learn to count the value of quarters up to a dollar.

"Twenty-five, fifty, seventy-five, a dollar,
If you love Kindergarten, stand up and holler!"

As students chant, have them place quarters in a pile.

Small Group Activities

Textbook Page 189

▲ Shop

Materials: 3 quarters, 3 dimes, 3 nickels, 10 pennies, priced objects similar to those pictured on textbook page 181 that are less than 95¢ and in multiples of 5 and 10

One student begins with 3 quarters, 3 dimes, and 3 nickels. His partner is the shopkeeper. The student spends his coins on items with a price tag until he is out of money. Then they switch roles.

To extend, add pennies to the **Shop** activity and price items to 99 cents.

▲ What Makes 100¢?

Materials: Quarters, dimes, nickels, pennies, Hundred Chart (BLM)

Using a Hundred Chart (BLM), if necessary, students determine what combinations of coins make 100¢.

Examples:

- 10 dimes make 100¢
- 20 nickels make 100¢
- 100 pennies make 100¢
- 4 quarters make 100¢
- 3 quarters, 2 dimes, and 1 nickel make 100¢

Exercise 5 • page 159

Extend

★ Greatest Value

Materials: Quarters, dimes, nickels, pennies, Hundred Chart (BLM)

Students each close their eyes and choose 4 coins from a tub of quarters, dimes, nickels, and pennies. Each student determines the value of the coins grabbed. Students can use a Hundred Chart (BLM) to help keep count.

The student with the greatest value wins a point for the round. Students return the coins to the tub and play continues.

To further extend, have students grab a handful of coins.

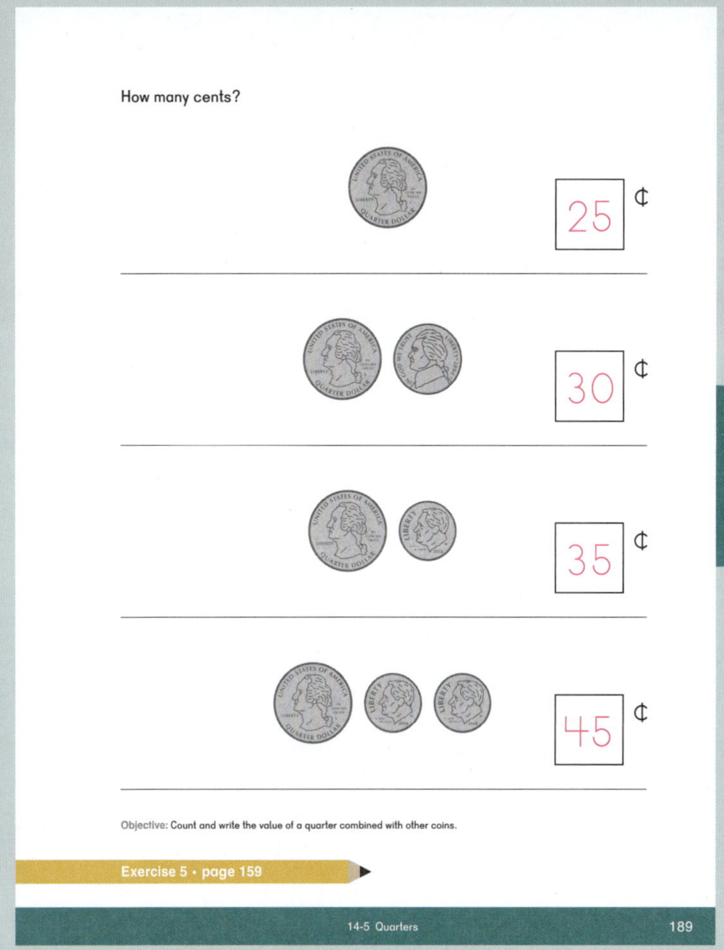
How many cents?

25 ¢

30 ¢

35 ¢

45 ¢

Objective: Count and write the value of a quarter combined with other coins.

Exercise 5 • page 159

14-5 Quarters 189

189

Lesson 6 Practice

Objective

- Practice skills from the chapter.

Practice lessons are designed for further practice and assessment as needed.

Students can complete the textbook pages and workbook pages as practice and/or as assessment.

Use activities and extensions from the chapter for additional review and practice.

190

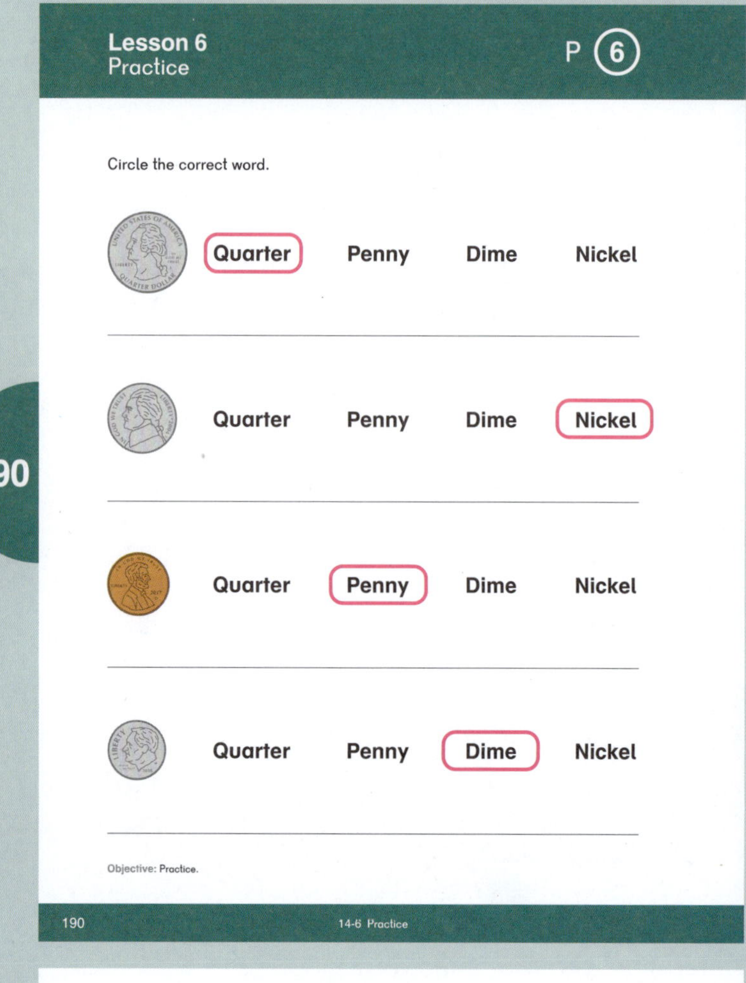

191

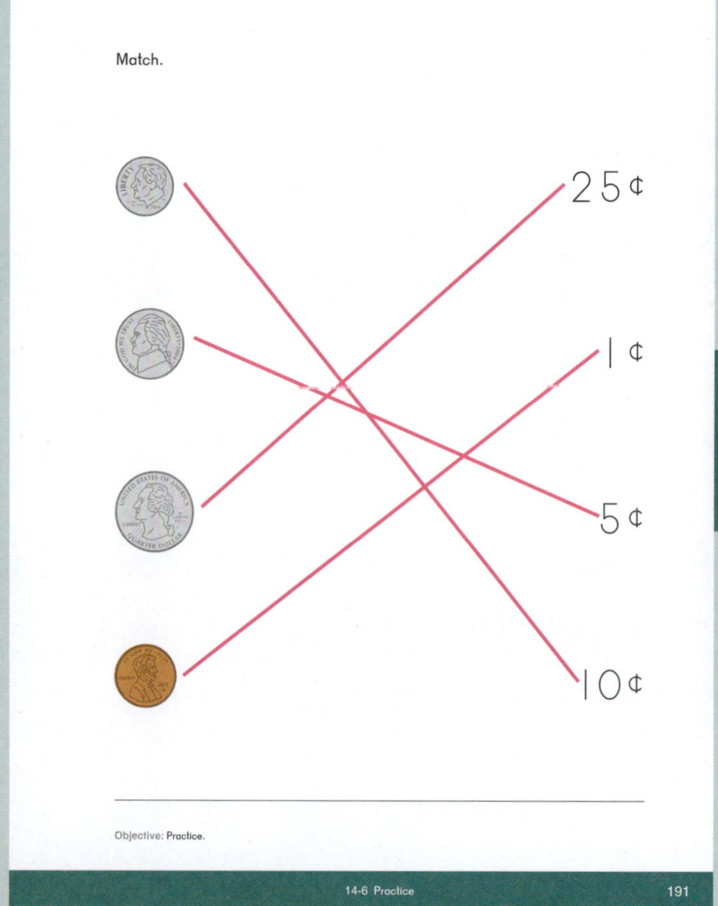

Exercise 6 • page 161

Extend

★ Clear the Board

Materials: 25 coins made up of a combination of pennies, nickels, dimes, and quarters for each player, 1 Blank 5 × 5 Grid (BLM) per player, die with modified sides: 1¢, 1¢, 5¢, 5¢, 10¢, 25¢

Students cover their Blank 5 × 5 Grid (BLM) randomly with the coins, 1 coin in each square.

Player 1 rolls the die and chooses a coin or coins to remove from his grid based on the roll. Player 2 rolls and also removes a coin or coins.

Players continue taking turns until a player clears his board. If a coin cannot be removed, that player loses his turn.

192

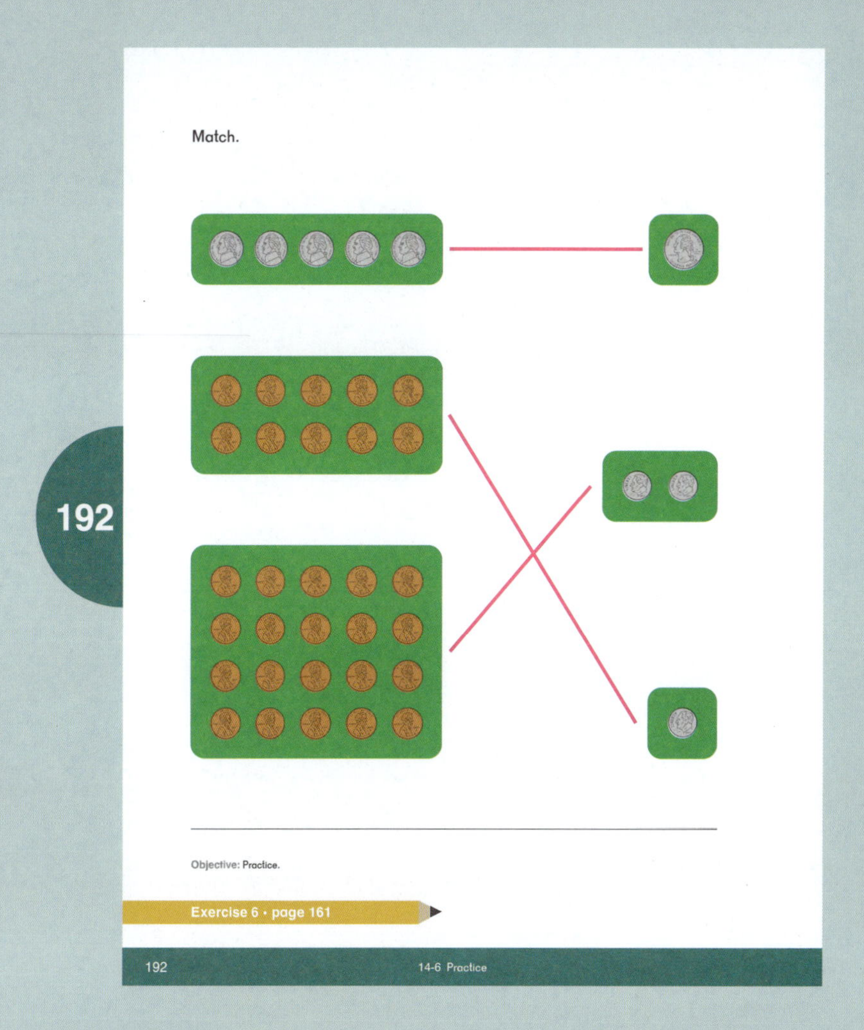

Exercise 1 • pages 151–152

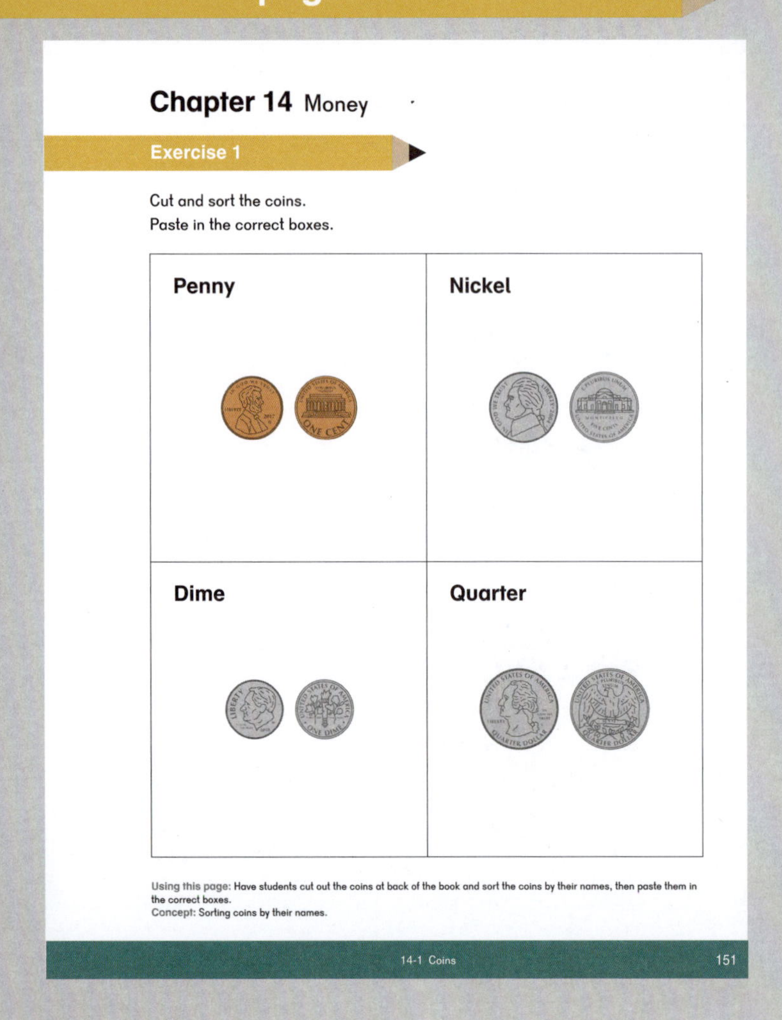

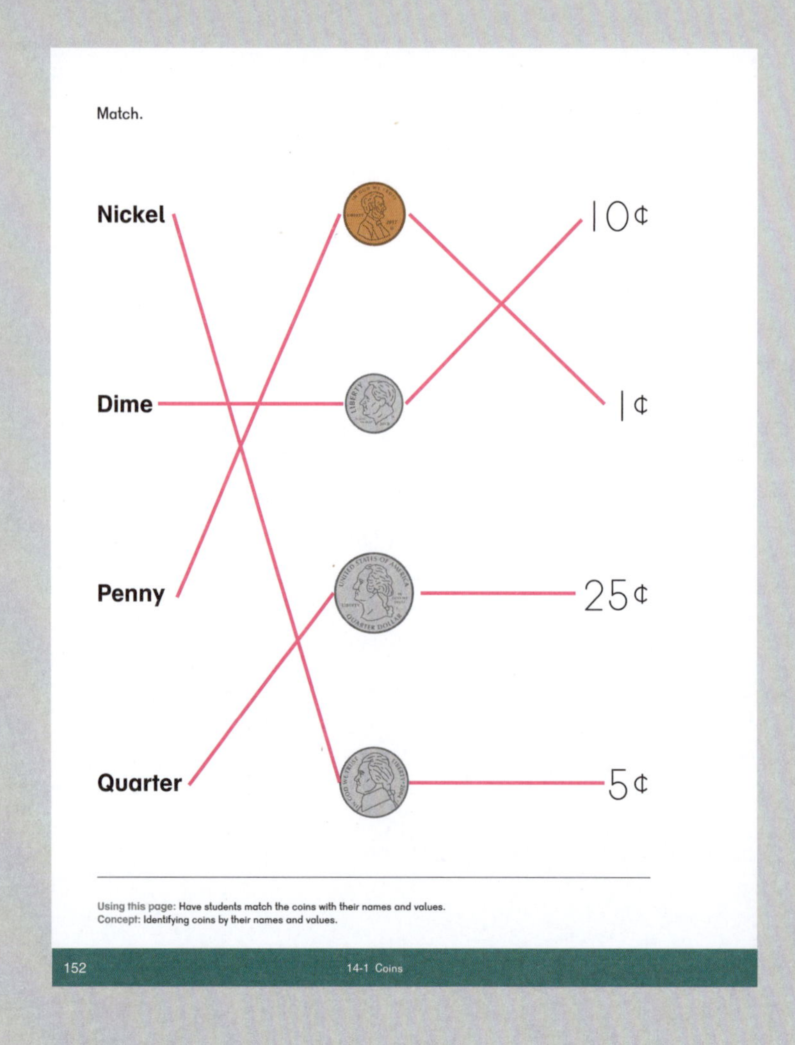

Exercise 2 • pages 153–154

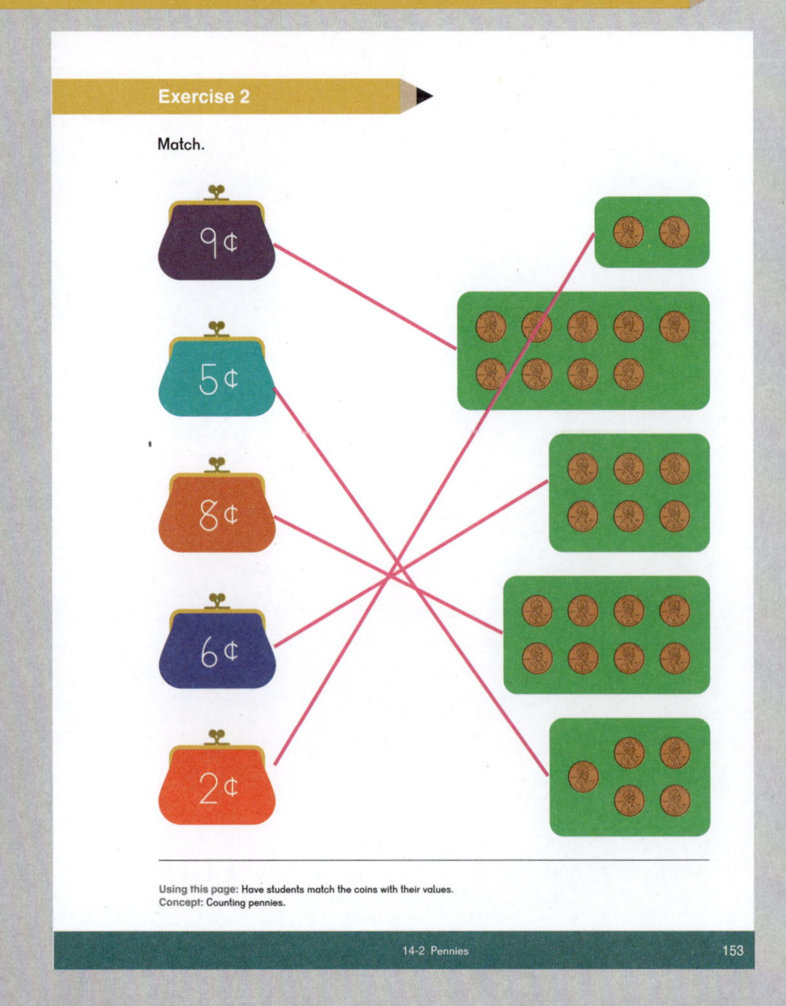

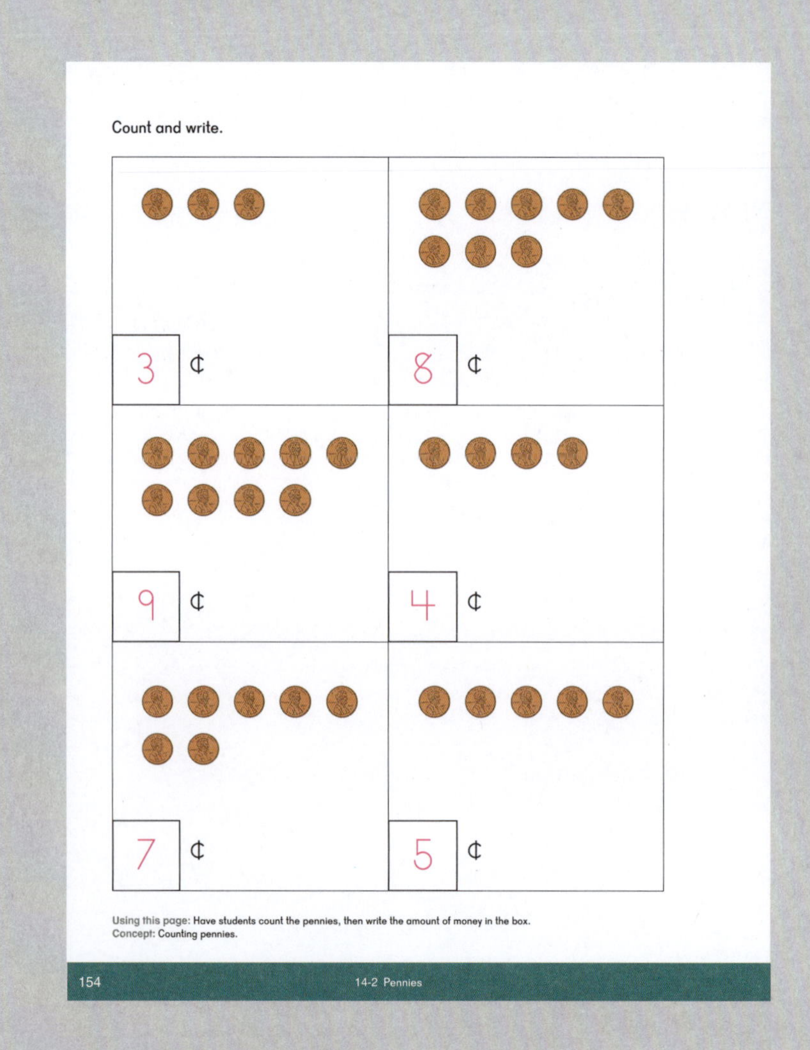

Exercise 3 • pages 155–156

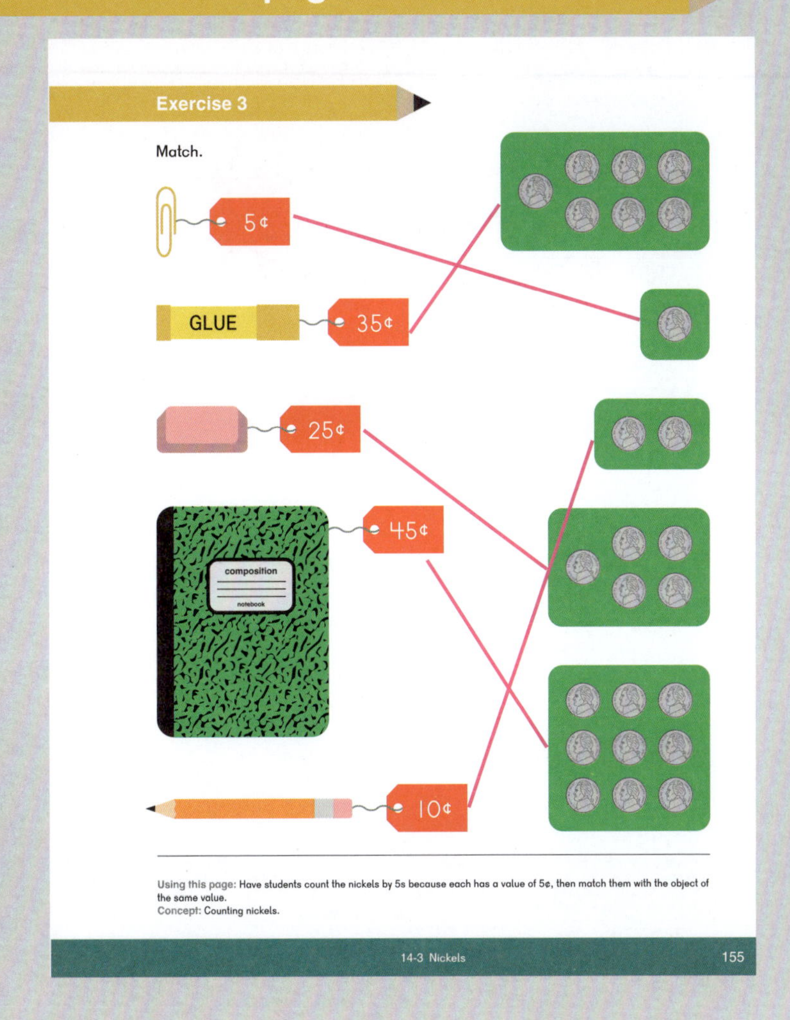

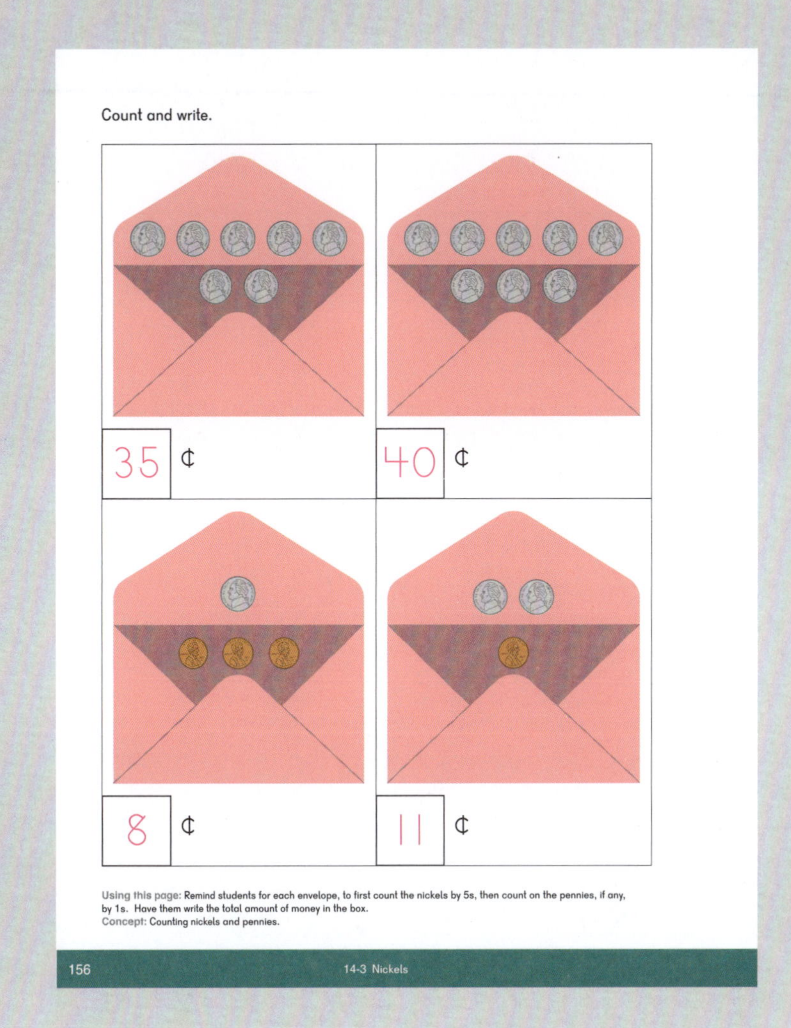

Exercise 4 • pages 157–158

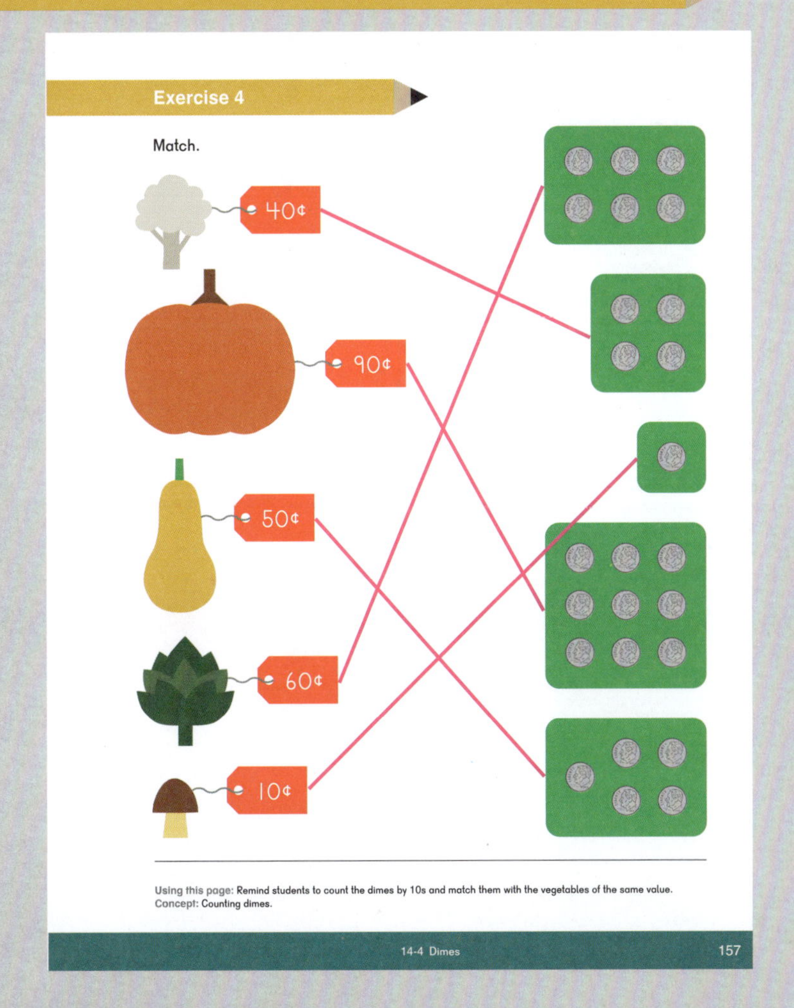

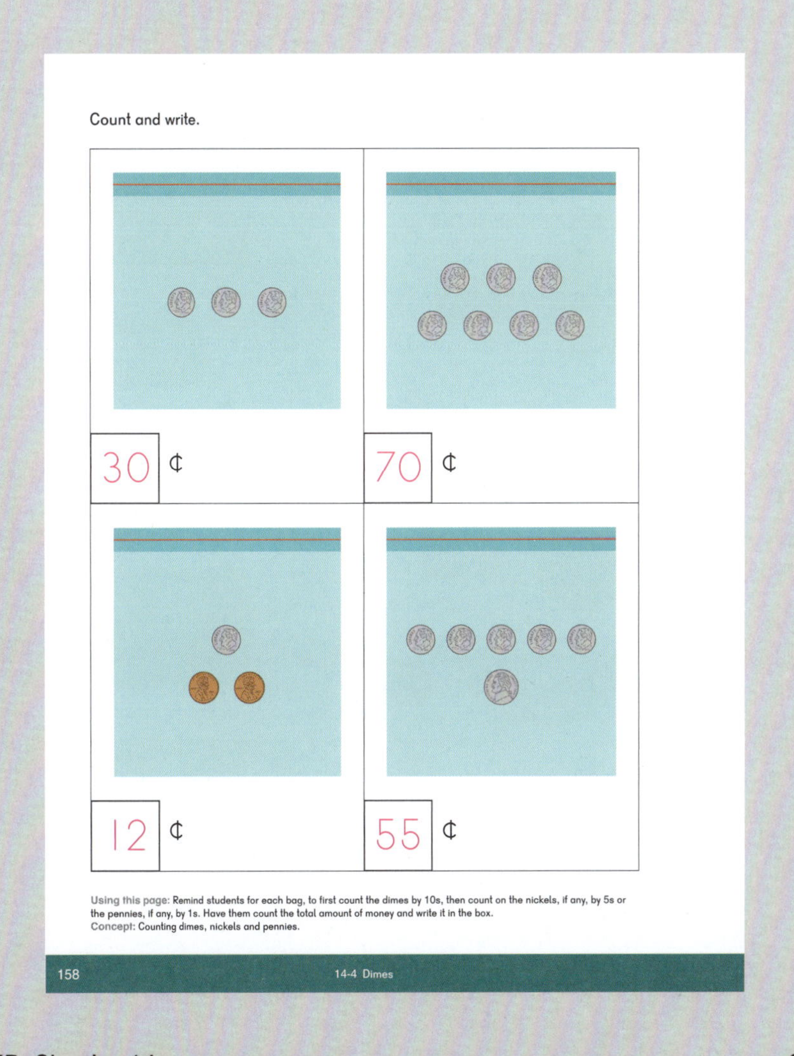

Exercise 5 • pages 159–160

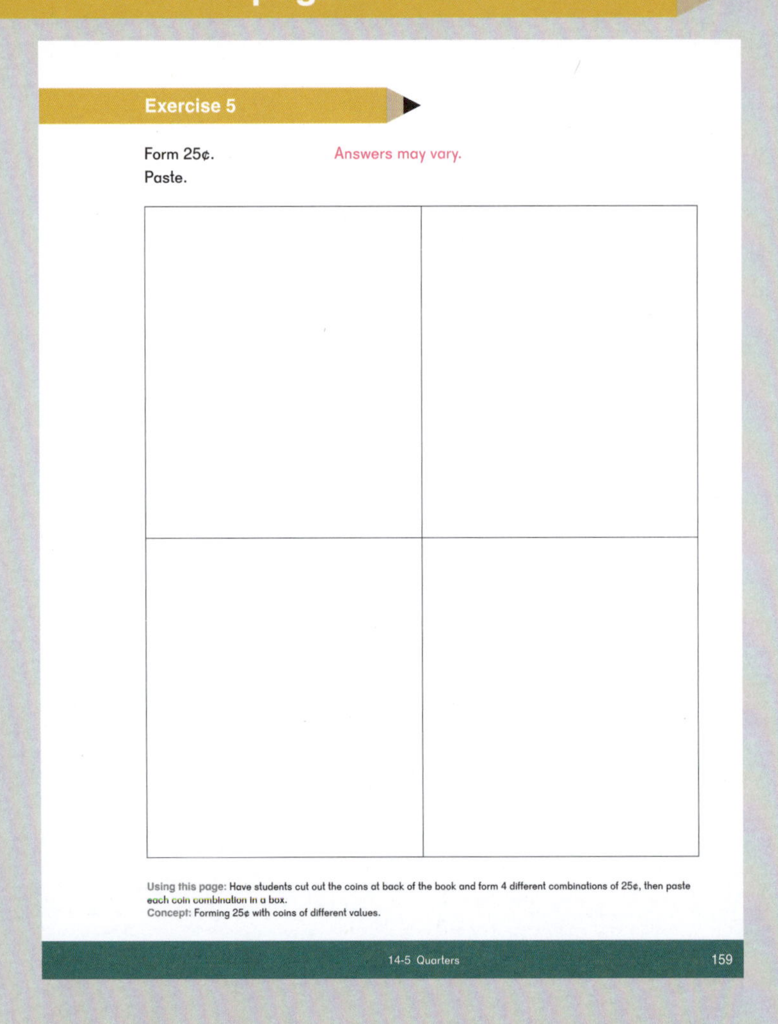

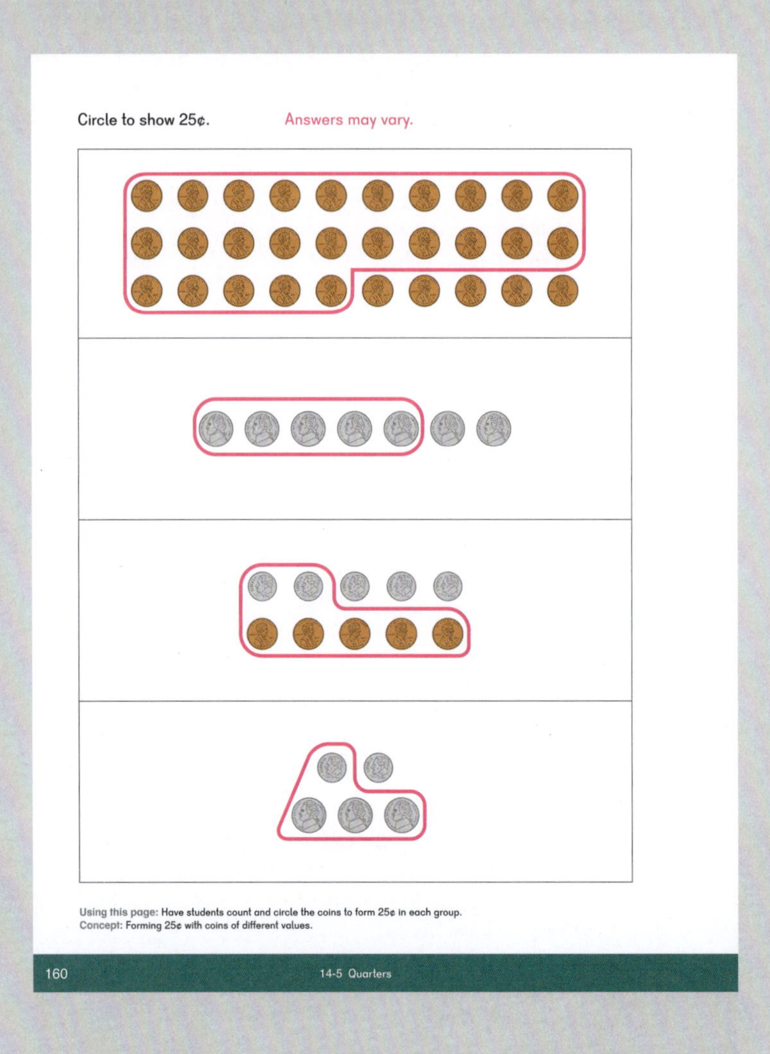

Exercise 6 • pages 161–162

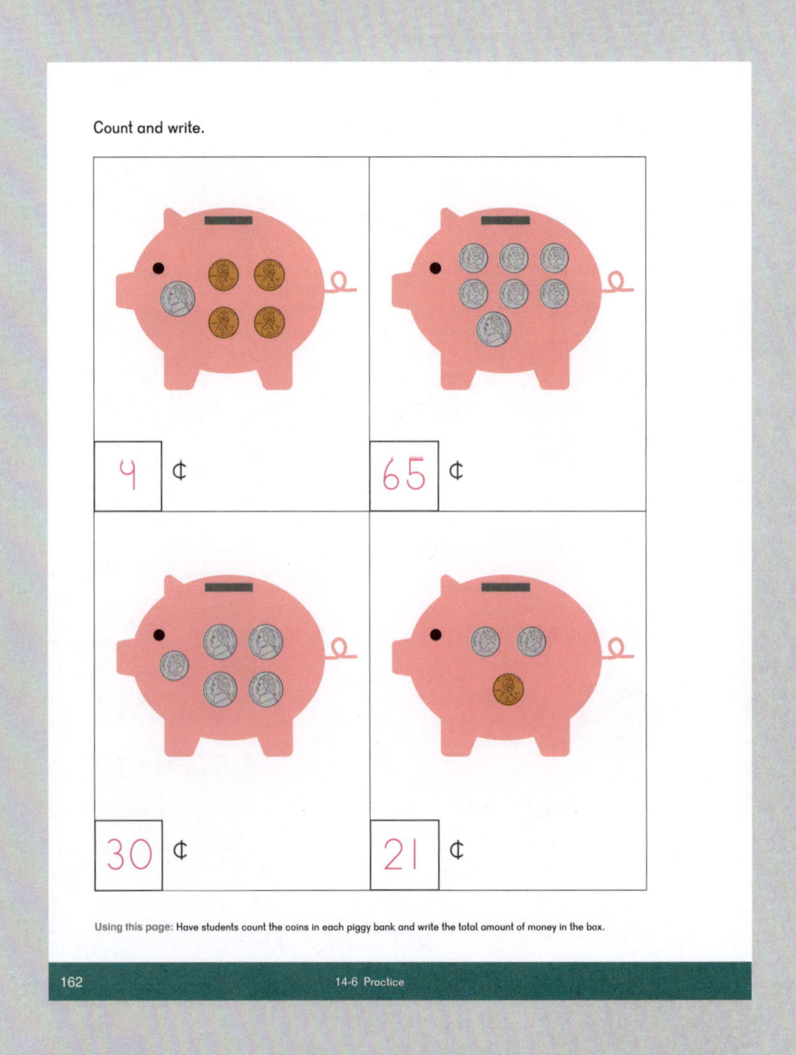

Blackline Masters for KB

All Blackline Masters used in the guide can be downloaded from dimensionsmath.com.
This lists BLMs used in the **Explore** and **Learn** sections.
BLMs used in **Activities** are included in the Materials list within each chapter.

10 and More Number Word Cards	**Chapter 7:** Lesson 6, Lesson 7
Animal Counters	**Chapter 9:** Lesson 6
Blank Double Ten-frames	**Chapter 7:** Lesson 3, Lesson 4
Blank Number Bond Template	**Chapter 8:** Lesson 4, Lesson 5, Lesson 7, Lesson 8, Lesson 9, Lesson 10, Lesson 11 **Chapter 9:** Chapter Opener, Lesson 2, Lesson 8 **Chapter 10:** Lesson 4, Lesson 5, Lesson 7, Lesson 8 **Chapter 11:** Chapter Opener, Lesson 3
Blank Ten-frame	**Chapter 7:** Lesson 1, Lesson 2 **Chapter 8:** Lesson 11 **Chapter 12:** Lesson 3 **Chapter 14:** Lesson 2
Double Ten-frame Cards	**Chapter 7:** Lesson 4, Lesson 6, Lesson 7, Lesson 8, Lesson 9
Hundred Chart	**Chapter 12:** Lesson 7, Lesson 9 **Chapter 14:** Lesson 3, Lesson 4, Lesson 5
Number Bond Recording Sheet	**Chapter 8:** Lesson 3, Lesson 11
Number Cards	**Chapter 7:** Lesson 4, Lesson 5, Lesson 6, Lesson 7, Lesson 8, Lesson 9 **Chapter 12:** Lesson 6, Lesson 8 **Chapter 13:** Lesson 4
Number Cards — Large	**Chapter 9:** Lesson 5
Number Path	**Chapter 9:** Lesson 6 **Chapter 10:** Lesson 6, Lesson 7, Lesson 8
Number Word Cards	**Chapter 7:** Lesson 5, Lesson 6, Lesson 7
Numbers to 50 Chart	**Chapter 12:** Lesson 10
Subtraction Sentences	**Chapter 10:** Lesson 3

Ten-frame Cards	**Chapter 7:** Lesson 5, Lesson 6, Lesson 7, Lesson 8 **Chapter 12:** Lesson 6
Tens Number Cards	**Chapter 12:** Lesson 2
Tens Number Words Cards	**Chapter 12:** Lesson 2
Tens Ten-frame Cards	**Chapter 12:** Lesson 2